Electronic Delivery of Documents and Graphics

Electronic Delivery of Documents and Graphics

Daniel M. Costigan

VNR VAN NOSTRAND REINHOLD COMPANY
NEW YORK CINCINNATI ATLANTA DALLAS SAN FRANCISCO
LONDON TORONTO MELBOURNE

NOTE REGARDING ORIGINAL MATERIAL CONTAINED IN THIS BOOK

Unless otherwise credited, all line drawings in this book are author originals (in some cases perhaps loosely based on similar drawings from other sources). Many of them are carried over from the author's previous book in this same subject area, *FAX-The Principles & Practice of Facsimile Communication*, published in 1971. Some of those earlier drawings have subsequently appeared, in perhaps slightly altered form, in various other publications. It should be noted that the contents of both this book and its predecessor are protected by copyright. None of the material, therefore, should be used in other publications without the express permission of the author or publisher.

Van Nostrand Reinhold Company Regional Offices:
New York   Cincinnati   Atlanta   Dallas   San Francisco

Van Nostrand Reinhold Company International Offices:
London   Toronto   Melbourne

Copyright © 1978 by Daniel M. Costigan

Library of Congress Catalog Card Number: 78-8166
ISBN: 0-442-80036-3

All rights reserved. No part of this work covered by the copyright hereon may be reproduced or used in any form or by any means—graphic, electronic, or mechanical, including photocopying, recording, taping, or information storage and retrieval systems—without permission of the author.

Manufactured in the United States of America

Published by Van Nostrand Reinhold Company
135 West 50th Street, New York, N.Y. 10020

Published simultaneously in Canada by Van Nostrand Reinhold Ltd.

15 14 13 12 11 10 9 8 7 6 5 4 3 2 1

Library of Congress Cataloging in Publication Data

Costigan, Daniel M.
   Electronic delivery of documents and graphics.

   Includes index.
   1. Facsimile transmission.   I. Title.
TK6710.C66   621.38'0414   78-8166
ISBN 0-442-80036-3

This book is dedicated to A. G. Cooley,
Captain Finch, the late H. E. Ives, G. H.
Ridings, the late John V. L. Hogan, the
RCA group, and all of those other now
largely forgotten pioneers who, more than
a generation ago, had already raised the
fax art to an impressive level of refinement.

PREFACE

There are undoubtedly many people who would like to be able, regularly and at reasonable cost, to dispatch documents and graphics from place to place electronically in a matter of minutes or seconds per page.

But it's a safe bet that only a relatively small number are aware that hardware offering that capability is in fact widely available.

How many people, for example, know that, for less than $30 a month, anyone who desires can rent a compact, desk-top terminal by which documents and graphics of all sorts can be sent and received electronically—via any existing telephone set—to and from compatible terminals anywhere in the country? How many know that similar systems have been the standard means for distributing official national weather maps since the 1930s, or that about a third of the photographs published each day in a typical metropolitan newspaper are received by wire and radio from all over the world? How many realize the extent to which branch banks regularly verify depositors' signatures via electronic dispatch from a central file, or the extent to which fingerprint records are electronically exchanged among law enforcement agencies?

These are not isolated examples of unique communication systems built around specially engineered, custom-built hardware. They are widely applied operations using production equipment from perhaps twenty different manufacturers throughout the world—many of them with well known names.

The "space-age magic" of which such systems seem capable is usually accomplished by a process that has in fact been around for more than a century—longer than the telephone and almost as long as Morse's telegraph—and in regular use in one form or another for at least the past 50 years. It is called *facsimile communication* ("*fax*" for short), and, reinforced with cumulative technological improvements, it has remained the predominant means by which existing documents and graphics are exchanged electronically between geographically separated points.

In the simplest of terms, fax is a system of slowed-down TV. Apart from speed, it differs from conventional TV in two important respects: (1) barring unusually bad transmission conditions, it can faithfully reproduce the characters, symbols, diagrams, etc., that constitute virtually every sort of document normally sent by mail, and (2) it receives the transmitted information not as tenuous images on a screen, but as substantially identical and permanent, tangible copies such as one might expect to obtain locally from an office copier. A terminal can range in price from about $1000 for the simplest, phone-coupled, desk-top transceiver to more than 100 times that amount for a high-speed, high-precision unit by which whole newspaper printing plates (which often include pictures) are remotely produced. At last count, there were well over 100,000 fax terminals of various sorts in regular use in the U.S. alone, and several hundred thousand throughout the world.

While fax has the graphics, or "copying," aspect of electronic "record" communications virtually to itself, there *are* alternative means of dispatching documents and graphics electronically. There are handwriting terminals, for example, with which handwritten messages and hand-drawn sketches can be exchanged over great distances via telephone circuits *in real time* (the time it takes the sender to form them). These devices cannot, however, transmit *existing* documents—unless of course, they are recopied by hand. By the best available estimates, there are, at present, perhaps 100,000 such devices in regular use throughout the world.

Then there are video systems—closed-circuit and slow-scan TV—which, at the present time, are in very limited use for record communications. There are communicating word processors for the combined generation and automated distribution of purely alphanumeric documents (possibly an emerging trend); and there are several relatively new component devices, each of which could form the nucleus of an effective record communication system. (Optical character reading devices and plasma panels are but two examples.)

This book will concentrate on fax as the most readily available, flexible, and economically viable medium for dispatching documents and graphics (as opposed to "messages") by wire and radio. It will examine the alternatives only to the extent deemed appropriate. (Even here it must draw the line at manual teletypewriter message systems, but will treat systems that automatically encode, store, and process alphanumeric characters for the re-creation of "finished" documents at a remote site.)

At this point one might ask why, if fax is all that these introductory remarks portray it to be, it has not caught on to a greater extent. Why only a few hundred thousand terminals at this late date in its long history and in a world that supports hundreds of *millions* of telephones? The reasons are both practical and psychological. But perhaps mainly it has been a matter of insufficient promotion and some persistent misconceptions.

The practical part is that, as with any communications medium (including the

PREFACE

postal service), fax may prove unsuitable for a given situation. In the business and professional communities especially, it has to be able to prove that it is more than just a luxury. Even for as little as $29 a month per transceive terminal (plus material costs and telephone tolls), one has to be able to justify the need to dispatch a written communication or picture in minutes rather than the day or two it would normally take by mail.

The practical part also includes the effects of certain regulatory obstacles and technical incompatibilities, which, in some instances, have made it impractical to adopt fax on a broad scale. For example, at this writing (mid-1977), one still cannot legally transmit fax signals overseas from the U.S. via the public telephone network. In other countries, there are similar regulatory restrictions on *domestic* communication. To a degree the situation is further aggravated by differences in signalling standards between terminals, both domestically and internationally. Fortunately, these obstacles are gradually being cleared away.

There is also a facet of the practical part that is really psychological. Fax has a way of suggesting possibilities of the sort that set people dreaming, and it has therefore had its share of entrepreneurs who have lost whole fortunes in the pursuit of a vision. The ill-fated fax broadcast venture of the 1930s and late '40s (Chapter 1 has the details) is but one case in point. That single failure has unfairly stigmatized fax to this day.

But the real psychological obstacle derives from the fact that people have become spoiled by the speed of technological advances in other, somewhat overlapping spheres of communication. When told, for example, that a modern fax terminal can send a document page in a minute or less, one tends to think of the instantaneity of TV and dismisses fax as a slow and relatively expensive medium. Or, having seen a piece of fax copy received via long-distance phone lines that contains some evidence of line impairments (similar faults would probably go unnoticed in a phone conversation over the same circuit), one tends to compare it with the output of a modern office copier and to judge the fax system's merits accordingly.

The fact that both these comparisons are technically inappropriate does not seem to matter. In today's world it is only the end result that counts, and not the technological miracle by which it was produced.

Ironically, it is an older generation of decision-makers who have perhaps contributed most to fax's bad press. Familiar with what fax *was* ten or more years ago, they are still dismissing it as too slow, too expensive, incapable of sufficient copy quality, and, in general, too unsophisticated for today's world. Unfortunately there has not been sufficient publicity to correct these misconceptions in the light of recent events.

Now, however, new forces are at work. In recent years, the communicating world in general—and the business community in particular—has been given cause to reflect on the practicality of continuing to move documents physically when there are electronic means available for doing it more efficiently. It is refreshing to see the emergence of a new generation of dynamic and perceptive young executives of successful companies putting things in perspective and making unbiased assessments of the "new fax."

It is one of the principal aims of this book to provide this new generation with the basic knowledge needed to make intelligent decisions in this increasingly important area.

DANIEL M. COSTIGAN

ACKNOWLEDGMENTS

The author wishes to thank all of the many equipment vendors who generously supplied product information and pictures for this book. In addition, the following institutions and individuals deserve special mention for their cooperation:

ANPA Research Institute, the Associated Press, Bill Austad, Don Avedon, Bell Telephone Labs, Bethlehem Steel, Mike Berwick, Bob Bishop, Ivar Blackberg, Jacques Cheminade (French Commercial Counselor in N.Y.), Dr. J. S. Courtney-Pratt, Carl Clough, Claude Clouthier, Council on Library Resources, Larry Dobrin, Dow Jones & Co., Hans Engelke, Oral Evans, Larry Farrington, Capt. W. G. H. Finch (U.S.N. Ret.), Frank Fitzgerald, Julie Frank, Hank Frey, D. E. Garner, Jr., Fred Gordon, Mike Gorman, John Hopf, Dewey Houck, the IEEE, the International Telecommunications Union, Dick Jenifer, John L. Jones, Win Kelker, Bob Krallinger, Ed Lewyt, Doug Linsley, Bill Lyon, Mike MacNaughton, Ken McConnell, Bob Meltzer, Harry A. Miller, Dick Murray, the National Weather Service, H. Lee Nicol, Wayne Owen, Joseph Paglia, the Philadelphia Police Department, Phillips Petroleum, John Porterfield, M. D. Post, Bill Rolph, John Sheehan, Akira Shimizu, John Shonnard, Fred Simpkins, Ian H. H. Smith, Southern Railway System, Glen Southworth, Carl Spaulding, George Stafford, George Stamps, M. Stephenson (Brit. P.O.), Nick Szabo, UPI, the U.S. Department of Defense, Bruno Vieri, P. Waddell (Univ. of Strathclyde, Glasgow, Scotland), L. C. Weekly, Ed Wern, Western Michigan University, Western Union, Gary Winkler.

INTRODUCTION

If you happen to be familiar with my book *FAX-The Principles and Practice of Facsimile Communication*, which was published in 1971, you will find that this one not only follows closely the lines of that earlier text but also borrows extensively from it. Yet in addition to updating the earlier book (a great deal has happened in facsimile communication since 1971), this one has expanded its scope, covering—along with fax—video systems, communicating word processors, and various other existing and prospective approaches to the electronic dispatching of documents and graphics.

The slant of this book (the intended audience and the kinds of information the book seeks to convey to that audience) remains relatively unchanged from that of its predecessor. If you are an electrical engineer seeking guidance in the design of electronic graphic communication equipment, you will no doubt find portions of this book useful to you, but, frankly, it was not written with you in mind. If you are accustomed to the level of technical book that abounds in equations, you will find this book, for the most part, comparatively simplistic. If, on the other hand, you are

- a *systems analyst* or *consultant*—or a communications planner or manager in government, industry, or the professional community—whose task it is to plan whole systems or to expand and improve existing ones, or
- a *technician* (or student technician) interested in electronic document or graphic communication systems and eager to learn all you can about them without getting hopelessly entangled in engineering particulars, or
- a *student* of the evolution of inventions, or
- a *business person* who feels that such systems may have a place in your office or plant, and who wants to learn more about them before talking to a salesperson—

then this book should be of value to you.

You will naturally be expected to have some familiarity with mechanics and electronics; the level of knowledge required will become evident as you begin to read. The book would obviously have to be three times its size if it were to attempt to teach the facsimile and related arts to a lay person whose only previous experience with communications equipment consisted of the twisting of knobs and the pushing of buttons.

Among the things the book will do is acquaint you with facsimile's history, survey some of the many ways in which facsimile and related technologies are being applied today, and explain (in reasonable detail) how these systems work and why they sometimes do not work as they should. It will also discuss standards, specific commercial systems, the variety of systems available and what they cost to buy or rent, and the inevitable trade-offs among cost, quality, and speed within a system.

Finally, it will examine current trends and speculate on what the future holds for fax, video systems, and for some of the newer media that are now just emerging as competitors in the race to provide electronic alternatives to the physical dissemination of documents and graphics.

–D.M.C

CONTENTS

Preface	vii
Acknowledgments	xi
Introduction	xiii

1 HISTORY — 1

Evolution	2
Newspapers by Radio	10
Miscellaneous Advances	14
A New Generation	15
Continuing Evolution	20

2 FAX TODAY — 22

Current Applications	22
Virtues	41
Limitations	47
Trends	51

3 HOW IT WORKS — 65

Scanning	65
Recording	76
Amplification	88
Phasing and Synchronization	91
Index of Cooperation	98

4 TRANSMISSION — 103

- D.C. and Subaudio — 103
- Modulation — 105
- Transmission Impairments — 115
- Other Facilities — 123
- Couplers and Modems — 131
- Redundancy Reduction — 139
- Transmission Sequence — 152
- Transmission Options — 154

5 QUALITY — 160

- Resolution — 160
- Contrast, Gray Scale, and Polarity — 174
- Impairments Due to Digitizing — 179
- Legibility Criteria — 183
- Pictorial Criteria — 191

6 ECONOMICS — 199

- Cost Factors — 199
- Trade-offs — 207
- Cost Per Copy — 209
- Economic Trends — 217

7 STANDARDS — 221

- Domestic Standards — 222
- International Standards — 231
- Intersystem Compatibility — 240

8 OTHER DOCUMENT AND GRAPHIC COMMUNICATION SYSTEMS — 245

- Manual Graphic Terminals — 246
- Video Systems — 253
- Communicating Word Processors — 261
- Miscellaneous Schemes — 263

9 THE FUTURE 267

Traditional Role; New Trends 267
Merging of Previously Separate Technologies 268
New Trends in the Way Documentation Originates 271
Microminiaturization: Its Economic Impact 272
Fax in the Home 273
Speed 276
Color 277
The Question of Paper 278

Current Information 281

Appendix 283

Index 323

Electronic Delivery of Documents and Graphics

1
HISTORY

Two significant events in the field of telecommunications took place early in 1926. One was the first public demonstration of television—by John Baird, in a little laboratory in London's Soho district in January. The other was the inauguration of commercial transatlantic radio facsimile service (for the transmission of news photos) by the Radio Corporation of America three months later.

That the two events occurred within a few months of each other is perhaps not in itself significant. What is interesting is the relative states of development of these two loosely related picture transmission media at that time. They were technologically similar, both employing essentially the same electromechanical techniques of dissecting and reconstructing pictures. But while TV remained confined to the laboratory bench, its static counterpart (static in the sense that it was concerned with still, as opposed to moving, images) already had several decades of practical application behind it, in the course of which it had steadily progressed toward a high state of refinement.

Although TV was to enjoy a somewhat more rapid technological growth than facsimile in the ensuing years, it was not until two decades later it finally caught up in the sense of becoming a full-fledged commercial reality. And it was not until the advent of communications satellites in the 1960s that it was able to span the seas on even a limited commercial basis.

As late as 1949 it was still believed in some quarters that facsimile could compete with TV as a home news and entertainment medium. Although that dream has not materialized, facsimile has managed to maintain and even gradually reinforce its position as an indispensable telecommunications medium.

Inasmuch as this book is devoted primarily to the subject of facsimile communication, it might be well at this point to clarify what *facsimile* means as a telecommunications term. There are conflicting definitions.

To some, the word is confined to describing what is essentially a "message" medium through which written or printed messages and sketches are exchanged via wire or radio and are visibly recorded by *non*photographic means, i.e., by electrochemical or electromechanical means not requiring further processing.

By this definition, the systems customarily used for transmission of news photographs are excluded.

To others, *facsimile* refers to any system by which printed *or* pictorial matter (graphics material in general) is transmitted electrically from one place to another and a reasonably faithful copy permanently recorded at the receiving end in any one of several forms.

This book takes the latter view, lumping together the news photo and so-called message or document facsimile systems simply as variations of facsimile recording technique, which is, in fact, the only genuine distinction existing between them.

EVOLUTION

Curiously, the origin of facsimile—or *fax*, as it is often referred to today—was not so much the result of a conscious quest for a new telecommunications medium as it was an offshoot of the invention of the electric clock.

In 1842, a Scottish inventor, Alexander Bain,[1] fashioned an electrically controlled pendulum mechanism that formed the basis for a master-slave system of interconnected synchronized clocks, a later refinement of which is still widely used in offices, schools, and other institutions. From this novel time-keeping system evolved Bain's pioneering device for the transmission of visible symbols over telegraph lines.

The inventor had shrewdly observed that since each of the synchronized pendulums of this clock system was in the same relative position at a given moment of time, if the master pendulum could somehow be rigged to trace a pattern of intermittent electrical contacts in its travel, the pattern could simultaneously be reproduced (by a means that had still to be devised) at one or more locations remote from the sending device.

The result of the ensuing experiments was an "automatic electrochemical recording telegraph," on which Bain received British Patent 9745 in 1843.

Basically, three modifications had been necessary to convert the synchronized clock to a graphics transmission device. One was the addition of a stylus to the pendulum of each device. The second was the addition of a clockwork-actuated "message block," and the third was the development of a suitable electrosensitive recording medium.

In the SEND mode, the added stylus functioned as a brush contact arranged to sweep over a message block containing a group of words formed by raised metallic letters. Each sweep of the pendulum intermittently disrupted and completed a circuit, producing, in effect, a series of pulses corresponding to the metal protrusions the stylus contacted within the breadth of the pendulum's stroke. With each full swing of the pendulum, the message board was caused

HISTORY

to drop down a notch so that, on the next swing, the stylus would contact a fresh segment of the message.

The resulting serial stream of d.c. pulses was sent by wire to the receiving device, where, in place of raised metal letters, the message board accommodated a sheet of electrosensitive paper. As the RECEIVE stylus swept over the face of the recording paper, it formed a dark mark wherever a pulse occurred. After several sweeps the dark marks formed a facsimile of the raised metal letters at the SEND terminal.

Synchronization was achieved by an automatic correcting action in which a pendulum tending to lead its remote mate would be electromagnetically "locked" at the end of a swing cycle, and then released by a momentary opening of the circuit when the lagging pendulum had caught up (Figs. 1.1 and 1.2).

Bain's device enjoyed a moderate success but was unable to compete with the electromechanical printing telegraph, which had meanwhile been perfected and which surpassed the Bain system in speed and reliability.

Since facsimile originated in Europe, it was only natural that its early commercial applications centered on Britain and on the continent. The first commercial system on record was established in France in 1865 by an Italian expatriate, Giovanni Caselli, who had patented an improved version of Bain's device. The system, which connected Paris with several other French cities, remained in operation for about five years.

Thereafter, the French themselves began to contribute to the advance of the technology (which remained quite slow). For example, in 1869, Ludovic d'Arlincourt devised a synchronization system using tuning forks as speed governors for spring-driven mechanisms. About 15 years later, Clement Ader, the noted French aeronautics pioneer, contributed one of the first electromagnetic light valves for photographic recording (similar in principle to one later perfected by E. C. Wente in the United States). Beginning around the turn of the century, Edouard Belin contributed a variety of innovations. He was, in fact, to be among the first to transmit a photograph across the Atlantic by telegraphy (1921).

Meanwhile, in England in 1850, Frederick Bakewell had successfully demonstrated the forerunner of the "cylinder-and-screw" arrangement that was eventually to replace Bain's pendulum and to reign for many years as the standard fax mechanism (Fig. 1.3). Many present-day fax systems are of the cylinder, or "drum," type.

The next major advance did not come until shortly after the turn of the century. Building upon a suggestion by Britain's Shelford Bidwell some years earlier regarding possible uses for the material selenium, Germany's Dr. Arthur Korn set about to perfect photoelectric scanning. For nearly 60 years facsimile had been tied to Bain's contact scanning technique with, at best, a few subtle

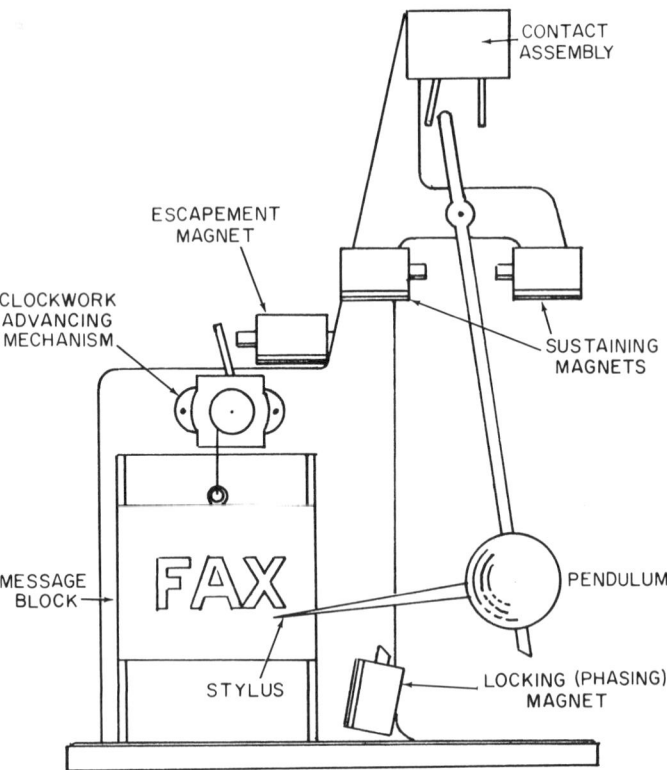

Fig. 1.1. Basic mechanics of Bain's electrochemical recording telegraph. In the SEND mode, the message block contained raised metallic letters, and in the RECEIVE mode, a piece of electrolytic paper. An electrically actuated escapement lowered the message board one notch for each swing of the pendulum, thus effecting line-by-line scanning by the oscillating stylus. (Original sketch by the author, based loosely on an old patent drawing.)

variations. In 1902 Korn publicly demonstrated the first practical photoelectric fax system for the transmission and reproduction of photographs. Five years later he established a commercial picture transmission system, which was at first confined to Germany, but which, by 1910, linked Berlin with London and Paris.

In 1922 Korn successfully transmitted a picture to America by radio. The picture, a photograph of Pope Pius XI, was sent from Rome and received at Bar Harbor, Maine. It was published in the New York *World* on the same day it was transmitted: June 11, 1922.[2]

Inspired by Korn's successes, a number of British and continental groups set about establishing long-distance picture transmission facilities of their own.

HISTORY

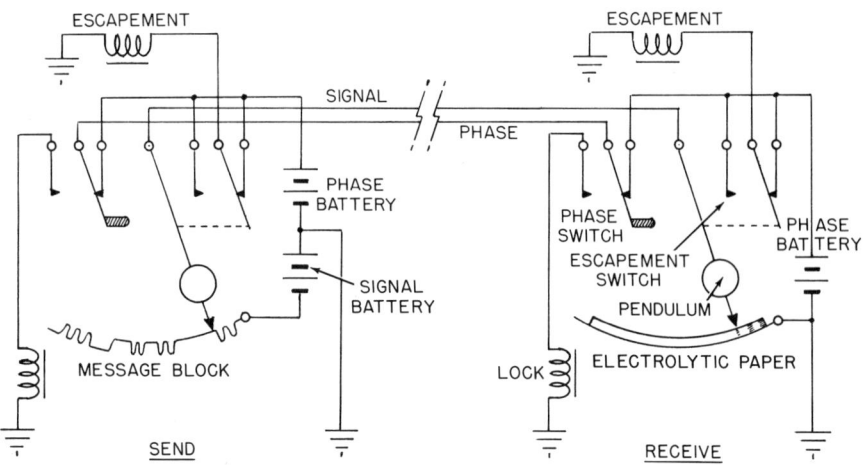

Fig. 1.2. Schematic representation of Bain's telegraph, illustrating (a) the basic scheme for visual electrolytic recording of a d.c. pulse train, and (b) the scheme for synchronizing, or "phasing," two pendulums by holding a leading pendulum at the "start" position while its lagging mate catches up. For simplicity, pendulum-sustaining circuitry has been omitted, and signalling and phasing are shown as isolated functions. (Original sketch—and circuit—by the author.)

Chief among them were Belin in France, the Telefunken Works in Germany, and the Marconi Company in Britain. But perhaps the most unusual system was one that actually preceded Korn's transatlantic triumph by about two years. It was the so-called Bartlane system developed by two Britons, H. G. Bartholomew and M. L. MacFarlane, for purely telegraphic picture transmission via the Atlantic cable. It converted picture elements to perforations in paper tape in a manner anticipating the present-day PCM (pulse code modulation) system of digitally encoding discrete amplitude levels of an analog signal.

In America there had been some sporadic developments beginning around the 1870s, notably Sawyer and deHondt's system of stop-start synchronization (c. 1875) and, closer to the turn of the century, the work of Amstutz, Flynn, Hummel, Palmer, and Mills in picture transmission. But serious fax experimentation (in the commercial sense) did not begin until the 1920s, after Korn had equipped the art with the faculty of sight by photoelectric means. Three telecommunication giants—American Telephone and Telegraph, RCA, and Western Union—set about developing picture transmission systems, chiefly for press use (see Fig. 1.4). All three separately developed systems came to fruition in 1924.

In May of that year, AT&T transmitted a 5 × 7-inch picture from Cleveland to New York (a distance of over 500 miles) in $4\frac{1}{2}$ minutes. And in July RCA

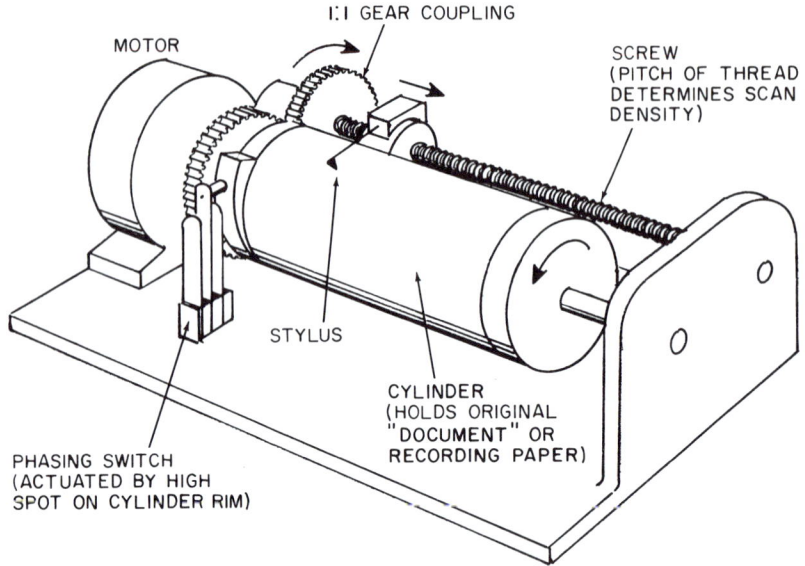

Fig. 1.3. Successor to Bain's pendulum was the still-familiar "cylinder-and-screw" mechanism (c. 1850) usually attributed to Bakewell. Scanning and recording were done lathe fashion, with the stylus (or light beam in a later innovation) progressing slowly along the screw, tracing a continuous helical scan around the face of the revolving cylinder. (Original sketch by the author.)

successfully sent a picture by radio to England and back. Meanwhile, Western Union established a limited commercial service called "Telepix," which, however, it was shortly to abandon for lack of customer interest.

It is an interesting sidelight that the AT&T "telephoto" system was developed by Dr. Herbert Ives,[3] a Western Electric Company (and, later, Bell Labs) engineer, whose father, Frederic Ives, had helped pioneer the halftone photoengraving processes in 1878. The elder Ives's innovations had figured prominently in a number of earlier (prephotoelectric) fax experiments, both in the United States and in Europe, by having provided the necessary means for transmitting photographs by contact scanning. The younger Ives was later to conduct the first public demonstration of television in America—in New York City on April 7, 1927.

The AT&T domestic system went commercial in 1925, and RCA's international radiophoto system followed suit a year later. Chief developer of the RCA system was Captain Richard H. Ranger, who was then a young assistant to the noted radio broadcast pioneer Dr. E. F. W. Alexanderson and who was later to distinguish himself in the field of electronic music.

At about this time, Western Union, while continuing to develop circuits and equipment for phototransmission, began to explore fax's potential as a message-

HISTORY

Fig. 1.4. Evolution of a newspicture, 1925. Original picture (upper left) was transmitted from Washington, D.C.; received simultaneously in New York, Chicago, and San Francisco (lower left); and published in the New York *Sun*, Wednesday, March 4, 1925 (lower right). (Courtesy Bell Telephone Labs.)

sending medium to supplement the firm's telegraph service. In the process, it developed a new, nonelectrolytic direct recording paper called *Teledeltos*,* which contributed notably to the success of the automatic facsimile telegraph system the firm inaugurated in the early 1930s. The development of this system culminated, in 1948, in introduction of the *Desk-Fax* (Fig. 1.5), a compact, desk-top fax message transmitter, thousands of units of which have since been installed.

Another important departure from the newspicture application was fax's adoption, in 1930, as a means of distributing weather data. Earlier experimental short-range transmissions (by C. F. Jenkins and others) had culminated, that year, in RCA's pioneering transmissions of weather maps to ships at sea via radiofacsimile. Since then fax's role in weather forecasting has grown steadily.

After having developed its picture transmission system to a high degree of refinement (at a cost of about two million dollars), AT&T sought to interest a major news agency in taking over the operations aspect of it. The Associated Press did so in 1934, dubbing it "Wirephoto," and within a year had made such a success of it that a mad scramble ensued among other newspaper interests to develop picture transmission systems of their own. AT&T continued to provide

---

*Teledeltos is a registered trademark of the Western Union Telegraph Co.

Fig. 1.5. Western Union's Desk-Fax, one of the most compact fax transceivers ever developed. Some 50,000 were produced. (Courtesy Western Union)

circuits—and, for a time, terminal equipment as well—to implement the private newspicture networks operated by the AP and others.

Out of the scramble emerged at least three new fax equipment manufacturers: Acme (manufacturing offspring of Acme Newspictures), Times Telephoto Equip-

(a)

Fig. 1.6. Some classic newspicture machines. The Western Electric photographic receiver (a) was the backbone of the AP Wirephoto net during the 1930s and '40s. The Times Facsimile Model CF portable transmitter (b) was introduced about 1940 and remained a popular newspicture machine into the 1960s. The Muirhead D-649 electrolytic receiver (c) was the mainstay of the AP network from the 1950s to the mid-1970s.

HISTORY

(b)

(c)

Fig. 1.6. (*Continued*).

ment (later renamed Times Facsimile), and Finch Telecommunications. Along with RCA, Western Union, and another pioneering firm called Radio Inventions, Inc. (founded in 1929 by radio and television pioneer John V. L. Hogan),[4] these firms constituted the nucleus of the American facsimile industry until the 1950s (see Fig. 1.6). The Bell System (specifically Bell Labs and the Western Electric Company) had meanwhile phased itself out of the development and manufacture of fax equipment. However, AT&T, the parent firm, has remained quite active to this day in the development and leasing of fax transmission facilities.

NEWSPAPERS BY RADIO

As early as 1926, Austin Cooley (who was later to propel his Times Facsimile Company to a dominant position among manufacturers of photofacsimile equipment) had experimentally broadcast printed news reports via commercial radio channels. His "Ray Photo" system, however, had failed to convince broadcasters that there was any future in this novel visual supplement to the transmission of speech and music. But in the minds of a small but growing fraternity of imaginative technologists, a vision was beginning to form of a day when a fax receiver would be a household fixture no less common than a radio set—to which, in fact, it would be an indispensable appendage.

About the same time (1926), in Austria, Captain Otho Fulton[5] was developing his "Fultograph" system for fax broadcasting. In 1928, the system went into experimental operation for about two years in at least four major European cities, including Berlin, where the receivers were made. The postcard-size pages were received on specially formulated electrolytic paper, and the RECEIVE mechanism, like the record players of the day, was spring driven.

In 1929, Charles J. Young and Maurice Artzt of the General Electric Company in the U.S. demonstrated a radio facsimile receiver using carbon paper as the medium for impact printing the received pictures onto plain paper. The device subsequently accompanied the two men to RCA, where it was further refined for use in fax broadcast experiments (see Fig. 1.7). Eventually, several hundred of the improved receivers were built for experimental home reception.

In 1934, John V. L. Hogan demonstrated his "Radio Pen" system of fax broadcasting in Milwaukee via radio station WTMJ. The receiver used an ink stylus for recording, as in some present-day laboratory chart recorders, except that it was magnetically actuated to respond to signal pulses. The resulting copy quality must have been marginal, and probably for that reason, the demonstration did not create much of a stir.

The real thrust in fax broadcasting was not to come until 1937. In that year, station KSTP in St. Paul, Minnesota, successfully delivered a special edition newspaper by radio facsimile, and at least four other stations quickly followed suit. One of the four was a non-commercial station in New York City, W2XBF, which was the first to be granted an FCC experimental fax broadcasting license. It was owned and operated by W. G. H. Finch, who had designed the transmitters and receivers that were to be used in most of the other trials (see Fig. 1.8).

This was to be fax's long-awaited bid for public recognition. The Federal Communications Commission required that receivers be placed in at least 50 homes within the broadcast area of each specially licensed station. Thus, by

HISTORY

Fig. 1.7. Sample of broadcast fax copy produced in 1937 by the RCA carbon paper process. (Courtesy RCA)

the close of the 1930s, with the number of experimental fax broadcasters having grown substantially, there were at least 1500 radio fax receivers installed in homes around the country. Most of them were attachments to existing broadcast receivers. Arranged to function like nocturnal robots, they would

Fig. 1.8. Fax pioneer W. G. H. Finch monitoring a radio newspaper broadcast, c. 1941. (Courtesy W. G. H. Finch)

automatically go into operation after the household retired for the night, printing out a complete radio-dispatched newspaper that could be read at the breakfast table the following morning.

The idea caught on to the extent that, by late 1940, there were some 40 commercial stations regularly broadcasting fax newspapers (experimentally), and one supplier alone—Crosley Radio Corporation—had sold some 10,000 Finch-designed radio fax attachments (Fig. 1.9). This was in addition to the several hundred RCA units that had been field installed.

Unfortunately, public demonstrations and trials had left much to be desired, and by the time fax joined the war effort in 1941, interest had flagged considerably. It was rekindled after the war, however, and, on June 9, 1948, the FCC officially authorized commercial fax broadcasting.

Even before the official authorization, four major newspapers—the Miami *Herald*, the Chicago *Tribune*, the Philadelphia *Inquirer*, and the *New York Times*—had experimentally broadcast special fax editions via their respective FM radio outlets. (Under the FCC authorization, fax had gone from an AM to an FM medium.) Of these pioneering papers, the Miami *Herald* took the venture most seriously, gambling considerable money and developmental effort on the belief that the fax newspaper would catch on. At the height of the postwar drive, it was transmitting *five editions daily* and was leasing receivers to hotels in the Miami area.

But the idea did not have time to take hold. For a rival with appreciably

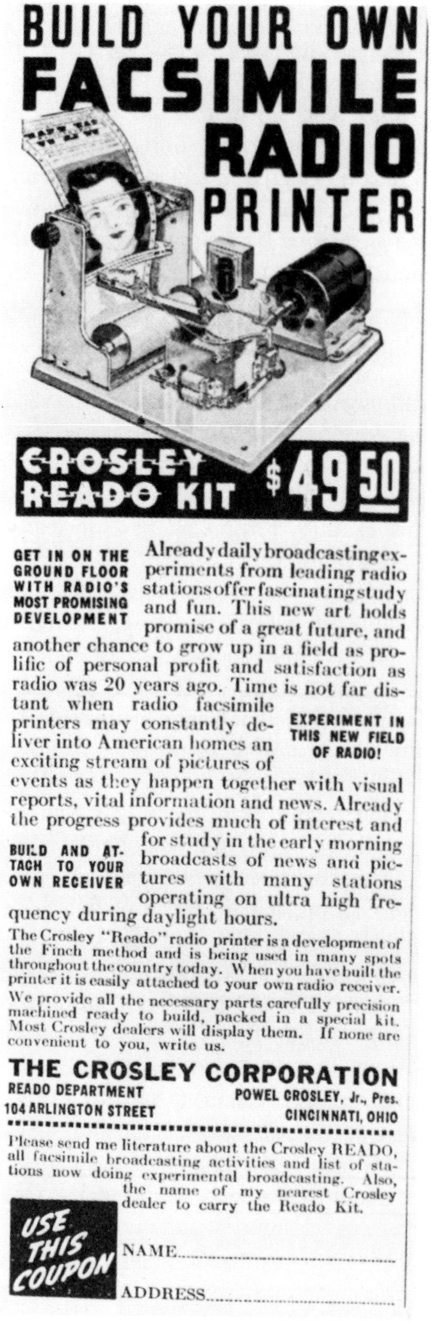

HISTORY

Fig. 1.9. Ad reproduced from a 1939 issue of *Radio News*. The kit was engineered by Finch; manufactured and marketed by Crosley. The finished unit had only to be connected to the speaker terminals of any standard receiver for reception of a broadcast fax newspaper. Some 10,000 of the units were sold.

greater popular appeal had entered the scene: commercial TV. For the fax venture to have succeeded under the circumstances, the recorders for home use would have had to be marketed at less than cost. Even then it is doubtful that a public preoccupied with the long-awaited debut of TV would have accepted fax service with sufficient enthusiasm to make it pay.

So fax quietly retreated to its behind-the-scenes role as a strictly utilitarian communications medium—a role in which it has continued to grow in stature and in a variety of applications despite strong competition from younger and more sophisticated systems.

MISCELLANEOUS ADVANCES

Although newspicture, weather chart and radio newspaper applications dominated the developmental activity during fax's formative years, it would be an oversight not to mention some of the more or less random (but nonetheless impressive) individual technological achievements during the first four decades of this century. For example, the early 1930s saw two innovations that were considerably ahead of their time. One was the all-electronic flying spot scanner patented by Manfried Von Ardenne in Germany in 1931, and the other was a modulated ink-jet recording method developed by C. W. Hansell of RCA at about the same time.

Von Ardenne's work made possible the later experiments of England's TV pioneer John Baird with high-speed TV-facsimile. (In 1944, Baird claimed the ability to transmit as many as 25 pages a second with his system via relatively wideband radio channels.) Hansell's ink-jet system was to remain essentially dormant until quite recently when the advent of the computer stimulated its revival as a means of increasing the speed of character printers.

While Baird was developing high-speed electronic fax in England, a system for the transmission of fax in color was being demonstrated by Finch in America. (Finch, incidentally, had also patented a high-speed cathode ray tube scanner in 1937, and had experimentally "simulcast" fax and music via FM multiplex the following year.)

Also in the 1930s, P. T. Farnsworth, V. K. Zworykin, and others developed the photomultiplier tubes that were to result in significant improvements in the design of fax scanners, and G. H. Ridings and his associates at Western Union developed a variety of automated fax machines.

No history of facsimile communication would be complete without acknowledgment of the work of America's C. Francis Jenkins. After inventing the first practical motion picture projector in 1894, Jenkins devoted most of the remainder of his career to the kindred technologies of fax and mechanical TV. His innovations were numerous and varied, and his aggregate contribution to both technologies was significant.

HISTORY

Since 1950, there have been continuing advances, some of which are just now beginning to influence the still-evolving fax art. Among the more important are the development of redundancy reduction (white-space skipping) techniques by Kretzmer, Hochman, and others; electrostatic recording (Xerography, RCA's *Electrofax*,* etc.); laser scanning (Bell Labs, MIT, Singer, Xerox, and others); and solid-state scanning—notably Bell Labs' invention of the charge-coupled device (CCD).

A NEW GENERATION

Several of the names that once formed the nucleus of the fax industry have since vanished, and several new—and in some cases more familiar—ones have appeared.

The assets of Acme Telectronix, heir to the fax manufacturing activities of the old NEA-Acme news organization, were purchased by the Fairchild Camera & Instrument Corporation in the early 1960s. Prior to that, in 1951, Acme Newspictures had been acquired by the United Press Associations, which subsequently (1959) were absorbed by International News Service to form the present United Press International. Fairchild has since divested itself of its facsimile product line, but UPI has remained active in the development of newspicture transmission equipment for internal use in its Telephoto system.

Times Facsimile Corporation, with fax pioneer Austin Cooley at the helm, was acquired by Litton Industries in 1959. Some of the Times-developed apparatus remained available briefly under the name "Westrex" (a Litton division, formerly the Western Electric Export Company). More recently, Litton's entire facsimile line, including separate systems for eleven different applications— some of the equipment embodying original Times Facsimile designs, some of it brand new, and some of it of foreign manufacture—bore the name *Litcom*. Late in 1970, the fax line was transferred to Litton's Datalog Division and renamed accordingly.

Finch Telecommunications, which had banked heavily on the success of the ill-fated fax broadcasting venture, changed hands around 1950 and went out of business shortly thereafter. (Finch himself had left the firm in the 1940s to serve as a U.S. Navy officer, and had returned to it briefly just before its demise to guide it back to a civilian product line.)

Radio Inventions was reorganized in the early 1950s as Hogan Laboratories Incorporated, which, with its "Faximile" manufacturing division, was subsequently acquired by Telautograph, Incorporated. Along with the handwriting machines for which it is perhaps best known, the latter currently produces a line of fax equipment, while its Hogan Faximile Division concerns itself primarily

---

**Electrofax* is a registered trademark of the RCA Corp.

with supplying its patented electrolytic *FAXpaper** to the news wire services for the direct recording of transmitted photographs.

Both RCA and Western Union have remained somewhat active in the fax field, though not as producers of apparatus for the general market. Both firms maintain facilities for fax transmission and both have engineered a number of novel systems, such as RCA's system for sending fax signals over commercial TV channels by utilizing the blanking period between frames. Western Union appears pretty much to have phased out fax service *per se*, as well as the design and production of fax terminals—two areas in which it had once been extremely active. It still, however, leases lines to fax customers, as does the Bell Telephone System.

Of the names that currently lead the roster of manufacturers of fax equipment for the general market, that of Alden[6] is perhaps the best established. The Alden Company was a maker of radio parts when it entered the fax field during World War II. Since then it has matured steadily in the broad field of electronic graphics recording and is today one of the leading producers of specialized fax equipment.

RCA, General Electric, and the Stewart-Warner Corporation were conspicuous among the handful of industrial giants who had briefly joined the regular fax manufacturers (notably Alden, Finch, and Radio Inventions) in the design and building of experimental broadcast fax equipment during the late 1930s and 1940s. Except for some newspicture receivers it built for the Associated Press a few years ago, that activity has been pretty much the extent of G. E.'s involvement with fax. Stewart-Warner, however, returned to fax manufacture in 1956, when it took out a license agreement with Hogan Laboratories and acquired the facsimile business of the Allen D. Cardwell Electronics Production Corporation. The Stewart-Warner *Datafax* equipment has ranked high in the document fax field for the past several years (Fig. 1.10). (For a brief period in the 1960s, the Datafax equipment was marketed also by the Dictaphone Corporation under the name *Dictafax*.)

The 1960s saw a sudden rebirth of interest in fax, due largely to its having gained access to the nationwide dial telephone network as a transmission medium. That access, and the introduction of new machines designed to take advantage of it made fax a viable alternative to voice and teletype communication within the business community. The Magnavox Company pioneered the commercial introduction of the desk-top, telephone-coupled "convenience" transceiver, which was to set the pace for fax's new thrust. (In the interest of historical completeness, it must be noted here that Finch Telecommunications had produced and marketed a relatively compact, desk-top business fax transceiver prior to World War II—but for use on leased private lines only. Similarly, the

---

**FAXpaper* is a registered trademark of Hogan Faximile.

HISTORY 17

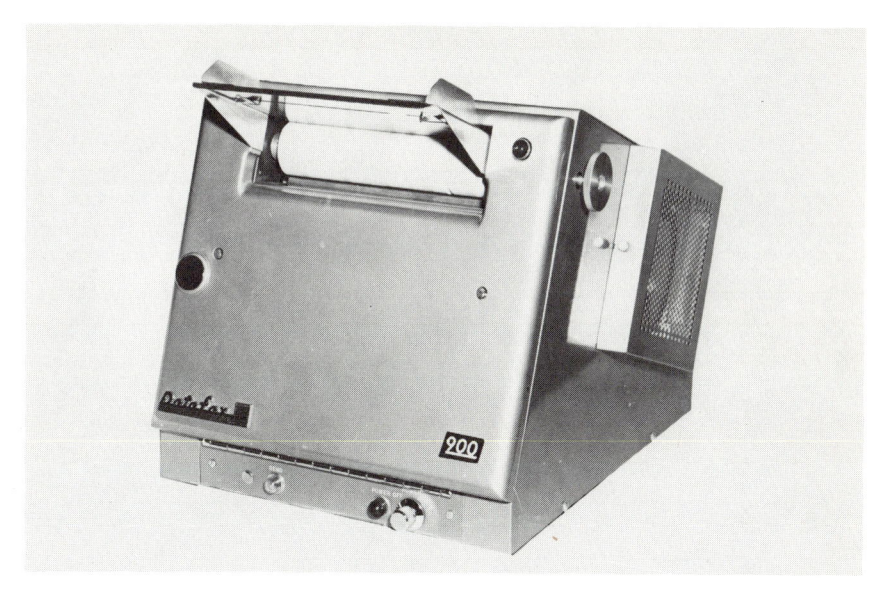

Fig. 1.10. An early Stewart-Warner *Datafax* transmitter, c. 1958. (Courtesy Stewart-Warner)

Western Union *Desk-Fax* system, introduced in 1948, featured an extremely compact transceiver, which was, however, designed to operate via a special, closed network.)

Through a marketing arrangement, the original Magnavox machine (the Magnafax 840) became the first of a series of *Xerox Telecopiers*,* the Xerox-registered name that was to become synonymous with this new breed of fax terminal (Fig. 1.11). Xerox subsequently began to manufacture and market its own machines, and Magnavox became a principal competitor with its virtually identical *Magnafax*** line. Among the others to join the phone-coupled transceiver market during this period were Stewart-Warner and a relatively new firm, Graphic Sciences, Incorporated.

Late in 1972, to the surprise of many, Magnavox decided to quit the fax business, and, in June of the following year, sold that segment of its product line to the 3M Company. The latter had already been in the fax business for about a year, marketing a Japanese-built transceiver.

Meanwhile, a handful of other American firms, large and small, had made brief appearances on the fax scene with systems that had definite potential but were somehow unable to make the grade. A. B. Dick, for example, had

---

*Telecopier is a registered trademark of the Xerox Corp.
**Magnafax is now a registered trademark of the 3M Co.

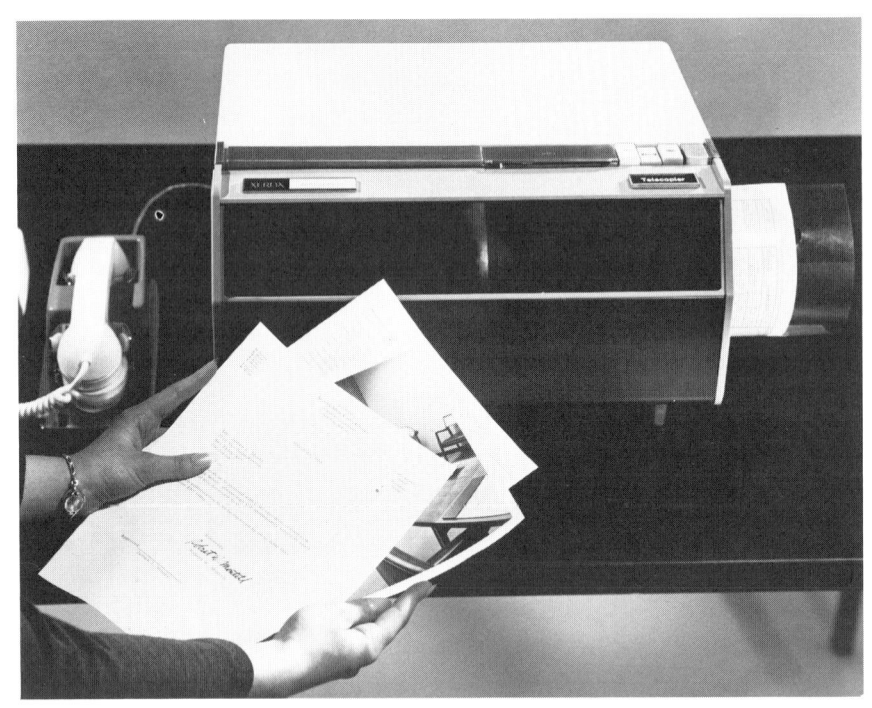

Fig. 1.11. The original Xerox *Telecopier*,* produced by the Magnavox Company, c. 1966. (Courtesy Xerox Corp.)

briefly marketed its highly advanced, all-electronic *Videograph* system for use on broadband circuits (to compete with Xerox's similar *LDX* system). A firm named Shintron in Massachusetts had brought out a very compact phone-coupled transceiver called the *Qix*, and a small New Jersey firm, Graphic Transmission Systems, had been formed just to produce and market its *Bandcom* system, which featured a very basic bandwidth compression scheme to permit faster transmission over phone lines. The latter was taken over by EG&G Incorporated (Bedord, Mass.) and subsequently dissolved. EG&G, however, has since undertaken a few specialized fax ventures of its own, notably its current involvement with the UPI's *Unifax II* electrostatic newspicture system.

Another comparatively large firm (already briefly mentioned) in and out of the fax business during the 1960s is the Fairchild Camera & Instrument Corporation, which introduced the first fully transistorized fax system (*Scan-a-Fax*) about 1963, and subsequently sold it to a small new company, International Scanatron. A new line of fax terminals introduced by Scanatron embodied some

---

*Telecopier is a registered trademark of the Xerox Corp.

HISTORY

of the Fairchild innovations along with a 2:1 bandwidth compression scheme invented by Scanatron co-founder John H. Clark. In 1972, Scanatron became Victor Graphic Systems, a subsidiary of Victor Comptometer, which had already been marketing a handwriting machine in competition with Telautograph. In September, 1976, the entire Victor Graphics line of graphic communication terminals was acquired by a newly formed Chicago company, Infolink Incorporated.

Two of the most recent additions to the list of companies in the convenience fax business are Exxon (yes, the oil people!), and Burroughs Corporation. Early in 1973, Exxon Enterprises Incorporated, a venture capital operation within the Exxon organization, announced development of a new, low-cost fax transceiver, and subsequently established Qwip Systems Incorporated to produce and market it. The *QWIP** machine's chief claim to fame is its price breakthrough: it rents for $29 a month per machine as compared with approximately double that amount for the typical phone-coupled transceiver.

Burroughs Corporation annexed Graphic Sciences, Incorporated as a subsidiary in February, 1975. GSI's *dex*** line of machines currently holds second place in the total U.S. domestic fax terminal market, trailing Xerox and just a pace ahead of 3M.

But, at this writing, the very latest addition to the roster of names is Panafax Corporation, a new American-based company created as a joint venture of the Matsushita Electric Company of Japan (maker of *Panasonic* products) and Visual Sciences, Incorporated, a New York engineering firm. Both firms were previously co-involved in 3M's entry into the fax field. The *Panafax* name will adorn a complete line of analog and digital fax terminals.

One of the more recent newcomers to fax in the specialized equipment department is the Harris Corporation—formerly Harris Intertype—whose Electronic Systems Division—formerly Radiation, Inc.—of Melbourne, Florida, produces the *Laserfax* line of analog fax terminals. *Laserfax* is the commercial counterpart of *Laserphoto*, the laser-facsimile system developed at the Massachusetts Institute of Technology and manufactured by Harris for the AP *Wirephoto* newspicture network.

Two other large American firms that have been rumored as being potential new entrants into the fax field are IBM and Pitney-Bowes. (The latter is, in fact, already involved in an "electronic mail" project for the U.S. Postal Service.) At this writing, however, there are no details as to the plans of either company with regard to a specific commercial product.

The fax business is by no means monopolized by U.S.-headquartered companies. For example, just about any large Japanese electronics firm that comes to mind is likely to be involved in some aspect of facsimile technology. The one per-

QWIP is a registered trademark of Exxon Corp.
**dex* is a registered trademark of Graphic Sciences, Inc.

haps best known to the American market in this regard is Matsushita. But Fujitsu, Mitsubishi, Nippon Electric, Toshiba, and Hitachi also manufacture fax equipment, as do a number of companies with less familiar names (e.g., Oki, Juki, Japan Radio, Nagano, Taiyo Musen.) In Europe, at least three countries— England, France, and West Germany—each has two or more native corporations actively producing fax systems. The European fax system names best known to Americans are Hell, Muirhead, Plessey, and Siemens.

CONTINUING EVOLUTION

The next generation of fax machines—after the desk-top transceiver—began to emerge about 1970 in the form of sophisticated terminals using computer technology to improve transmission efficiency. Addressograph-Multigraph had initially set the scene with the announcement of its highly advanced *Telikon* system—which, however, was never to get beyond the prototype stage. The first such devices to make it to the marketplace were those developed by two relatively small electronic firms, Dacom Incorporated and ComputerPix (later Comfax). After a couple of false starts, the *Dacom* and related *Rapifax* transceivers became available in 1974. For a brief period, Savin Business Machines and the Columbia Broadcasting System were also involved in that venture, along with Ricoh, Ltd. of Japan (which now owns and controls the Dacom/ Rapifax operation). Meanwhile, the Comfax-developed machine had been launched commercially as the EAI *FAX I* system, produced and marketed by Electronic Associates Incorporated.

Magnavox had developed a similar "data compression" system before deciding to get out of the fax business, and nearly all of the relatively large companies currently in the business have talked about introducing such systems. (At this writing, 3M has, in fact, already unveiled its new "Express 9600" system, Stewart-Warner, and Burroughs/GSI have systems nearly ready for introduction, England's Muirhead has already completed development of one for the British Post Office, and at least four Japanese firms are preparing such systems for marketing.)

Among the very latest generations of fax systems just beginning to emerge in the mid-1970s are those that combine improved transmission efficiency with laser scanning (or recording) and those in which the receive machine can be programmed to function as either a fax or a high-speed teleprinter terminal. As of this writing, the availability of combined data compression/laser terminals is pretty much confined to the periodical publishing sphere for transmission of page masters to satellite printing plants. Before very long, however, the wire-dispatching of newspictures is expected to be streamlined by the adoption of these combined technologies.

HISTORY

Combined fax-teleprinter terminals are, at this writing, scarcely beyond the prototype stage. Their introduction, however, is indicative of a trend toward the marriage of graphic and character transmission systems—the first step toward a feasible system of "electronic mail."

FOOTNOTES

1. Not to be confused with the noted Scottish philosopher-psychologist, with whom inventor Bain was contemporary.
2. Korn died in Jersey City, N.J., in 1945.
3. For a comprehensive account of Ives' work in picture transmission, see Ives, Horton, et al., "The Transmission of Pictures Over Telephone Lines," *The Bell System Technical Journal*, April 1925, pp. 187–214.
4. See *Proceedings of the IRE*, March 1956, p. 296.
5. Fulton was a Briton. Subsequently he moved his fax business to New York City.
6. Alden Electronic & Impulse Recording Equipment Co., Westborough, Mass.

2
FAX TODAY

There is virtually no limit to the variety of applications to which facsimile uniquely lends itself as a communications medium. In the spectrum of communications techniques, it bridges the gap between mail and so-called "data" transmission—or, at a higher level of graphic complexity, between mail and closed-circuit TV.

When the day or more it takes for a document to be delivered by mail is unacceptable, fax stands ready to transmit it in a matter of minutes—or seconds. It costs more, but the economic difference has been gradually narrowing.

As for its speedier and more sophisticated competitors, fax is preferable to data transmission from the standpoints of error rate and required operator training, and to closed-circuit TV from the economic standpoint. (This does not necessarily include slow-scan TV systems, or limited-range systems intended for referencing remotely stored documents.) Moreover, fax has a decided edge on pure digital data (nonfacsimile) techniques when it comes to the faithful—and economic—reproduction of complex graphics such as photographs, engineering drawings, signatures, and fingerprints.

Some current fax applications are briefly examined in the paragraphs that follow.

CURRENT APPLICATIONS
Newspictures

Historically, the distribution of newspictures ranks as one of the earliest fax applications (see Chapter 1). It is also probably the only one that has assumed an independent identity, namely, as *telephotography*,[1] the generic name by which the newspicture process is known technically in the United States (*phototelegraphy* in Europe). But while telephotography is often spoken of as a technique distinct from facsimile, it is in fact merely a branch of the fax art.

Through a complex communications network consisting of perhaps 40,000 miles of wire circuits, the two big American wire services—the Associated Press

FAX TODAY

(AP) and United Press International (UPI)—furnish member newspapers throughout the United States with about 250 fax-transmitted news photos daily. The AP alone has more than 1000 recorders installed and in regular use.

In addition to wire circuits, both services exchange pictures with the rest of the world via transoceanic cable and radio facilities. The public telephone network is also used on occasion for special transmissions (or retransmissions) to individual papers.

A typical newspaper installation consists of one photographic recorder for each of the wire services to which the paper subscribes (typically, two machines— one for AP and one for UPI). These are usually located in an area convenient to the darkroom in the paper's picture section. Larger papers may have two machines per service to ensure that there is always one ready to record an incoming picture.

Pictures to be transmitted are usually briefly described over a loudspeaker via an auxiliary voice link with the wire service control desk. The loudspeaker may be placed near the picture editor's desk to enable him to decide in advance whether or not to press the button that starts the recorder (or, with the older photographic process machines, whether to bother loading the recorder with a sheet of sensitized paper). Pictures are transmitted complete with captions.

Typically, although the paper may record and process 80 to 100 pictures a day, only a small fraction of these will actually be used. Still, roughly a third of all the pictures used in one edition of a large metropolitan daily are wire service pictures. Certain periods of the day are set aside for limited-range transmission of pictures that are of only local or regional interest.

Direct fax recorders, which require no chemical processing, have come into fairly wide use for the reception of newspictures because they eliminate the photo processing step, together with its material costs. The received picture emerges completely processed and ready to use. Both the AP and UPI are, phasing out the older photo-process machines in favor of direct recording types. The latest-generation AP machine uses laser technology to record on a special "dry silver" paper. The latent image is developed (automatically within the machine) by application of heat. UPI has introduced its new electrostatic machine, in which image development is by application of "toner"—a kind of electrostatic ink. It is a process similar to that used in some office copiers.

Within the next few years, it is expected, newspicture systems will be further modernized to permit computer storage and automatic cropping and enhancement of pictures. Realization of these and other "exotic" capabilities will require embodiment of computer and cathode ray display technologies within the newspicture receiving terminal. Allowances for eventual adoption of these technologies, along with higher transmission speeds, have already been made to some extent in the new equipment currently being introduced.

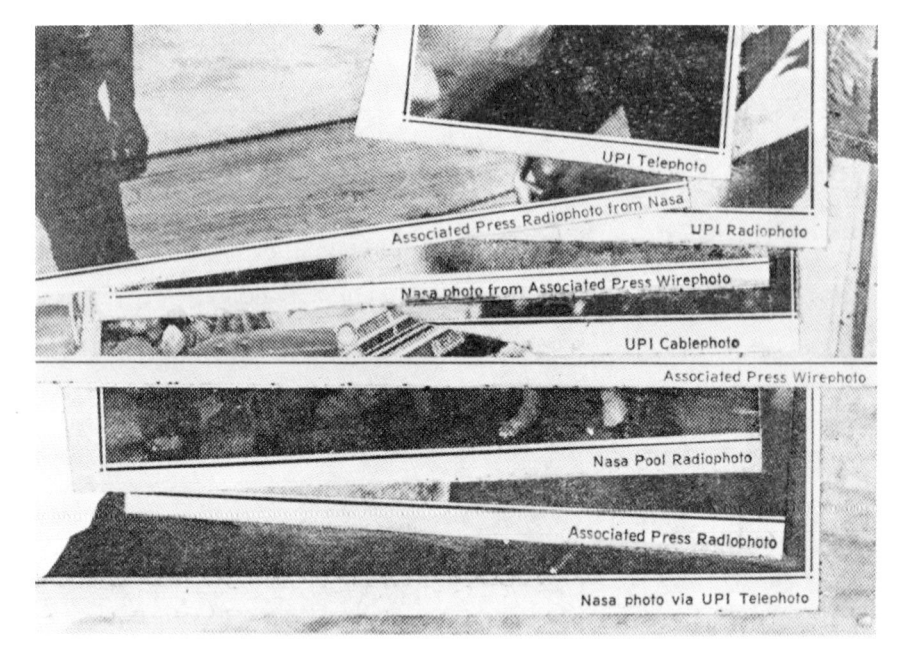

Fig. 2.1. The variety of credit lines for fax-transmitted pictures appearing in a single issue of a major daily newspaper should dispel any doubts as to the influence of facsimile in the field of journalism. (Source: New York *Daily News*)

Weather Forecasting

Radio weather map service to ships at sea via facsimile was initiated experimentally in this country in 1930. From this important beginning evolved today's worldwide facsimile weather network.

Within the continental United States there are three principal facsimile weather map networks: the National Network (largest of the three; see Fig. 2.2); the National & Aviation Meteorological Network (NAMFAX), which provides both national (NAFAX) and upper atmosphere weather data for high altitude aircraft operations; and the Forecast Office Network (FOFAX), which distributes supplemental data from a variety of sources, including earth-orbiting weather satellites. This composite of networks ties in, via shortwave radio, with ships at sea and with similar networks throughout the world.

At the hub of the three U.S. national fax networks is the National Meteorological Center (NMC) located just outside Washington, D.C. Into the NMC pours a daily volume of some 30,000 encoded reports representing raw data gathered by a vast complex of observation stations. The "stations" range

FAX TODAY

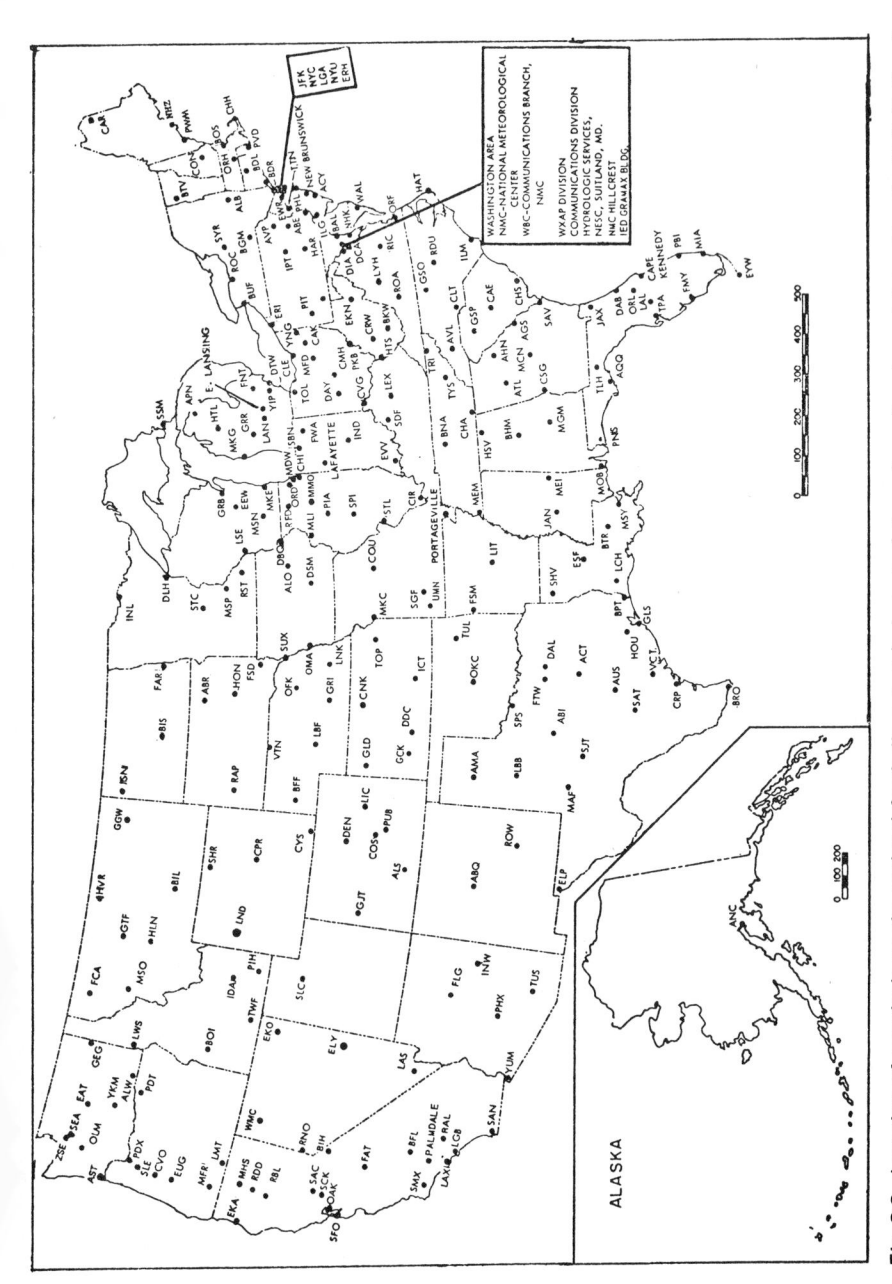

Fig. 2.2. Location of terminals on the national facsimile network, largest of three fax networks maintained by the U.S. National Weather Service. (Courtesy National Weather Service)

from fixed land installations (the source of most of the reports) to ships at sea, commercial aircraft in flight, and satellites.

The data collected by the NMC is methodically decoded, analyzed, and computer-processed for use in chart (weather map) production. Until recently, charts were drafted manually or by mechanical plotters for input into the networks via more or less conventional fax scanners. Now they are largely computer-produced (see Fig. 2.3). For the most part, there is no paper original; the computer outputs its automatically merged data directly onto the various networks, the signal having been formatted for compatibility with existing fax recorders in the field.

Principal recipients of the fax transmissions are local Weather Service offices and Forecast Centers all over the United States. In addition, certain maps and satellite pictures are dispatched by HF radio facsimile, and by various special circuits, to ships at sea and to foreign countries. At the Forecast Centers, the facsimile maps are supplemented with additional data, and forecasts are prepared

Fig. 2.3. Weather chart transmission room at the National Meteorological Center in Suitland, Maryland. From this room, the output of a chart-generating computer goes directly on-line to fax receivers all over the country. The recorders at the lower right monitor the outgoing signal. (Courtesy National Weather Service)

for the various news media, for airports (and planes in flight), and for ocean-going vessels.

Near real-time distribution of infrared and visual cloud cover pictures from geostationary meteorological satellites is accomplished by a special system, designated GOESFAX, in which the received satellite signals are computer-processed (for gridding and superimposing of national boundaries, among other things), and transmitted via specially conditioned voice-grade circuits to the Forecast Centers. The pictures are received on high-resolution photo-facsimile recorders. As an additional service to TV stations, short motion picture film loops are prepared from a succession of fax-recorded cloud cover pictures.

In yet another application of fax to the communication of weather information, it is possible for a fax receiver location to dial a special number and receive, by telephone, a paper reproduction of a weather radar display for a given area.

An exact count of the number of fax terminals in use within and around the United States (including ships at sea) for weather reporting purposes is difficult to obtain, but it can be reasonably estimated that there are a few thousand. The National Network alone serves more than 1000 locations.

Supplementing the work of the NMC is the Combined National Weather Service and World Meteorological Organization's Regional Center for Tropical Meteorology in Miami. Through its tropical regional analyses (TROPRAN) fax circuit, it provides tropical weather data for distribution to forecast offices in the southern states and HF radio broadcasts of appropriate fax charts to the Caribbean and South America.

Means of speeding the distribution of weather graphics and of cutting transmission costs are constantly being explored by the U.S. National Weather Service. As early as 1969, bandwidth compression techniques were adopted to double the speed of map transmission via ordinary telephone lines on the FOFAX network. More recently, coded digital techniques have been used to gain even more speed. In addition, when the NAMFAX network went into operation in 1972, it introduced an automatic reject capability wherein only the maps that satisfy a given station's needs are received. This latter innovation has made it possible to cut out some of the redundancy in transmission circuit mileage.

Commercial

Facsimile has established itself in a variety of commercial applications, its capacity for improving communication efficiency having been demonstrated in at least three general areas: (1) the shipping of goods, (2) the expediting of customer orders, and (3) the expediting of monetary transactions.

In the shipping of goods, fax is used effectively to transmit waybills, bills of lading, trucking manifests and the like from the point of origin to freight destinations along extensive rail, air, and highway shipping routes, or to central control locations.

At least one major freight-carrying airline fax-dispatches all of its air freight waybills from originating and receiving points within a limited geographic area to a main operations center, thus making it easier for shippers to track down cargoes. One advantage of this procedure is that the operations center has an exact copy on hand, so that it can not only answer shippers' questions but also make additions or corrections when necessary. Another is that no special training (as would be required of a teletype operator) is needed to operate the fax equipment. In between waybills, the fax transmitters at the air terminals can be used to send periodic reports of operations to the center.

Similarly, railroads have found fax useful for the dispatching of "train consists"—lists of rolling stock constituting the various trains, formerly sent by mail or teletype. When mail was used, shipments frequently arrived ahead of the

Fig. 2.4. Bank of high-speed fax recorders at a rail freight control center. Waybills are dispatched here via microwave radio from key points along a 10,000-mile rail network. (Courtesy Xerox Corporation)

FAX TODAY								29

Fig. 2.5. Customer orders are received via fax at the main office of a big steel firm from scattered district offices. (Courtesy Bethlehem Steel)

"consists," resulting in delay. And while teletype is fast and relatively inexpensive, fax eliminates the need for data conversion and thereby minimizes the chance for error in conversion and transmission. One U.S. railroad operates a waybill-dispatching fax system covering over 10,000 rail miles, and another uses a number of special high-speed terminals (ten documents a minute) for the same purpose in a private microwave network covering 900 miles (see Fig. 2.4).

As for the expediting of customer orders, a number of manufacturing and service firms whose operations are by nature decentralized have found fax systems very useful. One major eastern steel firm, for example, maintains 140 fax terminals to link its 72 sales offices and plant locations to its general offices (see Fig. 2.5). Orders on the customers' own letterheads are fax-dispatched by the sales offices to the main office, along with forms from the sales offices containing pertinent supplementary information. Within $4\frac{1}{2}$ minutes of insertion of an order into the fax scanner at the sales office, processing of the order can begin at the main office. Errors are virtually eliminated.

The system also permits more effective communication of inquiries and order changes, which formerly consumed the time of two people engaging in a long-distance phone conversation in which the risk of errors ran comparatively high.

This system is more or less typical of similar ones in use in other industries, but fax is also finding considerable use in the expediting of orders for services, as opposed to commodities. Utilities, in particular, have discovered that in some ways it is more practical to relay customer service orders by fax than by the more conventional method of teletyping them to the various departments concerned. For one thing it eliminates the need for trained teletype operators, and for another it minimizes errors that interfere with the satisfactory completion of service obligations.

Legend has it that as far back as the late 1920s, a million-dollar transatlantic business deal was saved by the successful transmission of some vital signatures from New York to London via radiofacsimile. Since then the use of facsimile in the expediting of monetary transactions has become almost commonplace.

A number of banks use fax regularly today to verify signatures on withdrawal slips or other transaction forms where comparatively large sums of money are involved (see Fig. 2.6). Verification can be made while the customer waits,

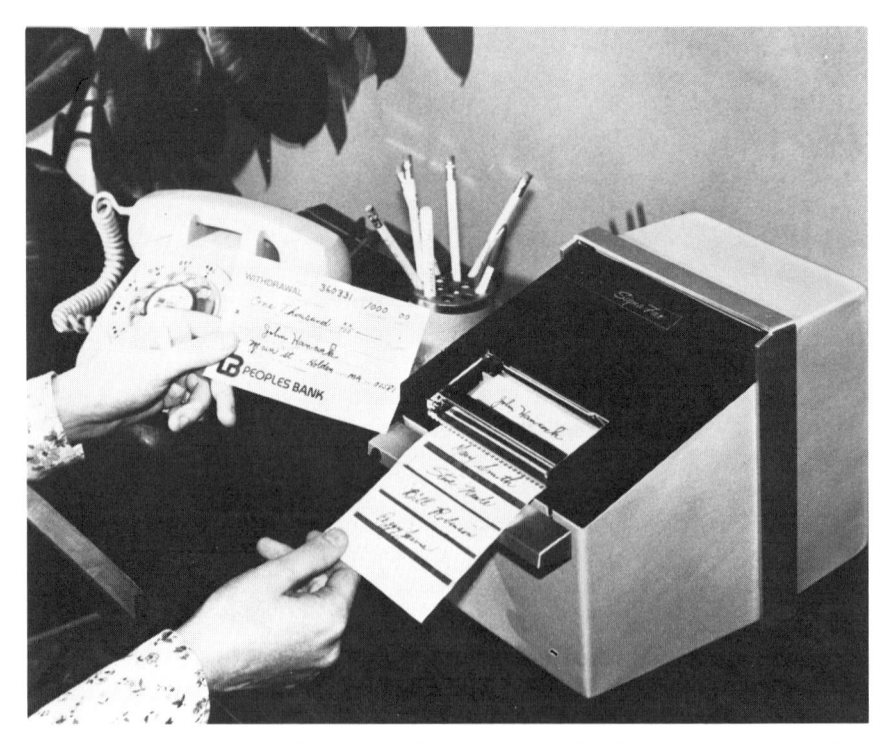

Fig. 2.6. Fax receiver specially designed for signature verification at branch banks. In a matter of seconds after initiating a request, the branch teller receives an exact copy of the customer's signature, as centrally filed at the main bank. (Courtesy Alden Research)

even though the depositors' signature file may be located several miles away. The special fax equipment developed for this purpose is sold commercially.

Fax lends itself to a variety of other banking applications as well. A Chicago bank, for example, uses it to dispatch stop-payment notices on checks from the bookkeeping office where the notices are received to the main office in another part of the city. In New York one of the nation's large banks maintains a network of 14 digital data compression fax terminals (capable of sending a document in a matter of seconds over ordinary telephone lines) for the intracity dispatching of securities purchases, money transfer cables received from overseas, and foreign exchange buy/sell orders. Much of this communication had previously been done by voice over the telephone, with the risk of introducing very costly errors.

Publishing

The publishing industry uses facsimile in essentially three ways: (1) to expedite graphic communication between editorial offices and printing facilities, (2) to dispatch news copy from satellite offices or bureaus to a paper's main news room, and (3) to eliminate duplication of typesetting by separate printing facilities.

A number of periodical publishers have found fax well suited for the interchange of material between geographically separated departments. Typically, manuscripts ready to be set in type are transmitted from the editorial offices to the printer, who in turn wires back proofs ("galleys," page proofs, etc.) of the typeset material for proofreading. If there are corrections or revisions to be made, and if the fax system is capable of producing a readable second-generation facsimile of its own output, the marked proofs can then be fax-transmitted back to the printer.

By using fax rather than mail, a publisher can save several days of production time—a saving than can prove vital when last-minute revisions have to be made in the face of a fast-approaching deadline. Where the only alternative is to put a person in a car or on a plane to deliver the material, a good case can usually be made for cost savings through the adoption of fax, even if the system is used only sporadically.

Besides connecting editorial offices with printing facilities, fax is also used to some extent to link a newspaper's main news room with its satellite offices. The *Wall Street Journal*, for example, uses fax to dispatch copy from its Washington Bureau to the news room in New York. The advent of portable fax machines with built-in telephone handset couplers has enabled reporters at remote news sites to file their hand-written or rough-typed stories by phone with the home office.

Fig. 2.7. Fax transmitter for full-size newspaper pages. A companion recorder at a satellite printing plant reproduces the transmitted page proof as a photographic printing master. (Courtesy Muirhead)

One of the more specialized fax applications in the publishing industry—but one that appears to be catching on quite rapidly—allows large periodical publishers to operate regional printing plants. Again, the *Wall Street Journal* is a case in point. It maintains regional plants for its eastern and Pacific Coast editions. To avoid having to duplicate all of the various steps associated with letterpress printing, the main and branch plants in each region are linked by a special fax system that permits final page proofs from the main plant's presses to be transmitted to the branch plant and recorded as photographic masters for a less involved printing process (see Fig. 2.7). That way, typesetting is necessary at only one plant, and page proofs from that plant can be sent simultaneously to the branch plants. The *Journal's* two fax-linked West Coast plants are separated by about 400 miles, while the 3-point East Coast network consists of links approximately 150 and 1100 air miles in length. The longest of these is the newest—from Chicopee, Massachusetts (the paper's principal plant) to Orlando, Florida—and is unique in that it interconnects via a channel on Western

FAX TODAY   33

Union's WESTAR I communications satellite in geostationary orbit 22,300 miles above the equator. In fact, the estimate of 1100 air miles given above for the Chicopee-Orlando link needs qualifying: the fax signals actually travel some 50,000 miles between terminals. The shorter links in the system are terrestrial microwave.

A Japanese newspaper, *Asahi Shimbun*, pioneered the use of such systems, having established the first one in 1959, tying its main Tokyo plant to a branch plant in Sapporo 500 miles away. The equipment was developed and produced by Muirhead of England. Since then other Asian papers and several in Europe and other parts of the world have adopted the system, and at least four other fax manufacturers have developed (and are presently producing) equipment to

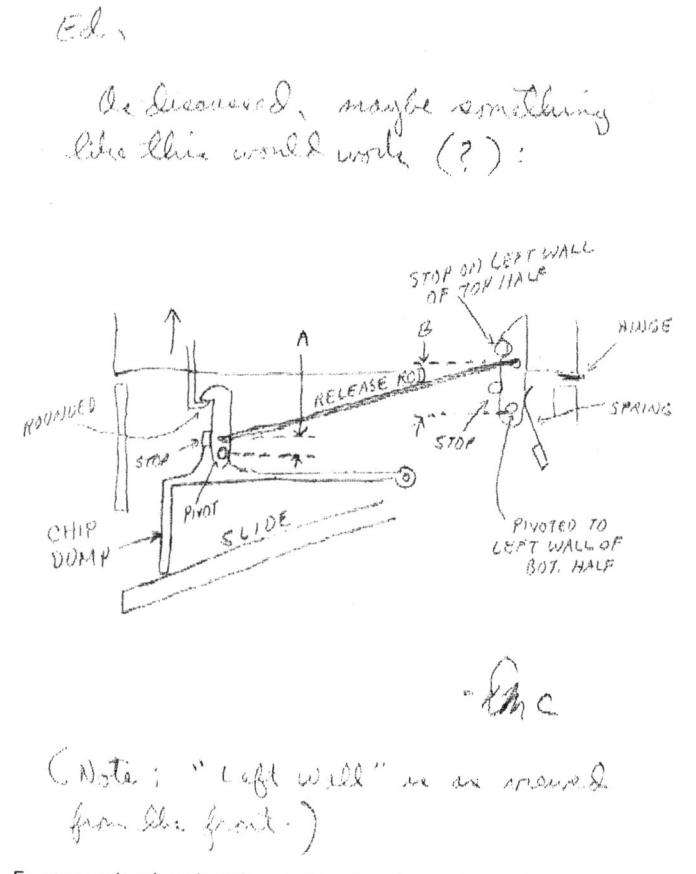

Fig. 2.8. Fax-transmitted engineering sketch. Imagine trying to describe even this relatively simple sketch orally by phone.

implement it. Digital data compression (redundancy reduction) devices have recently been introduced to make the normally broad bandwidth, high-resolution transmissions more economical over long distances.

Among the major American periodicals besides the *Wall Street Journal* currently being published by facsimile (at least in part) are *Business Week*, the *Christian Science Monitor*, and two St. Louis papers: the *Post-Dispatch* and the *Globe-Democrat*. The *New York Times*, an early experimenter with facsimile, plans to add its prestigious name to that list sometime in 1978.

Engineering and Manufacturing

Liaison between a manufacturer's engineering and production facilities, or between separated engineering facilities working on different aspects of the same project, has been greatly enhanced by the use of facsimile. An aircraft concern in California, for example, exchanges engineering data via fax with subcontractors in two southern states and in Canada. The graphic data transmitted, such as engineering sketches, columns of figures, or curves representing test data, supplement telephone technical discussions between engineers at the separated facilities. The result is a far more complete exchange of complex information than would otherwise be possible.

Once a product is in production, or a plant is being tooled for production, fax can also be quite useful for dispatching engineering change orders from the firm's engineering department to the distant plant—particularly if the product is very complex.

Let us assume that the engineering department has discovered a flaw in the design of a new electronic device that is being mass-produced on a rush basis in a plant 1000 miles away. An engineering change has been worked out that will effectively eliminate the flaw. Meanwhile, the units are being turned out at the rate of ten an hour, and for each unit already produced, the change will cost about $50 (in manpower and wasted components).

The traditional dilemma is, do you stop production and leave an entire production unit idle for the time it takes to get the change order to the plant, or do you keep producing and spend several thousand additional dollars in modification costs? A fax system that can dispatch the necessary paperwork to the plant in a matter of minutes offers a new option.

Whether you choose that option will depend, of course, on the operating cost of the fax system as compared with the cost of occasional production delays that might result without it.

In the engineering realm, fax also comes into play in the drilling of offshore oil wells. A drilling foreman on a platform some 75 miles from shore sends drilling logs by radiofacsimile to a land-based geologist to determine whether to begin installing a casing in a well. Since the offshore rig, with its 45-man crew, costs the oil company an estimated $25,000 a day to operate, every minute saved is worth about $20. And a single oil company can have perhaps a dozen drilling platforms operating simultaneously.

Law Enforcement

Facsimile has proved quite effective for the exchange of fingerprints and "mug shots" among law enforcement agencies.

A typical system works something like this: a criminal suspect, upon being booked at a local precinct house, is photographed with a Polaroid camera and fingerprinted. Within minutes, the combined mug shot and fingerprint can be fax-dispatched to a central headquarters, where, with the aid of efficient file retrieval techniques, a quick check can be made to see whether the suspect has a record. By speeding the process, facsimile helps reduce detention time and expedites the administration of justice.

In Philadelphia some 24 Harris Laserfax terminals form the nucleus of a law enforcement network interconnecting ten police headquarters, plus the district attorney's office at City Hall. The system is capable of processing a document a minute at a scan resolution of 285 lines per inch. The New York State Police operate a statewide fax network whose more than 60 fax terminals process an average of 20,000 fingerprint records a month.

In England and Germany, police departments have experimented with a system in which shots of suspects are radioed from headquarters directly to patrol cars. Similar mobile fax systems experimented with in this country some 25 years ago were found impractical because of the state of the technology at that time. However, since then at least two fax equipment manufacturers, including Muirhead, have marketed commercial mobile fax systems for that purpose. Figure 2.9 illustrates how such systems might be used to help locate missing persons.

In addition, there are stationary facsimile systems being offered specifically for law enforcement applications— Litton-Datalog's *"Police-Fax,"** for example, and Dacom's *"Identfax."** These systems are designed to afford the comparatively high resolution needed for the faithful reproduction of fingerprints— about 200 scan lines to the inch—and therefore lend themselves to other applications where high-definition reproduction is required.

Messages

The term "message facsimile" probably originated with fax's adoption by the Western Union Telegraph Company in the 1930s as an automatic message-

---

**Police-Fax* is a registered trademark of Litton Industries' Datalog Division, and *Identfax* is a registered trademark of Dacom, Inc.

MISSING PERSON FAX DISPATCH

DATED February 20

NAME Ned Nagitsock

LAST SEEN (DATE) February 13
IN VICINITY OF Orangeburg & Blue Hill Roads
LAST SEEN WEARING white shirt, brown trousers

BUILD thin HAIR blonde
EYES blue HEIGHT 5' 8"
AGE 12 SEX M RACE cauc.

MISSING PERSON

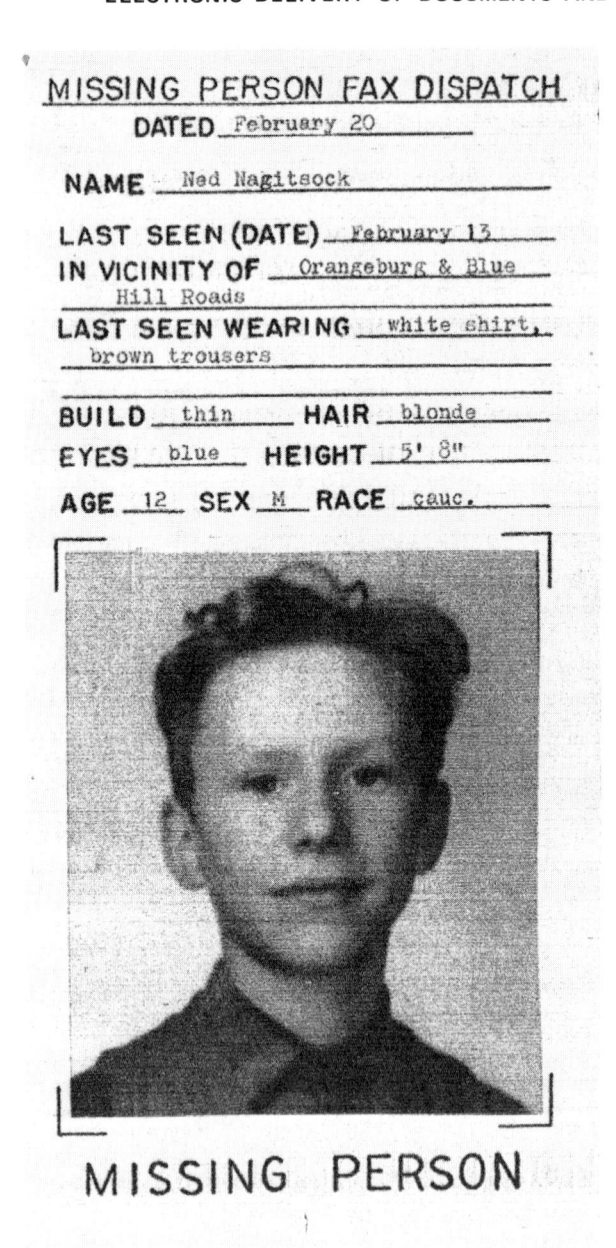

Fig. 2.9. A police application of facsimile. Reports such as this can be dispatched via radio to hundreds of patrol cars simultaneously and received automatically (with about the same quality as you see here) on compact, dashboard-mounted recorders. (The report illustrated is, of course, fictitious.)

sending medium. The object was to permit hand-written "messages" to be sent to the nearest WU office via scanners (some of then coin-operated) in bus terminals, lobbies, and other public places. Upon receipt at the telegraph office, the message could then go out over the wire in the usual manner. The idea of exchanging urgent messages by facsimile subsequently caught on in the business community, and today it ranks as one of the principal uses of the medium.

To say that the sending of messages is a principal use of facsimile is perhaps a rather equivocal statement to make in view of the difficulty of drawing a clear distinction between simple messages and commercial or technical communications. "Will arrive Penn Station at 5:30" and "try substituting these values: . . ." are both messages, the basic distinction being that one is personal and the other technical.

The author's concept of a true message facsimile system, as the term is generally applied in business and industry today, is a system that is used for graphic communications *in general* and does not have any one specific function, like the dispatching of weather maps or customer orders. A great many decentralized businesses today maintain fax systems for the exchange of everything from budget reports to engineering sketches. The term "convenience fax" is sometimes applied to such systems, particularly if they are used only casually.

Western Union's original automatic message facsimile system enjoyed a measure of success before it was discontinued several years ago (see Fig. 2.10). One of the problems with it, according to G. H. Ridings, who helped develop it, was the vulnerability of slot-type transmitters to abuse of various sorts. Besides their convenience as receptacles for gum and candy wrappers, hairpins, and similar refuse, Ridings points out, the units "made it easy for hoodlums to shock our central office personnel by sending sketches and poetry of the type found on public washroom walls."[2]

Desk-Fax, a later Western Union development that made it possible for a businessperson to send short messages via a telegraph central office direct from his or her own desk, has been eminently successful, but it is evidently in the process of being "retired" by attrition. As of August 1970 there were nearly 30,000 of the compact Desk-Fax transceivers in operation throughout the United States. By January 1, 1977, that number had diminished to some 2000.

In Japan, a telegraph message system employing conventional fax scanners has been used by the Nippon Telegraph & Telephone Corporation for relaying telegrams between collecting offices and main stations.

Libraries

The idea of facilitating interlibrary loans by use of facsimile has great appeal. As of mid-1976, at least 10 such systems were in regular—and evidently successful—operation around the country, and about a half dozen more were planned.

Fig. 2.10. Receive and transmit units for an automatic facsimile telegraph system. (Courtesy Western Union)

The two largest existing systems are the 22-point Michigan State Library Facsimile Communications Network and the California State University and Colleges system, serving 17 member libraries throughout that state.

Unfortunately, this favorable picture is marred by the record of some 15 other such systems (some of them merely trials) that have failed, or proved only marginally successful.

One reason library fax systems—or telefacsimile systems, as library people prefer to call them—have disappointed is that a great deal of time must still be spent manually processing the requests at both ends of the system. In addition, where the average single request is for a number of pages (of a book, document, etc.), the problem of queuing is introduced, and a traffic jam ensues at the transmit station—especially where a straight voice-grade fax system is used and transmission is therefore relatively slow.

But probably the biggest obstacle to fax's adoption by libraries is economic. The University of California trial of March 1967 involving a high-speed fax system between two campus libraries 65 miles apart is a case in point. While the average turnaround time between a request and its fulfillment was reduced from six or seven days to about a day, the cost of each 10-page transaction at a volume of 500 transactions a month came to about $10. At 100 transactions a month, the cost per transaction would have been $40. A survey of users of the service indicated that the need for the transmitted material was not urgent enough to justify the high cost.

The situation apparently did not change appreciably in the ensuing eight years. Reporting on a 1975 survey of the use of fax in U.S. libraries, Hans Engelke of Western Michigan University notes that, in many of the experiments that have failed, the interlibrary loan traffic did not warrant such a system and therefore it could not be economically justified.[3] However, the same report mentions a couple of factors that could point to an eventual brightening of the picture for fax in the library environment. One is a rise in the cost of TWX teleprinter service, fax's principal rival in the expediting of the interlibrary loan process. The other is the obvious fact that the attractiveness of fax to the library community is more or less directly related to the speed advantage it can offer over competing media. In view of this and the technological advances that have recently occurred—and those yet to come (see below, Trends)—there would appear to be good prospects for the eventual economic justification of large-scale library telefacsimile networks.

Miscellaneous

Fax is also used in a variety of ways that are not sufficiently widespread to warrant general classification. Mention of a few of these applications will attest to the medium's versatility.

A big electric utility uses fax to relay trouble reports to its emergency service crews. After the original incoming report is written on a card by a telephone contact clerk, it is promptly inserted into the appropriate fax transmitter. The report reaches its destination in about the same time it would take to relay it by phone—but with less chance of error, and with the advantage that while it is being transmitted, the clerk is left free to handle other calls.

A commercial translation service in the nation's capital uses fax to improve turnaround time on client requests for translations of foreign correspondence, and a large chemical firm uses it to help gather, from its scattered branches, input for a monthly operating report to headquarters. A contracting firm gets payroll information via fax from job sites, enabling it to get pay checks out faster. Another actually faxes pay checks to its people working out of state. (The receiving machines at the work locations are loaded with pre-printed check forms for the purpose.) A noted political cartoonist uses it to dispatch his work from his home to the newspaper he works for 1200 miles away. Groups of scattered realtors use it to exchange descriptions and pictures of homes for sale for out-of-town prospective buyers, and some companies use it for routine direct-wire document communication among buildings of a plant complex. Still others use it to obtain—via telephone—copies of records stored in remote warehouses. Fax was reportedly used to dispatch the now-famous first Howard Hughes will—the one involving a Utah gas-station owner—to a former Hughes aide for verification.

A popular office machine called an electronic stencil cutter utilizes the fax principle to produce master stencils for inking-type duplicators (see Fig. 2.11). The machine is, in essence, a fax scanner and recorder, mechanically and electrically coupled in a single compact unit. The original item to be duplicated is scanned photoelectrically and reproduced on the special stencil paper by a kind of arcing process. Scanning and reproduction take place side by side. (It is often possible, incidentally, to produce masters for various duplicating processes on conventional fax recorders, in place of the usual "hard copy.")

Similar but more elaborate devices are used in the printing industry for the production of color separation plates. In fact, the plates for most of the color reproductions found in current periodicals are produced by the fax process on these so-called color scanners.

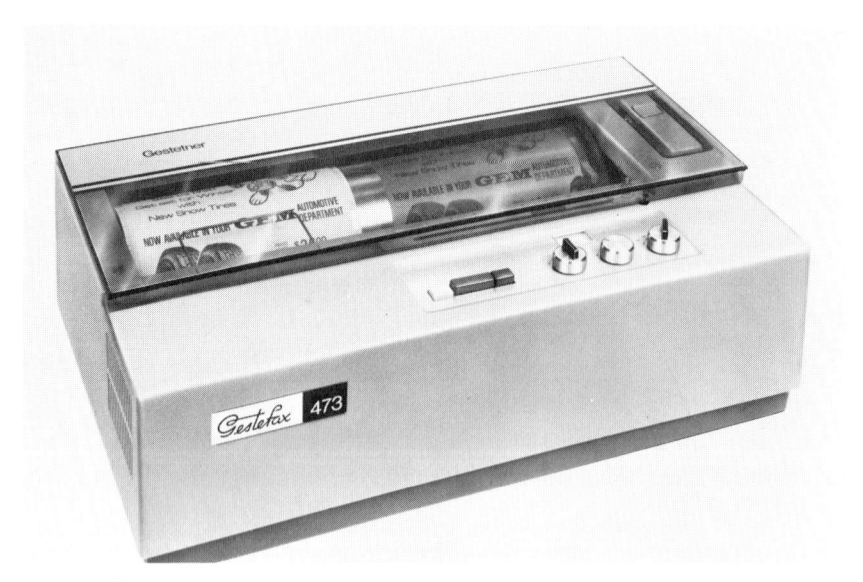

Fig. 2.11. Fax stencil maker. The scanner portion at the left "transmits" the original document to the recorder portion at the right, where it is reproduced as a duplication master. (Courtesy Gestetner Corp.)

There are also military applications. Polaroid photos of enemy positions, for example, can be radioed to command posts from fax transmitters mounted in reconnaissance vehicles. And, from central microfilm files in a protected area, drawings and parts lists can be remotely retrieved by fax, *as needed*, to facilitate the repair of aircraft at the fringes of a battle zone.

Fax has also been used to transmit X-ray photos from hospitals to analysis centers, to record the output of electron microscopes, to dispatch railroad passenger tickets from a main reservation office to local agencies, and to provide the crews of fishing boats with print-outs of fishing conditions and other pertinent information. Both technically and functionally, fax is a versatile tool whose potential for new and unusual applications remains practically limitless.

VIRTUES

Most of fax's advantages over competing media have already revealed themselves in the preceding pages. It might be useful, however, to recall each of these and examine it in more detail.

Faster Than Mail

One of the prime justifications for fax's existence is its ability to deliver a reasonable facsimile of a document or picture in minutes or seconds as compared with days. A quick glance down the list of specific applications described in the preceding section will reveal that better than 75 percent of them involve the need to expedite delivery of the items involved. It is hard to imagine newspictures or weather maps, for example, ever having been delivered by any means other than fax.

One might argue that if some kind of electrical transmission is inevitable, there are faster and possibly cheaper techniques than fax available. The argument has merit, but much depends on the nature of the material being sent—and that brings us to another of facsimile's prime attributes.

Input Flexibility

True, documents can be sent as binary-coded digital data (via keystroke-generated tape, for example) at the rate of more than 300 characters a second over ordinary telephone lines, which amounts to sending the contents of a typical one-page, single-spaced business letter in about four seconds. To send the same item, a fax system roughly comparable in cost and transmission bandwidth would require about *six minutes*.

So, for pure alphanumeric material we can say that, from the standpoint of speed, digital data techniques have an edge over fax. Let us be sure, though, to qualify this by pointing out that the speed advantage really applies only to machine-originated communication, e.g., telemetry or the output of a computer. Any time conversion to an encoded form is required prior to actual communication, the encoding time must be included in any speed comparison between fax and digital data techniques.

Human beings seldom, if ever, originate communication in coded form. Where an $8\frac{1}{2} \times 11$-inch page of information must be encoded on a manual keyboard for transmission, the time consumed is normally about twice that for sending the same material via one of the slower fax systems.

But it is in the area of graphic communication—the sending of maps, diagrams, photographs, etc.—that fax really proves its worth. Although digital data techniques have advanced to the point where they can be used to transmit simple graphics as well as alphanumerics with relative ease and economy, the general rule has been that, as the graphics increase in complexity, these techniques become sharply less economical and reliable. That is the reason fax has long been popular in the Far East. The pictorial nature of most Oriental written languages presents some obvious problems in the use of teleprinters for message transmission.

In the simplest digital data systems for the transmission of graphs, symbols must usually be rigidly standardized, limited in number, and inputted manually with a keyboard. More elaborate (and expensive) systems may employ Optical Character Recognition (OCR) or other highly sophisticated techniques to permit insertion of the original item directly into a scanner. But in the present state of the art, even these systems are limited in the variety of symbols they can recognize and accurately reproduce. (Redundancy reduction techniques, which are also usually digital, will be discussed later.)

By contrast, the same comparatively simple fax system that is used to exchange business correspondence and other types of alphanumeric material can, without alteration, faithfully reproduce a reasonably complex sketch, chart, map, or photograph as well. It is this intrinsic indifference to complexity or variety of input that, perhaps more than anything else, makes fax hard to beat as a graphic communications medium.

Error Immunity

There are two characteristics of the fax process that make it inherently less susceptible to error than competing digital communications media. One is that although it segments copy into lines in the process, it subsequently reconstructs its input as a unified whole; that is, as a camera or office copier would reproduce it, making no distinction between essential and nonessential details. And the

FAX TODAY

other, closely related to the first, is that it analyzes and reproduces its input on an element-by-element basis, and not as a sequence of codes representing the components of which the input consists. These characteristics generally apply whether the transmitted fax signal is analog or digital in nature (with the exception of systems in which "digitizing" is for the purpose of compressing the signal).

The distinction between elements and components is significant here. Component, in this instance, means a character (number or letter), which is, in turn, composed of a number of elements. In a conventional fax system having a scan density of 96 lines per inch, a $\frac{1}{8}$-inch-high letter A would be dissected by 12 scan lines, and would be represented in the transmitted signal by a total of about 22 elemental pulses (Fig. 2.12).

In a data system the same letter A would be translated into a binary code,

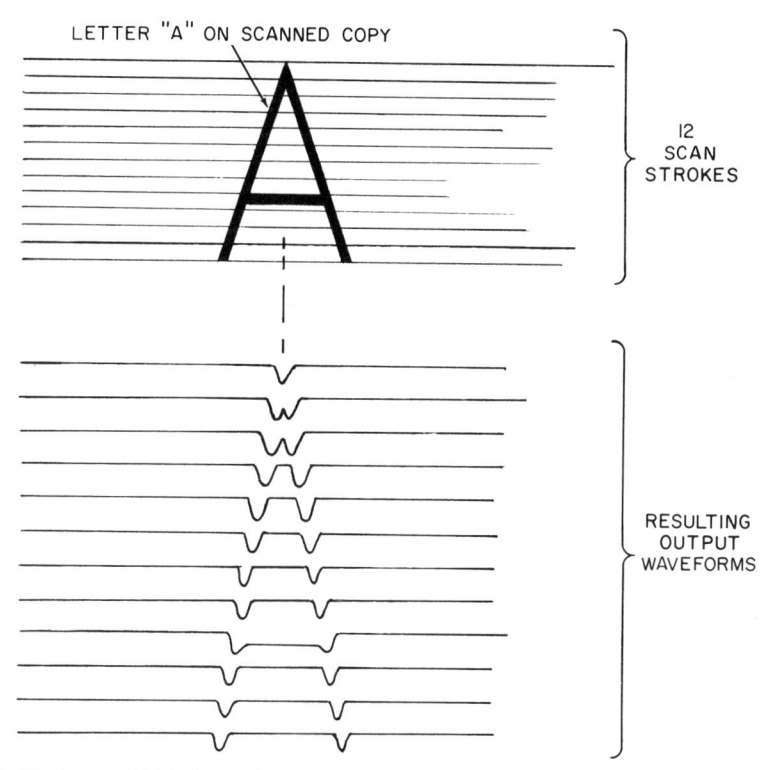

Fig. 2.12. Letter "A," dissected by 12 scan strokes, takes the form of some 22 electrical "blips" in the transmitted picture signal. The blips constituting a single character may be spread out over several seconds of time.

such as 1000001, which would be represented in the transmitted signal by only seven pulses (one for each of the *bits* in the code). If any of these binary pulses were lost in transmission, or if, as a result of external electrical interference, one or more pulses were incorrectly recognized, a different code might result and an entirely different character would be printed at the receiving end of the system.

However, if one, or even a few, of the 22 pulses representing the character in the *fax* signal were changed, the effect on the reproduced character would scarcely be detectable.

Although the data system has the capability of indicating when it has made an error, it usually cannot report the nature of the error. Thus, where errors are indicated, and they are significant, retransmission is generally necessary—a fact that must be considered in assessing the data system's speed advantage.

In a fax system, even when the transmitted signal is in digital form, with codes representing gray levels, as long as it is on an element-by-element basis, errors will not have a significant effect on output integrity. In fact, tests performed in connection with the trial of an all-digital newspaper page facsimile system, in February 1968, revealed that the effect of error rates as high as one in 100 may not even be perceptible in the output.[4]

No Special Training

There have been, and no doubt still are, facsimile machines that require a measure of skill to operate. The steady trend, however, has been toward simplicity of operation. Most of the fax terminals in use today can be operated by nonskilled personnel after only a few minutes of instruction.

This is an attribute that makes fax especially attractive in situations where the dispatching of high volumes of alphanumeric data is required, and where employee turnover makes it difficult to retain a staff of skilled operators of keyboard-type transmitting equipment. The problem of securing skilled personnel to perform comparatively menial tasks has become acute throughout industry. But at least in this one area—data communications—fax has proved a ready and relatively inexpensive solution.

Essentially Automatic

Fax's simplicity of operation is due largely to the fact that it is essentially an automatic process. Apart from the training required, operation of a keyboard-type data transmitter keeps an individual occupied throughout transmission (or throughout the preparation of a tape, which amounts to the same thing), whereas

FAX TODAY 45

operation of a fax transmitter requires only as much of the individual's time as is needed to insert the copy.

Many modern fax systems have been automated to the point where they can operate completely unattended. Recording paper is automatically fed from a roll long enough to produce a few hundred $8\frac{1}{2} \times 11$-inch copies, and an automatic feeding capability at the transmitter permits loading of a stack of documents for feeding into the scanner one at a time. (Figs. 2.13 and 2.14). In addition, most modern fax receivers feature automatic answer and "hang-up" capabilities.

Besides saving on labor, a fully automated system of this sort makes it possible to take advantage of the lower nighttime phone rates, to utilize leased lines more efficiently by keeping them in use at times when they would normally be idle, and to continue communicating with a branch office in a different time zone after one location or the other has closed down for the day.

Fig. 2.13. Recording paper fed from a continuous roll within the machine helps make possible the completely unattended operation of a fax receiver. (Courtesy 3M Co.)

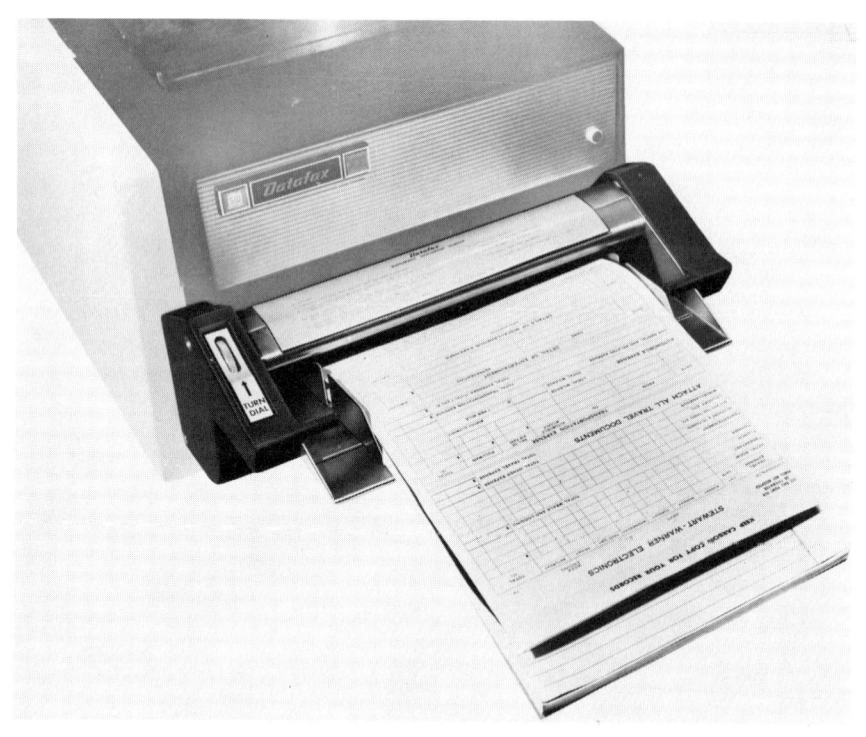

Fig. 2.14. An automatic feeding accessory attached to a fax scanner is usually all it takes to make a modern fax system completely automatic. (Courtesy Stewart-Warner)

Flexibility Through Trade-Offs

Facsimile is not a live medium as are voice communication and conventional TV. In other words, it is not committed to re-creating a sequence of changes with respect to a time frame. Like other so-called data communication systems, it can be operated as slowly as is tolerable or as rapidly as transmission and economic parameters will permit, so long as it preserves proper spatial relationships and reasonable fidelity of dark-light, or mark-space, transitions.

It is thus a comparatively flexible medium. If a fax user wants to utilize an existing phone line as the transmission medium, equipment can be obtained that is compatible with that choice. However, one must be willing to trade something off for the bandwidth economy gained. That "something" may be quality or speed (or both)—or, in the case of a sophisticated digital data compression system, it may be the extra dollars required for the special electronics to maximize transmission efficiency.

If, on the other hand, speed is desired, fax systems are available for a variety

FAX TODAY

of transmission speeds. Here again, some trading is necessary, and it may be either quality or bandwidth economy that is sacrificed. With sufficient funds to pay for the broad bandwidth required—or for the sophisticated electronics to make maximum use of available bandwidth—one can have a system that is both fast and capable of high resolution output.

Trade-offs will be discussed at greater length in subsequent chapters.

LIMITATIONS

So far in this chapter, fax has been represented as a find of paragon of perfection among communications media. Along with its many virtues, however, it has certain limitations. The important ones are examined in the next several paragraphs.

Redundancy

It is the nature of a conventional analog fax system to examine every square fraction of a millimeter of the physical item transmitted (with the possible exclusion of upper and lower margins), and to transmit an electrical analog of what it sees, even when it sees absolutely nothing. In other words, a large portion of the transmitted signal represents the blank space in margins and between characters, words, and lines of information.

In the transmission of a typical one-page business letter (about 250 words), for example, the proportion of the total signal representing the actual "marks" necessary to reconstruct the letter at the recorder is roughly five percent. (This is based on 9350 scan-inches for the entire $8\frac{1}{2} \times 11$-inch sheet, of which 1520 contain "pulse trains" whose "black" elements represent an estimated average of $\frac{10}{32}$ of each pulse train inch.) The proportion for a simple sketch would be even smaller.

It was pointed out that analog fax and voice communication differ basically with respect to the comparative significance of time as a transmission variable. However, from the standpoint of transmission efficiency, the two have much in common. Both demand considerably more channel space than they actually need for the amount of information they convey. In the information theory terminology, their signals are said to be *redundant*.

A number of techniques have been devised to reduce transmission redundancy or required bandwidth; but while these techniques improve efficiency from the standpoint of cutting transmission time or permitting use of lower-grade facilities, they usually impose a higher terminal cost. Thus, depending on the volume of messages transmitted, the use of these techniques can result in a net loss economically. Moreover, in most such systems, binary codes are substituted for all, or at least part, of the original analog signal, and the potential for error increases accordingly.

More will be said of redundancy reduction in subsequent sections.

Relative Slowness

Fax has capitalized on the fact that it is considerably faster and more direct than mail. But measured against our conditioned space-age expectations—particularly as they concern anything electronic—it often gives the impression of being a ponderously slow way of communicating electrically. The characteristic redundancy of its signaling process, and what this means in terms of bandwidth requirements, does, in fact, place fax among the more sluggish electrical communications media.

Its comparative slowness becomes apparent when we consider two hypothetical information retrieval systems in which the material transmitted is limited to the contents of a central store arranged for remote access by phone or by telegraphed retrieval codes. In one system, the central store consists of the actual documents containing the original data, an arrangement that makes facsimile the logical transmission medium. In the other system, the data has been extracted from the documents and stored in binary form on magnetic tape, which permits use of a straight digital transmission mode with high-speed teleprinters or CRT display units as the output hardware.

Both systems use regular telephone circuits as links between the inquiry stations and the central store. But in the fax system, the information contained on a single document page takes from three to six minutes to transmit, whereas in the digital system the same information is sent in a matter of seconds.

The above comparison assumes conventional analog, element-by-element fax transmission. The recent emergence of commercial digital data compression (redundancy reduction) fax terminals has altered this picture somewhat. The use of such terminals can reduce the fax transmission time to a minute or less per page via ordinary telephone circuits, thus narrowing the speed gap substantially.

Speed is, of course, only one consideration in choosing between fax and data systems. Other factors influencing the choice are the degree to which graphics are to be handled along with alphanumerics; the comparative terminal costs; and, in the case of a digital system, the costs associated with data conversion and storage.

Vulnerability to Transmission Impairments

When we hook an oscilloscope to the output of an audio system, we are likely to see a certain amount of "hash" and other imperfections that are not apparent to the ear. Likewise, transmission impairments that go essentially unnoticed in voice communication may have a drastic effect on the quality of a fax recording.

Much has been done in recent years to minimize the effects of transmission impairments on fax signals. FM has been adopted, for example, to reduce the effect of characteristic variations in transmission level in telephone circuits,[5] and *amplitude quantizing* techniques have been applied to permit reshaping of fax signals into discrete pulses that are unaffected by moderate amounts of "noise."

But in spite of these efforts, fax signals remain vulnerable to the effects of a variety of transmission phenomena, the more important of which will be explored in some detail in subsequent chapters.

Servicing Difficulties

Locating the source of trouble in an erratically operating facsimile system can be a problem, particularly where the system contains only two terminals. Typically, the terminals are miles apart, and the connecting link is controlled by someone other than the firm responsible for servicing the terminals. Relatively few of the troubles normally encountered in such a system can be predictably isolated to a given point.

Of course, this is not a unique problem; it is common to all electrical communications systems where two or more separated terminals are interconnected by wire or radio. In most cases, however, it is a problem that we (as customers) have come to take for granted. In the case of the telephone, for example, where servicing is the common carrier's problem, we tend to ignore the fact that a huge force of technicians is constantly at work keeping the system going. And in the case of commercial radio and TV, the customer himself can usually determine whether the trouble is at the transmit or receive ends of the system simply by a twist of the station selector knob.

Locating the source of trouble in a fax system is usually not that simple. True, some terminals lend themselves to "self-testing" by the ability to hook send and receive units back to back to help isolate the trouble. This is possible even with some transceivers. But, self-test features notwithstanding, it is usually advisable that a firm providing fax terminals for lease, or selling service on purchased gear, be prepared to provide *coordinated* service between geographically separated terminal locations.

While the situation just described is that which prevails in the U.S. at this writing, there is no sound reason to dismiss "end-to-end" fax service (by a single vendor) as a future possibility. Contrary to some popular misconceptions regarding government restrictions on the scope of common-carrier communication services, there is no sound basis for considering fax outside that scope. It is largely a question of whether the carrier views it as a sufficiently marketable service to be feasibly offered on a large scale. Let it suffice to say that fax's stature in that regard has been steadily improving.

Lack of Standards

Despite some improvement in recent years, several of the various makes of commercially marketed facsimile terminal units (transmitters, receivers, and transceivers) remain incompatible with one another. In fact, between the recent introduction of advanced digital terminals and the concurrent adoption of sophisticated automatic control signalling in analog systems, "interbrand" compatibility has taken a decided backward step.

Within large fax networks, certain basic standards have naturally had to be adopted. Networks for the gathering and distribution of weather data are an example. At least within each of the various aspects of the overall system—e.g., map distribution and the reception of cloud cover photos from satellites—there is standardization, and we find fax equipment of several different makes linked within a network.

Also, within a given newspicture network, there can be various makes and types of receivers in use, all of which, however, must conform to the standards established for that network.

But outside of these long-established and inherently widespread networks, the standardization situation is generally disappointing. Suppose, for example, the personnel at one terminal of a fax network within a decentralized business concern are dissatisfied with some characteristic of the recordings they have to handle. Perhaps there is a tendency for the copy to smudge, or it may have a chemical odor that some persons find objectionable. The range of satisfactory alternative pieces of equipment that will be fully compatible with the rest of the machines in the system is likely to be very limited—if, indeed, there is any alternative at all.

Similarly, when someone with a telephone-coupled fax terminal wants to communicate via the dial network with a party at another location, he first has to ascertain that the two terminals are compatible. This can be as much a problem within a company as between companies. Imagine how it would be if (apart from language barriers) a similar situation prevailed for voice communication by phone!

Progress itself can lead to problems. Whenever a fax manufacturer introduces a significant technological advance, such as an appreciable boost in transmission speed with existing communication facilities, it tends to make obsolete (in spirit, at least) a good many existing installations. Unquestionably, the stimulation of product improvement through competition is to be encouraged, but where it concerns the multiple terminals of a relatively large network, which, *collectively*, may be periodically rendered obsolete by new advances, it can confront system planners with an uncomfortably unstable situation.

Actually, considerable standards work *has* been done in the facsimile field by several organizations, and certain formal standards have been adopted and are strictly adhered to in various areas of fax communication. (More on this in Chapter 7.) It is in the broad area of general purpose, or so-called message or convenience, fax that standardization has been perhaps least effective.

The Electronic Industries Association's standards of 96 scan lines per inch (LPI) and 180 lines per minute (LPM) for message fax systems has been widely adhered to by manufacturers. But the problem often facing the customer or user is that while a transmitter and receiver—or two transceivers—of different makes may both conform to the 96/180 standard, each may use a different system of modulation (one may be FM and the other AM), or each may employ different phasing and synchronizing schemes. Any such variations will make the units incompatible, despite adherence to established resolution and scan-rate standards.

Further, there has been a notable trend in the voiceband analog market toward the adoption of scan rates higher than 180 LPM, usually accompanied by some degree of resolution degradation. Popular alternative speeds are 240 and 360 LPM, and scan resolutions of 90 and fewer lines per inch appear to be gaining in acceptance.

At this writing (mid-1977), there are positive indications that the recently finalized international recommendation for a dual-speed (180 and 360 LPM) message fax system with bandwidth compression capabilities will eventually become the new universal standard for voiceband/analog fax terminals. This so-called "CCITT Group 2" recommendation will be further discussed in Chapter 7. (Bandwidth compression will be explained in Chapter 4.)

TRENDS

In the first section of this chapter we examined some of the areas in which fax is actively utilized at the present time. Now, to round out the picture, it might be well to examine briefly some current and emerging trends, to get some idea of what new dimensions the medium is beginning to assume and in what new directions it appears to be headed.

Fax by Phone

The use of the public switched telephone network—the direct-distance dial (DDD) network—for facsimile communication is not entirely new. Phone couplers for the purpose have been around since the 1930s, although they were

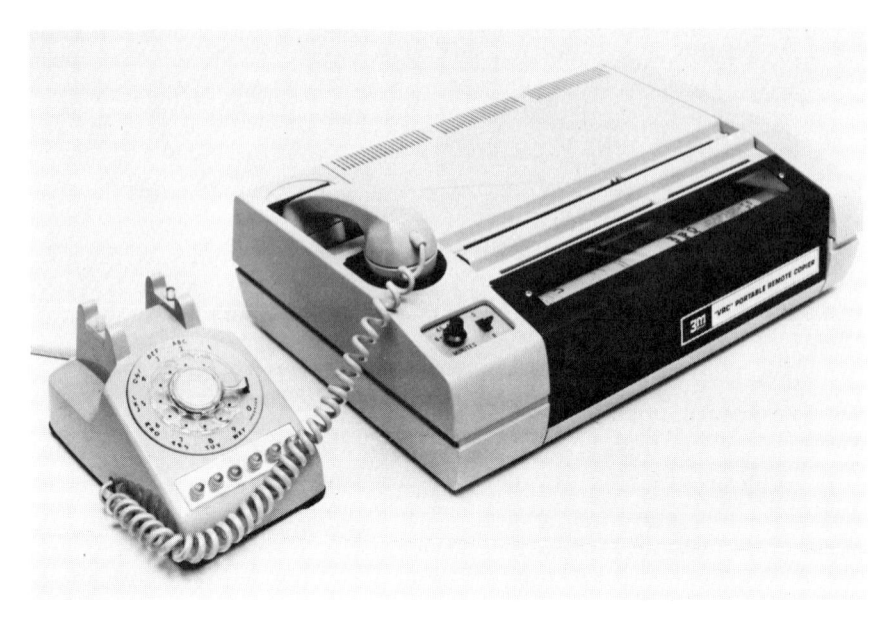

Fig. 2.15. Phone coupler, in this case integral to the terminal machine, permits use of any existing telephone for fax transmission and reception without direct hook up. (Courtesy 3M Co.)

seldom used until quite recently (see Fig. 2.15). In 1963, the common carriers made available devices to interface electrically with fax machines and condition their signals for transmission over the telephone network.

What had already begun to develop as a trend (facsimile via the dial network) was given fresh impetus in the middle and late 1960s by the emergence of a new breed of relatively compact, desk-top transceiver specially designed for the purpose. At the same time, restrictions on connection of "foreign" (noncommon-carrier) devices to telephone lines were eased, with the result that simpler, less expensive direct-hookup arrangements have become available and more of the signal conditioning is now integrated into the fax machine.

Continuation of the present trend should see a steadily expanding use of the telephone as a convenient coordinating device and interface for the selective communication of graphic materials via the dial network. The trend may be retarded somewhat by lack of compatibility between fax machines, but at the same time it may be nourished by a renewed emphasis on miniaturization and improved portability of the machines.

Supplementing the regular dial network are common-carrier switched *broadband* nets, permitting reductions in fax transmission time down to a few seconds per page (see Chapter 4). These facilities are in relatively limited use at the

FAX TODAY

present time, but are expected to grow in proportion to the increasing demand anticipated for such services.

Solid State

The fax equipment industry was at first comparatively slow to adapt to solid state electronics. Until about 1965, most new fax gear being marketed still used vacuum tube circuitry. By the end of the 1960s, however, the situation had pretty much been reversed. While no doubt there is, as of this writing (mid-1977), some vacuum tube equipment still in service, and some still being sold "off the shelf," it would be difficult to find fax gear currently in production that uses any kind of vacuum tube other than a photomultiplier or CRT.

In fax, as in everything electronic, solid state technology has permitted a substantial reduction in equipment bulk. Portability has already been mentioned as a feature that could well enhance the utility of phone-coupled fax transceivers. Along with its potential for miniaturization and weight reduction, solid state's inherent ruggedness has made possible the design of compact transceivers, about the size of an attaché case, that can be carried in the trunk of a car without being jounced out of commission (see Fig. 2.16).

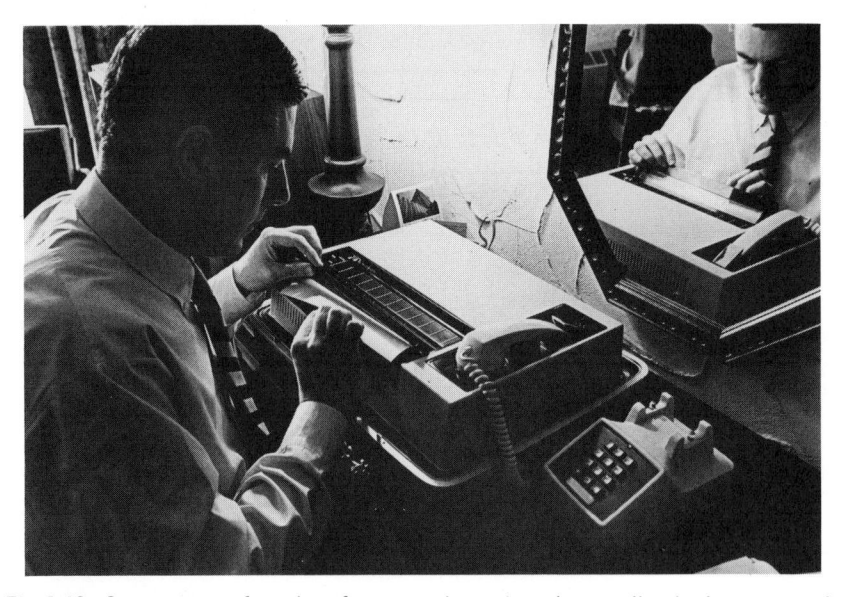

Fig. 2.16. Compactness of modern fax transceivers gives the traveling businessperson the capability to file daily reports to the home office from any convenient phone. (Courtesy Xerox Corp.)

Among the latest solid state advances influencing fax equipment design are large-scale integration (LSI)—the confining of multiple electronic functions to a single integrated circuit—and the strip photoconductor, which has permitted simplification of scanning mechanisms. Just emerging (and momentarily hindered by high cost) are charge-coupled devices (CCDs) and similar technologies that virtually eliminate the need for mechanical moving parts in fax scanners.

Cathode Ray and Laser Techniques

The continuing quest for higher speed in fax transmission has led naturally to the development of potentially faster scanning and recording techniques—notably to the adoption of cathode ray devices and lasers. Besides eliminating mechanical moving parts, cathode ray devices provide a scanning beam that can be made to move extremely fast and to undergo very rapid changes in speed and direction of movement without being subject to the effects of friction and inertia that might afflict a mechanical system under like conditions. Its extreme quickness of response makes it especially suitable for systems using white-space-skipping techniques for redundancy reduction.

As indicated in Chapter 1, cathode ray technology is not exactly new to fax. But its practical application had to await the advent of the TV-fax weather satellite (TIROS I) in 1960. The technology has since been used in at least three commercial fax systems and is at the heart of at least one other still under development at this writing.

As for the laser, it makes possible an extremely small scan spot of concentrated intensity from a source that offers speed, durability, and power-efficiency advantages over its more conventional counterparts, e.g., incandescent and gas-discharge lamps.[6] It has been used in a variety of experimental scanning and recording devices and is, at this writing, in use in perhaps three or four production fax systems, most of them of a specialized nature (the AP's new *Laserphoto* newspicture system, for example). IN 1975, Xerox unveiled its Telecopier* 200, the first commercially available laser fax system for general business applications.

Other features offered by these newer techniques are quieter operation, reduced mechanical wear and tear, and an improved ability to scan an item while it remains absolutely stationary. Various cathode ray and laser scanning and recording processes are discussed in the next chapter.

Data Compression

A trend that is rapidly gathering steam is the abandonment of straight, cycle-by-cycle signaling in fax transmission in favor of data compression—redundancy

---

*Telecopier is a registered trademark of the Xerox Corp.

reduction or "white-space-skipping"—schemes to permit speedier transmission over existing facilities (or transmission at the same speed over less costly facilities). Among the first such systems to be offered commercially were the Scanatron analog *Pacfax* system and the Dacom DFC-10 digital signal processor, both of which were introduced in the late 1960s—initially as separate interface units for existing private-wire fax terminals.

By 1973, there were three highly advanced, self-contained data compression fax systems commercially available for use on voiceband transmission circuits: the Dacom 400 series (Fig. 2.17), EAI FAX I, and Rapifax 100 transceivers. All three offer an average 6-to-1 speed advantage over conventional analog systems at a given resolution. In 1977, the 3M Company's competing transceiver became available and it is expected that before the end of 1978 additional data compression fax systems will have become available from other fax terminal manufacturers (Burroughs/GSI for one) and perhaps even from some companies entering the fax field for the first time.[7]

Redundancy reduction techniques have also been embraced in the latest-generation weather chart and newspaper page master systems and are expected eventually to be adopted by wire services for newspicture transmission as well. At this writing, at least three additional business fax systems embodying advanced white-space-skipping techniques are being readied for marketing in the U.S. One of them boasts a compression ratio approximately double the existing average 6:1.

The schemes vary widely in principle, although most of them have in common their reliance on digital techniques to modify the signal format. Among the drawbacks of such systems are the increased risk of error through the use of digital encoding, and the boost in terminal costs imposed by the increased complexity. On the other hand, continued technological advances, together with the economic effects of competition, should improve the outlook for the fax user.

In any event, from the standpoint of transmission efficiency, data compression is a welcome trend. Some of the more popular techniques are explored in Chapter 4.

Computer Input/Output

The perfection of digital data compression techniques has led to the adoption of fax terminal machines by the computer community for use as graphic input and output devices. The U.S. National Weather Service, for example, recently began using a digital fax scanner to input "backgrounding graphics" into a computer's memory, to be merged with the purely alphanumeric weather data that changes from hour to hour.

Similarly, there are microfilm-input scanners designed specifically for the

Fig. 2.17. A modern data compression fax terminal. The integrated circuit board (page 57) constitutes the heart of a digital microprocessor, which converts the scanner's output signal to a form that can be transmitted more efficiently via band-limited circuits. When the same terminal functions as a receiver, this combined "compressor/reconstructor" converts the incoming encoded signal to a form suitable for conventional recording. (Courtesy Dacom, Inc.)

conversion of film images to computer-compatible codes on magnetic tape or discs. To the micrographics community, such terminals are known as CIM (computer input microfilm) devices. They complement COM (computer *output* microfilm) devices, which perform the opposite function: producing microfilm from computer-output digital data. A COM device can be programmed to function as a microfilm-output fax receiver. (More on COM in Chapter 8.)

FAX TODAY   57

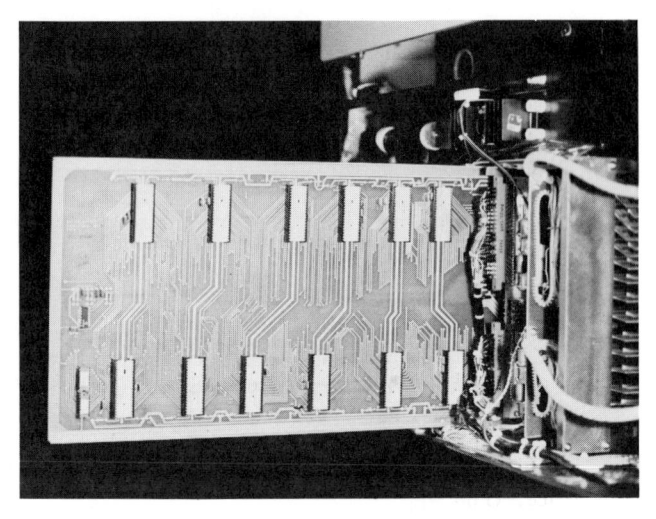

Fig. 2.17. (Continued).

Digital Transmission

Apart from their value as a means of compressing transmitted information, digital techniques offer significant advantages in the bulk handling of electronic communications of all sorts. For one thing, a digital signal, consisting of discrete pulses instead of complex analog waveforms, can be regenerated at intervals along a long-distance circuit with devices somewhat simpler and less expensive than the precision amplifiers required in analog circuits. Moreover, the signal can be regenerated any number of times without the risk of accumulating noise and distortion in the process, as is usually the fate of repeatedly amplified analog signals.

For these reasons digital transmission is expected eventually to replace analog as the norm for electronic communication. This does not mean that terminal devices such as telephone sets and economy-class fax machines (which are analog by nature) will become obsolete. Where such devices are used, the conversion to digital and back to analog will usually be the responsibility of the common carrier. In the case of fax and other visual systems, however, the customer may have the option of doing his own converting and in that way possibly gaining some transmission efficiency and reproduction fidelity in the process.

Many fax applications lend themselves to *two-level*, or binary (black-white), "digitizing." Alphanumeric material—text, tables, etc.—is a good example.

Other examples are weather maps, graphs, and engineering sketches. Once digital transmission facilities have been established, it should be possible to send two-level fax signals quite a bit faster and more economically than these signals can be sent over existing analog facilities.

Increasing Automation

There was a time when fax machines had to be manually phased before recording could commence. (Phasing is explained in the next chapter.) Happily, that day is past. Now the machines phase themselves, and a receiving machine will "answer a call" from a sending machine and "hang up" when transmission is completed.

As was mentioned earlier in this chapter, many (if not most) fax recorders are arranged for continuous roll-feeding, with enough paper on a roll in most cases to accommodate over 200 $8\frac{1}{2} \times 11$-inch documents. And to complement this continuous reception capability, many scanners can be equipped with automatic document feeders to permit uninterrupted transmission.

Other automation features currently available in commercial fax systems are automatic "equalization" (the ability of terminals to self-adapt to transmission circuit conditions), automatic speed switching in multispeed systems (one terminal automatically adjusting itself to the speed of the other), automatic "polling" (the ability of a receiver to "pull in" copy selectively from pre-loaded remote transmitters), and automatic "misdirection protection" (prevention of blind transmission to an unauthorized or inoperative terminal).

Microfacsimile

An allied trend certain to influence the further automation of facsimile systems is the increasing use of microfilm for the storage of records of all sorts. The tie-in with automation is that microfilm is often the first step toward total mechanization of a document or drawing file. Optical codes can be added to a reel of film, for example, to enable automatic selection of any one of over a thousand documents recorded on it. Similarly, frames of microfilm mounted in tab cards or in sheet form ("microfiche") can be automatically selected by punchings, or notches, in the code portion of each card or film sheet. In either system, the selected microimages, instead of merely being projected on a screen for viewing, can be scanned, and the resulting signal transmitted to a distant fax recorder. The output of such a system will ordinarily be an enlarged paper

FAX TODAY	59

reproduction of the scanned microimage, although special recorders can easily be designed to produce duplicate microfilm output if that is desired.

One fax equipment manufacturer, Alden, offers two microfacsimile transmitters, one of which features semi-automatic retrieval of microfilm images from 16mm roll film in cartridges (Fig. 2.18). Regrettably, this one firm appears—for the moment at least—to be the only supplier of ready-to-use microfacsimile terminals, and sales are not exactly brisk. Some three or four other companies have developed such equipment, some of it quite sophisticated. But for one reason or another, none of it has advanced beyond the prototype stage. As of this writing (mid-1977), the PRC Corporation of McLean, Virginia, and at least one other firm are preparing to announce commercial microfacsimile systems to fill the present gap.

Largely because of the scarcity of appropriate equipment and the resulting high cost of that which is available, there are many existing systems in which the

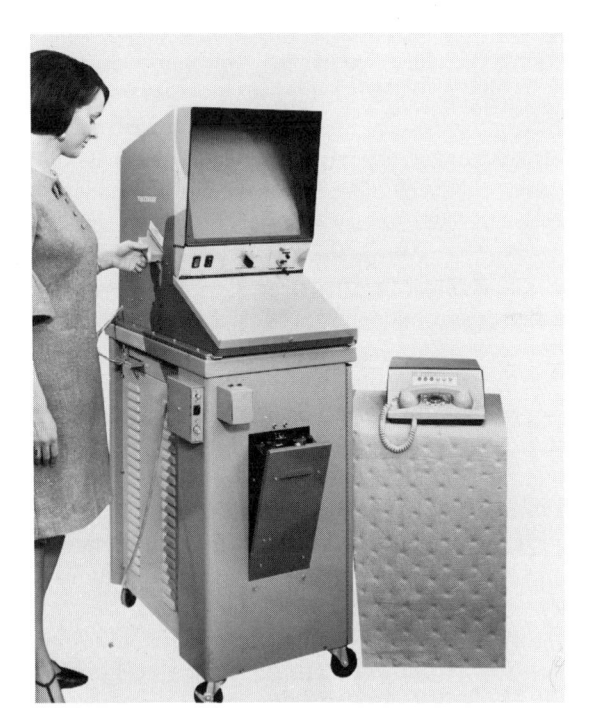

Fig. 2.18. Microfilm input fax scanner. From a film cartridge inserted into the side of a microfilm "reader," an image is selected, centered on the screen, and then transmitted by phone to a distant fax recorder. (Courtesy Alden Research)

true input is microfilm, but the film images are reproduced as paper prints for transmission by conventional fax terminals. Examples are library systems that use enlarged prints of microfilmed catalog cards as the vehicle for fax-dispatched interlibrary loan requests, and systems for the remote accessing of microfilmed records in isolated record warehouses.

But although to date the selection of commercial equipment to facilitate it has been limited, microfacsimile continues to loom as a concept with strong practical potential. Any central microfilm data bank containing material for which the need is generally urgent—a microfilmed newspaper "morgue" file, for example, or a central repository of microfilmed criminal records—is a candidate for microfacsimile. And the number of such microfilm "stores" is growing steadily.

Fax Networks

A somewhat false start was made a few years ago in the establishment of fax communication centers in major cities, between which documents of all sorts could be speedily interchanged for a fee. The operator of a network of such centers leased blocks of telephone lines between cities and bought or rented the necessary fax terminal units. Theoretically, by striking an optimum balance between the number of machines and the volume of business, the fax center could send material from city to city more economically than could a business concern with its own private system and a lower volume.

Few of these pioneering operations proved successful, partly because of awkward fee structures and billing procedures, but perhaps also because it was just so much more convenient—and private—for a business firm to have its own fax machines, even if they did not necessarily pay their own way. By and large, the fax network operations that *have* been successful are those that have specialized in a particular clientele—e.g., the legal and medical professions, pharmacies, trucking fleets.

One apparent exception is the New York-based Photophone Incorporated, a relative newcomer to the fax network scene. (It began operations in the spring of 1977.) It does not specialize, nor does it confine its operations to the domestic sphere. In addition to 13 major American cities, its service coverage extends to (at last count) 10 foreign cities, including London, Paris, and Rome.

In a public network recently established by the Mexican government, the concept of public fax service has been expanded to include the placing of terminals on the premises of business customers. In that respect it is similar to a system introduced by Western Union a few years ago in the United States, namely *Info-Fax 100*, which from all indications has not been widely implemented.

Similar public networks have been established in Europe, and have apparently met with some success—particularly in Great Britain. The European networks are by and large operated by the postal systems of the countries involved. Our own U.S. Postal Service has for several years been talking about establishing a similar operation.

Despite its previous failures and shortcomings, the fax network concept has remained very much alive in the U.S., thanks largely to a fresh approach pioneered by a new communications firm called Graphnet Incorporated. Using circuits leased from major common carriers like the Bell System, Graphnet intends to provide a computer-controlled digital network by which the subscribers' own fax terminals can communicate—point to point or on a "broadcast" basis—with other terminals (including those normally incompatible) anywhere in the country. Moreover, through some unique signal formatting capabilities, Graphnet will be able to permit a customer to receive, on a fax machine, a message that was input into the network via a teletypewriter or punched-tape reader. Also, by line concentration, central data-processing, store-and-forward, and various other sophisticated techniques, it promises to vastly enhance transmission efficiency.

One of the elements of Graphnet's system (now partially implemented) is a concept called "packet switching," in which blocks of digital data, virtually indistinguishable from one another except for their destination codes, are serially interspersed to make optimum use of the transmission channels on a time/bit-rate basis. Other companies, like Fujitsu of Japan, have developed similar systems for internal use by large decentralized corporations.

The apparent success of the packet-switched fax concept is evident in the recent announcements by ITT and Southern Pacific Communications Corporation (SPC) of hybrid fax/data networks similar to Graphnet's. ITT's proposed service is called *Fax-Pak*, and SPC's is called *Speedfax*. Both are expected to be operational by the time this book goes to press.

Graphnet recently enlarged the scope of its operations to include a computerized electronic message service called *FAXGRAM.** Suggestive of Western Union's popular *Mail-gram* system (an alphanumeric network service that uses teleprinters installed in post offices), FAXGRAM is reportedly arranged to work with fax terminals as well as teleprinters, thereby giving it a graphics capability.

Another fax-oriented network service, one that is suggestive (in certain respects) of both Mail-gram and FAXGRAM, is *Datapost*, which uses fax machines in combination with the U.S. Postal Service's Express Mail Service to dispatch priority mail electronically. It is operated by TDX Systems, Incorporated, and at last count, was serving 60 major business clients throughout the U.S.

**FAXGRAM is a registered trademark of Graphnet, Inc.

At this writing, it is still too early to predict how successful these new network fax services will be. But they represent a trend that is definitely worth watching.

Meanwhile, lurking on the sidelines is CATV—Community Antenna, or CAble TV—which for years now has been promising eventually to be the means by which mail, newspapers, printed program schedules, etc., are distributed within a community. With the impending relaxation of government restrictions on CATV's scope, that perennial promise may not be far from becoming a reality.

The "Total Data" Terminal

One of the features inherent in the newest trend in fax network configurations, namely the ability to interconnect dissimilar terminal devices, is also inherent in a newly emerging breed of fax terminal. At this writing, at least three fax manufacturers—Alden, Dacom, and Stewart-Warner—have developed voiceband fax receive terminals that can function alternatively as teleprinters. On receipt of a fax signal, the terminal processes the signal and reproduces the transmitted graphics line by line in the normal manner. Alternatively, upon receipt of an alphanumeric message in ASCII (the American standard teleprinter code), it automatically switches modes and converts the received characters into a matrixed raster format for recording by the normal fax process.

At 2400 bits per second (normally the lowest bit rate used on voiceband circuits), present configurations of the hybrid fax/data terminal can reproduce alphanumerics at the rate of 12,000 characters (about three $8\frac{1}{2} \times 11$-inch pages) per minute. Therefore, speeds of six to eight ASCII pages a minute via voiceband circuits is by no means beyond expectation within the present state of the art.

The development of such devices suggests the emergence of a "total data" or "alphagraphics" terminal—an all-purpose digital communications station for which fax is but one of perhaps several selectable modes of operation.

"Remote Servicing"

The sometimes difficult task of isolating the trouble in a malfunctioning fax system was cited earlier in this chapter as one of the medium's shortcomings. In an effort to minimize that problem, some of the larger fax vendors have established service centers that both their field technicians and customers can phone to help pinpoint the source of a system trouble. The proven economic success of the concept for those vendors who have already been actively applying it could very well signal an emerging trend.

The centers are staffed by a handful of expert technicians and equipped with

FAX TODAY

a variety of fax terminal machines and testing devices. They can be reached by a phone call—usually toll free—from any of that vendor's customer locations all across the country. The number to call is usually posted conspicuously in the vicinity of the customer's terminal. Through an exchange of questions and answers, and possibly some circuit testing and intercommunicating of fax messages, the center technicians are able to minimize the time needed by their field counterparts to make a repair. (A similar system has been in use by telephone companies for years.) Often, with the aid of telephoned instructions from the center, the customer can locate and rectify the trouble himself. In this way, the vendor is able to obviate a certain percentage of costly service calls.

Rising Mail Costs

Only a few years ago the suggestion that facsimile might someday be more economical than mail would have been ridiculed. But in fact by 1976 it had become possible to send documents relatively short distances by fax for the same amount it cost to mail them individually (13 cents a page). And fax, of course, has the bonus of getting a document to its destination within minutes of having been sent. The only catch is that realization of so low a unit cost requires that transmission be confined within a telephone exchange area and that it be maintained almost steadily for eight hours a day. (The example assumes a unidirectional document flow between two $42-a-month fax terminals interconnected via dedicated telephone sets within the same exchange area and transmitting about 75 copies a day, 22 days a month, at a material cost of 7 cents a page.)[8]

Of course, where a point-to-point system is concerned—which for fax is generally more economical at high volumes than a dial-up system—a valid comparison would have to take into account the possibility of using lower bulk mailing rates, thus favoring the unit-cost of mail. The fact remains, however, that as mail costs have risen, message fax terminal and transmission costs have generally declined. Moreover, technology is available today that permits the "bulk" transmission of fax signals as well by converting them to interleaved digital bit streams, together with address codes that automatically divert each message to its proper destination.

As already noted, the U.S. Postal Service has been considering the feasibility of dispatching mail by various electronic means, including facsimile. Meanwhile, the actual use of fax systems for that purpose continues to grow steadily within private industry.

These are the principal trends currently in evidence in fax's continuing evolution. Speculation on future developments is the subject of Chapter 9. The next chapter examines fax's "vital" organs and provides a comparative look at some of the many variations in technique that have evolved over the years.

FOOTNOTES

1. Not to be confused with the taking of pictures through a telescopic lens, which is also called telephotography.
2. Quoted from Ridings' article, "Facsimile Communications, Past, Present, Future," in the November 1962 issue of *Signal*, official journal of the Armed Forces Communications and Electronics Association.
3. Draft copy of "A Survey on Telefacsimile Use in Libraries in the United States," January 1976.
4. See the ANPA Research Institute's *R.I. Bulletin* #949 (March 18, 1968).
5. There is a continuing debate as to the merits of frequency versus amplitude modulation in fax transmission via voice channels. Some experts contend that, in view of improvements in the transmission quality of telephone circuits in the past decade, the use of FM no longer yields any particular advantage. (Modulation is discussed in Chapter 4.)
6. A clarification is needed with regard to laser power efficiency: Although by concentration they make more efficient use of the light emitted, present lasers are *less* efficient than conventional light sources in terms of input energy versus total output energy.
7. The 3M "Express 9600" was officially announced in April 1977.
8. An estimated $10-a-month basic service charge was applied for each of the two "dedicated" phones (regular dial service phones, but occupied totally with fax transmission). Chapter 6 explores the economics of a fax system in greater detail.

3
HOW FAX WORKS

The facsimile process consists of converting visual details of a diagram, document, etc., to an analogous electric current, conditioning the current for transmission by wire or radio to a receiver, restoring it to its original form at the receiver, and amplifying it sufficiently to drive a reproducer, which converts the current variations to a visual facsimile of the transmitted item.

Of course, it is not that simple. The transmission aspect is a study in itself, even without considering the need for perfect (or near perfect) synchronism of the scanning and recording mechanisms, which may be separated by thousands of miles.

This chapter will briefly explore all aspects of the basic process except the actual transmission of a picture signal from one terminal to another. The transmission aspect is covered separately in Chapter 4.

SCANNING

Modern commercial fax systems generally use electromechanical scanning techniques to convert visual tonal variations on the input copy to electrical variations for transmission to a receiver/recorder. With one or two possible exceptions, all fax scanners in use today are photoelectric and are arranged to bounce light from the input material ("subject copy," as it is officially defined)[1] to a photocell or photomultiplier tube for conversion to electricity.

The object of photoelectric scanning is to produce an electrical analog of the tonal variations that make up the image details to be transmitted. The scanner unravels the subject copy into a serial stream of light intensity variations, which it simultaneously converts to a varying electric current. The unraveling is accomplished by the movement of a spot of light over the copy surface—or, conversely, by the movement of the copy surface relative to a fixed spot of light—in such a way that the spot progressively covers the entire copy area. The subject copy is the mirror by which the source light is bounced to the photoelectric transducer (photocell, photomultiplier, or some form of solid state photoreceptor), and it is a mirror that varies constantly in reflectance throughout scanning. The reader should not infer from the above that fax scanners are exclusively bounced-light devices. There are scanners designed for use with transparencies as well—e.g., a microfilm scanner.

There are two basic requirements a photoelectric transducer must meet to qualify for use in fax scanning. One is that its electrical output at reasonable light levels be sufficient to overpower electrical "noise." (The electronics of a fax system will be discussed in more detail in subsequent sections.) The other is that its spectral sensitivity be broad enough to make the device responsive to all but the palest of colors, interpreting them as gray levels. It is also desirable that the gray-scale (or color) response be reasonably linear, although this is by no means a requirement and, in fact, might even prove a deterrent to the design of a practical digital fax system.

Scanning methods vary both optically and mechanically. Optically, the basic variation is between flood and spot projection, which have to do with the manner in which the optical path connects the light source to the transducer via the copy surface. Mechanically, there are three variations: (1) the subject copy alone may move relative to the scan spot; (2) both the subject copy and scan spot may move—at right angles to each other and at different speeds; or (3) the scan spot alone may move.

Spot Versus Flood

Spot projection in fax scanning is the optical projection of a tiny spot of light onto the surface of the subject copy, the bounced spot being picked up directly by the photoelectric transducer (Fig. 3.1). This approach is akin to the "reading" of an optical sound track in a motion picture projector. The only real difference is that in the projector, the spot is sent *through* the graphic medium instead of being bounced off it.

In flood projection (Fig. 3.2) the subject copy is illuminated by diffused light in the general area where it is being scanned, and the bounced light is optically projected through a tiny aperture onto the cathode of the photoelectric transducer.

Both approaches are used in modern fax scanners, and each has its advantages and disadvantages. Spot projection utilizes the available light more efficiently, but it requires a more sensitive photoelectric transducer than does flood projection. The chief drawback of flood projection is its need for a relatively high-intensity light source, which may impose restrictions on the physical compactness of the scanner. With spot projection, the scan head usually embodies the light source and transducer in a fixed relationship and in proximity

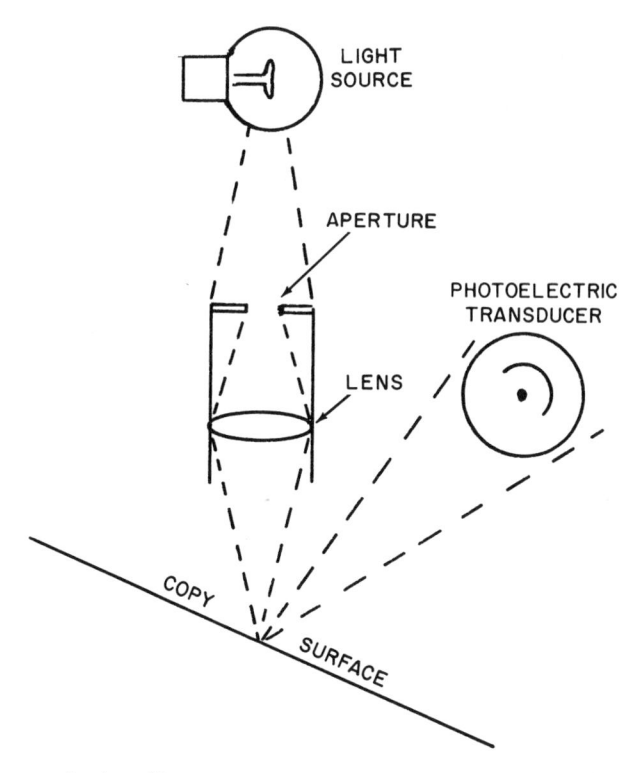

Fig. 3.1. Spot projection. Illuminated aperture is optically projected onto the copy surface, from which the image is bounced to the transducer.

to each other. Flood projection allows the illuminating lamps to be separate from the scan head, although they need not be.

Scanning mechanisms fall into two general categories with regard to the manner in which the subject copy interfaces with the scanning beam: *cylinder* and *flat bed*.

Cylinder Scanning

Cylinder (drum) scanning dates back more than a century, but it is still widely used in both newspicture and document transmission applications. It requires that the subject copy be wrapped around a cylinder and rotated to effect a continuous helical scan of the entire copy. Usually the scan head, containing the photoelectric transducer (and possibly the light source as well), advances slowly along the axis of the cylinder, like the cutting tool of a lathe, as the

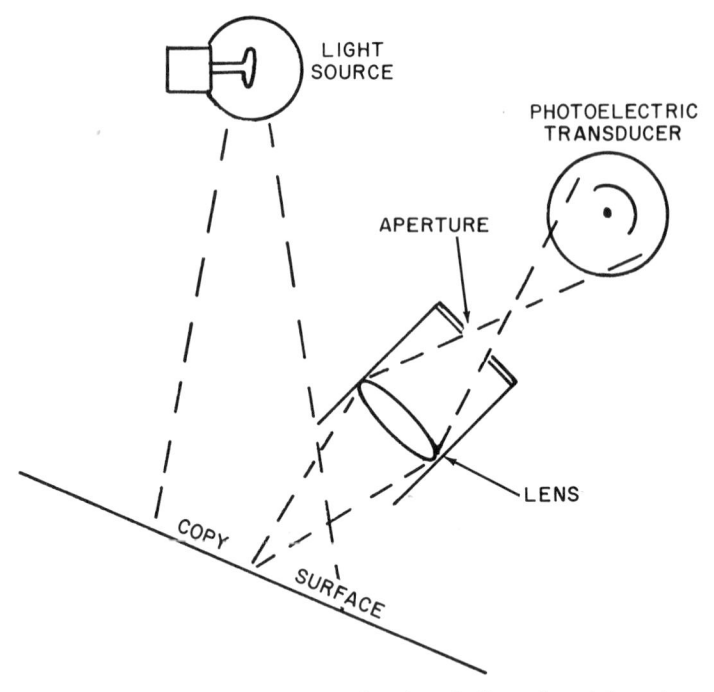

Fig. 3.2. Flood projection. Floodlit copy surface is optically projected through an aperture to the transducer.

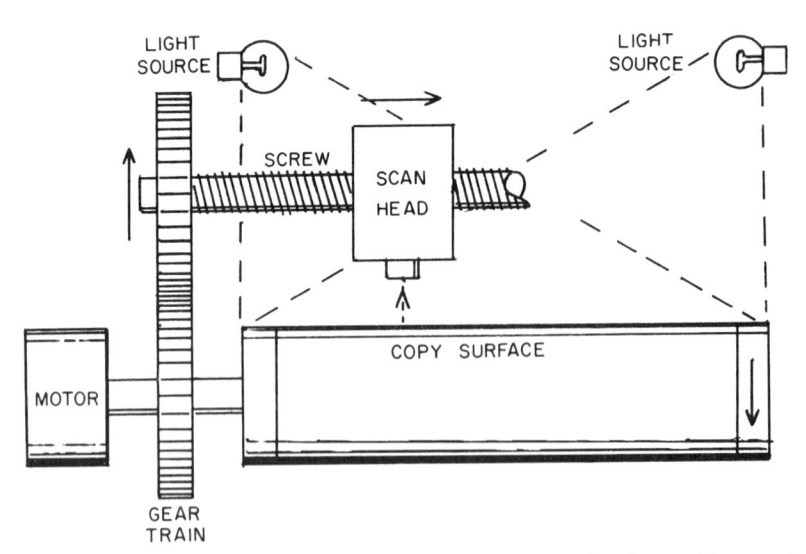

Fig. 3.3. "Lathe-style" scanning. Copy is mounted on a rotating drum. The scan head moves laterally the space of one scan interval (or spot width) for each rotation of the drum.

HOW FAX WORKS

cylinder spins (Fig. 3.3). However, the cylinder can do all the moving—radial and axial—while the scan head remains stationary, or the scan head may even rotate on a fixed plane *within* a cylinder (or semicylinder) while the copy is caused to move laterally across the scan grain. The latter two schemes are illustrated in Figs. 3.4 and 3.5, respectively.

Traditionally the chief drawback of cylinder scanning has been the bother of having to affix the copy to the cylinder. However, methods have been devised to facilitate this procedure, and it is no longer a factor in comparative operating convenience. Two current examples of commercial drum-type scanners in which the copy is manually inserted but automatically "wrapped" around the drum are the Litton/Datalog *Messagefax* transmitter and the *dex** (*de*cision *ex*pediter) transceivers produced by Graphic Sciences, Incorporated. In addition there is at least one *totally* automatic drum-loading machine currently available, the Xerox *Telecopier 410*,* in which the copy is hopper-fed onto the drum for scanning and released into a return hopper at the end of transmission.

Flat-Bed Scanning

Flat-bed scanning permits insertion of the subject copy into a slot, in which a feed mechanism takes over. In most flat-bed scanners, the inserted copy can

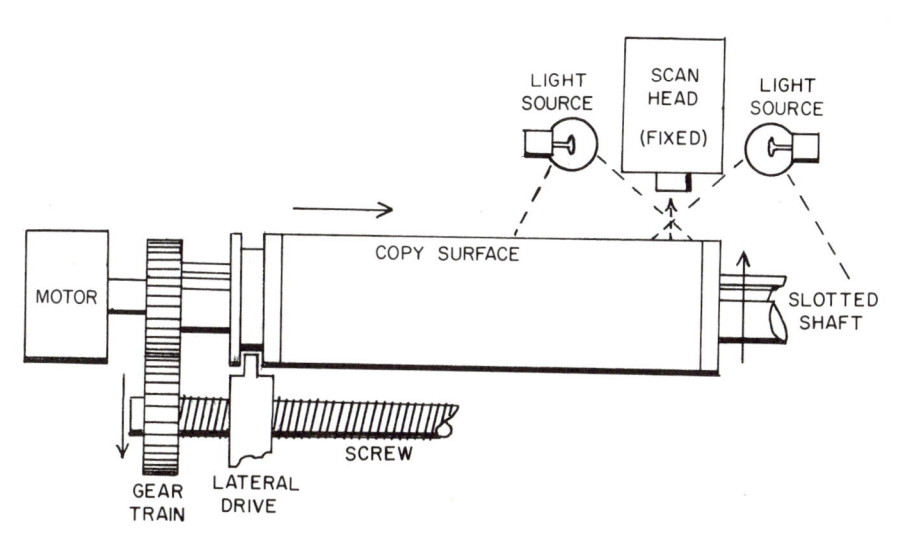

Fig. 3.4. Drum scanner in which the copy moves both radially and axially while the scan head remains stationary.

---

*dex is a registered trademark of Graphic Sciences, Inc.; *Telecopier* is a registered trademark of Xerox Corp.

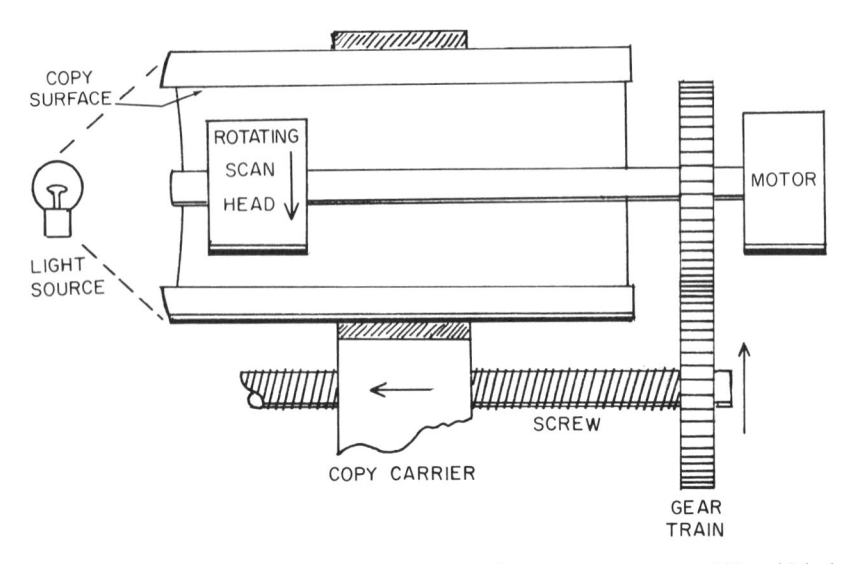

Fig. 3.5. Laterally moving copy has been curved to form a semicylinder, within which the scan head rotates.

be observed slowly being swallowed as it advances over the scanning area within the machine (Fig. 3.6), eventually to be released through a return slot. Figure 3.6 also illustrates one of the more common scanning techniques used in flatbed scanners. The lateral scanning action is achieved by rotation of a spiral aperture relative to a fixed slit.

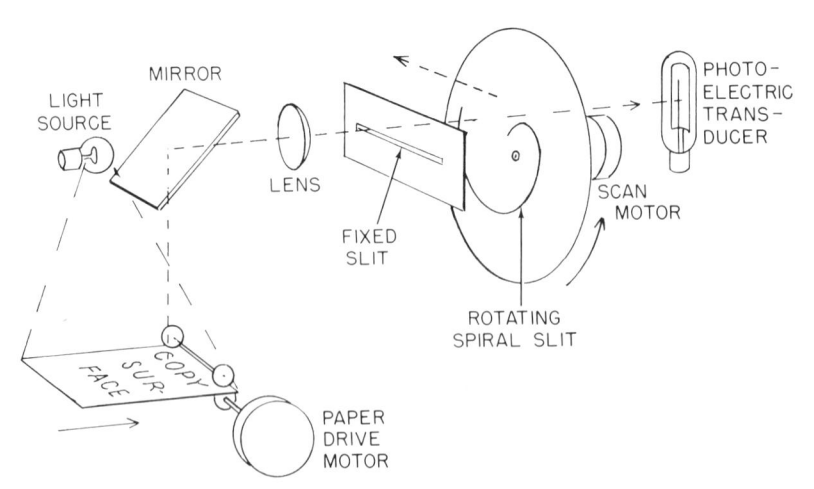

Fig. 3.6. Flat-bed scanning by the fixed/spiral slit technique. An image of the slowly advancing copy is optically projected onto the fixed slit. The spiral slit effects a lateral scan in the direction indicated by the dotted arrow.

HOW FAX WORKS

Another means of converting rotary to lateral motion in a scanner is by "piping" the light from the flat subject copy to a rotary scan head via optical fibers. Nippon Electric of Japan is one firm that has successfully employed this technique in commercial fax gear.

In other flat-bed scanners, particularly those employing electronic *flying-spot* scanning, the copy is merely laid on a platen, where it remains at rest throughout the scanning process while the scan spot does all the moving. Figure 3.7 shows a typical flat-bed scheme using a cathode ray tube (CRT), and Figure 3.8 shows one using a laser as the light source.

The term "flat bed," incidentally, should not be misconstrued as implying that the copy remains perfectly flat in its progress through the scanner. In some cases it may, but most feed mechanisms are such that the copy must undergo some degree of bending between the input and return slots of the machine.

The Electronics of Scanning

Historically the generation of a fax signal has been an electronic process ever since scanning first embodied photoelectric principles around the beginning of this century. Before that it was strictly an electromechanical, interrupted-contact process—a form of telegraphy. With the more recent advent of CRT (Fig. 3.7) and solid state scanning, however, the term "*electronic* scanning" has taken on a new meaning.

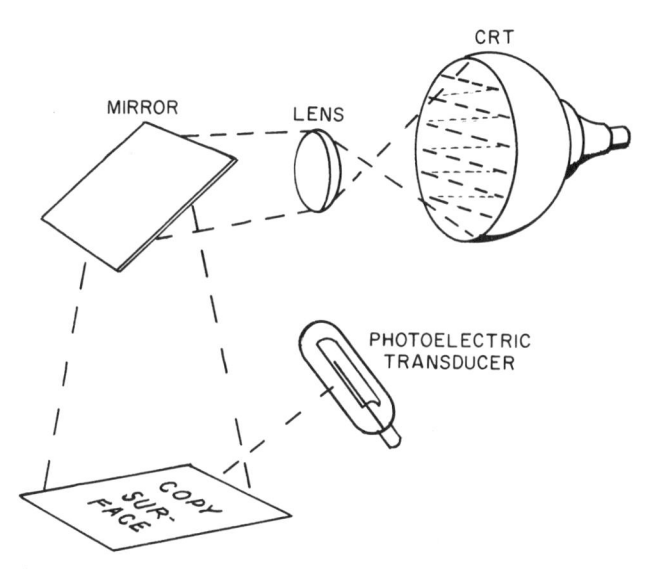

Fig. 3.7. Stationary copy is scanned by an optically projected "flying spot" from the CRT screen.

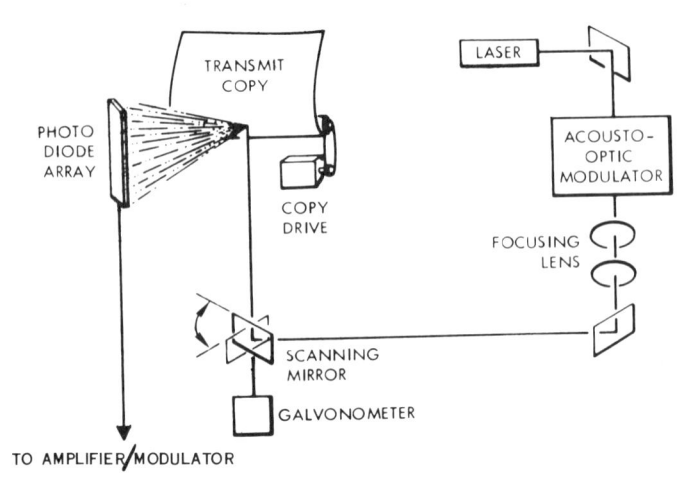

Fig. 3.8. A typical laser scanner. A mirror affixed to a galvanometer movement produces, by its precision oscillations, lateral movement of the scanning laser beam. The acousto-optic modulator has no purpose in the scanning process, but is provided in this particular configuration so that the same laser can be used for recording as well. (See Fig. 3.20.) (Courtesy Associated Press)

Conventional scanning is a mechanical process only up to the point where the scan beam is intercepted by the photoelectric transducer. Thereafter it becomes electronic. The transducer may be a photocell, a photomultiplier tube, or a photodiode. The principal difference among the three lies in the relative strengths of their outputs at a given scan-spot intensity.

The current in the output circuit of a typical photocell is in the microampere range—or a maximum of about one volt across a high-resistance load—and therefore requires substantial amplification. The photomultiplier, on the other hand, contains a number of elements called *dynodes*, whose surfaces have the property of emitting more electrons than they receive from the adjacent emitting element. An ascending series of positive—or descending series of negative—potentials from one dynode to the next instigates a relaying of electrons from the cathode to the anode, with the stream being reinforced at each step of the way. The result is an output current that may be as much as 100,000 times greater than that from a simple photocell. Figures 3.9 and 3.10 show basic circuits for the two types of transducers.

One apparent disadvantage of the photomultiplier is its higher cost, although this is largely compensated for by the reduction it affords in total amplifier circuitry. Its high voltage requirements might also be viewed as a limitation on its use but will usually not pose a problem if the equipment has been properly designed.

HOW FAX WORKS 73

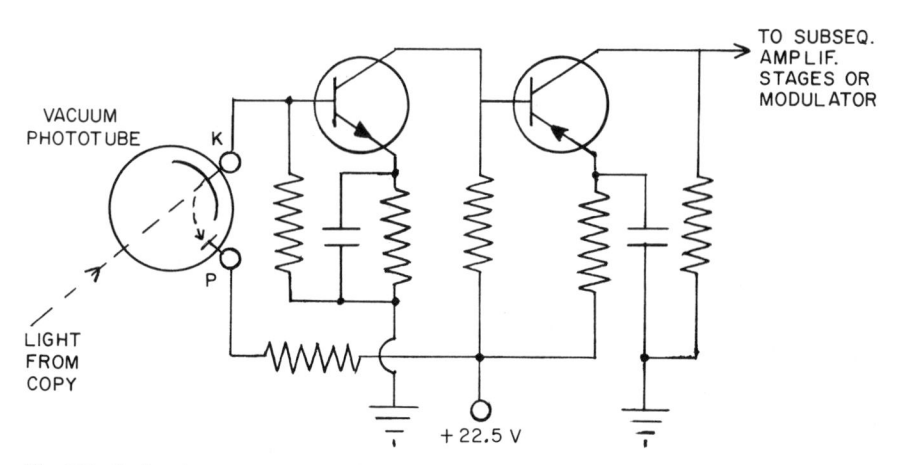

Fig. 3.9. Basic photocell scanner circuit. Amplifier stages are direct-coupled to permit passage of d.c. components characteristic of fax baseband signals.

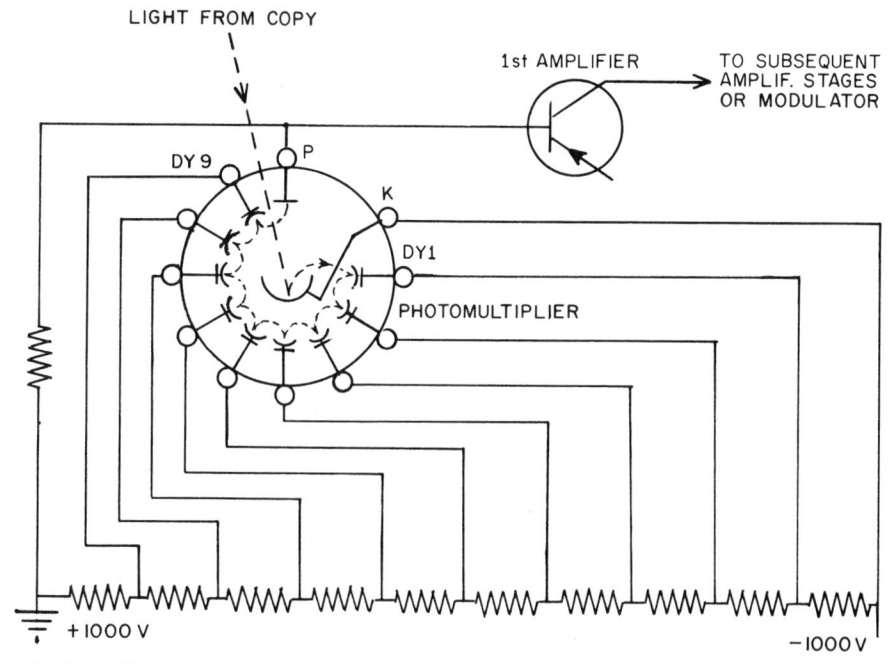

Fig. 3.10. Basic photomultiplier scanner circuit. Dotted arcs indicate the path of the multiplying electron flow from cathode (K), through the dynode chain (DY 1-9), to anode (P). High voltage is required because each dynode from 1 to 9 must be progressively less negative, in increments of about 100 volts.

Photodiodes, depending on the type, may have less or greater sensitivity than the more traditional types of photodetectors just described. Typically they are current-regulating junction devices having silicon as the base material. Their sensitivity is governed largely by the efficiency with which the received light can be concentrated at the actual junction. But apart from sensitivity, they are inherently compact, stable, and fast-responding. In a *strip* or *line* photodiode, the p*n* junction point is elongated so that light striking any point along the length of the device will have equal effect. The advantage of this latter device is the design simplification it offers by reducing the need for complex optics in a scanner.

Some of the more recently developed devices, such as the *pin* photodiode (so called because of the sandwiching of an *i*ntrinsic, effectively neutral, material between the active *p* and *n* materials), are able to approach the sensitivity of photomultiplier tubes. They also cost less, offer a possibly wider spectral response, and are considerably more compact.

The circuitry associated with the photodiode is similar to that shown in Fig. 3.9.

It should be noted that whether the transducer is a photomultiplier or one of the simpler devices, its output contains a d.c. component and therefore cannot, without prior alteration, be processed through a conventional capacitor or transformer-coupled amplifier. This will be discussed further in a later section.

In electronic flying-spot scanning, the moving spot on the face of the CRT is merely the source of illumination, which, upon being modified by density variations in the subject copy, impacts on the cathode of the photoelectric transducer, just as in conventional scanning. The electronics associated with the CRT consists of power and sweep circuits virtually identical to those in a standard TV set. This scan technique is particularly suitable where the subject copy is in the form of a transparency, as in the case of a microfilm input fax system.

With the recent advent of self-scanning, integrated-circuit image sensors—notably the charge-coupled device (CCD) introduced by Bell Labs in 1970—electronic (nonmechanical) scanning ceased to be the exclusive province of cathode ray technology and became possible on a solid state basis as well. Basically a silicon memory device, the CCD can function as a photoelectric transducer with a high-speed scanning capability accomplished by application of shift register action, as in an electronic calculator. Variations in light intensity occurring along a linear array of tiny photodetectors can thereby be sequentially "read out," just as the varying charge levels are read out by electron-beam scanning from the mosaic of a conventional TV camera tube. (See Fig. 3.11.)

HOW FAX WORKS

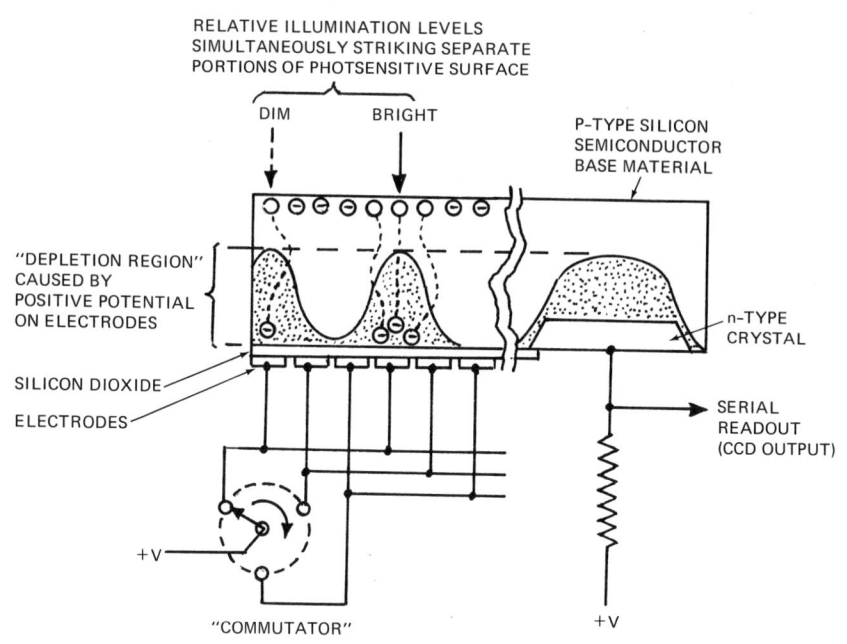

Fig. 3.11. Principle of the charge-coupled device (CCD), used as a self-scanning photoreceptor. Surface electrons are displaced by light bombardment into the depletion region, where they can be step-shifted to the output tap (right) by sequentially switching the positive potential from one electrode to the next. For illustration purposes, the diagram shows a mechanical commutator performing the step-switching function. In the actual device, this commutator action would be achieved with an integral *shift register* circuit.

Besides CCDs there are also self-scanning photodiode arrays, such as the familiar *Reticon* device, which perform the same function in a slightly different manner.

At this writing, these devices are still relatively scarce and expensive and therefore are just now beginning to show up in commercial fax machines.

Yet another aspect of scanner electronics is one that applies only to digital, or "binary" (two-level), systems and is technically a step beyond the scanning process per se. That is the conversion of the scanner output from its natural analog form to a pulse format. The process may or may not involve amplification, which is covered in a separate section further along in this chapter. At its simplest it may consist of applying the transducer output to a basic transistor circuit arranged to conduct whenever the scanner output reaches or exceeds a certain level. The resulting output would be the generation of d.c. pulses (of varying width) representing the black or white peaks of the copy.

In practice the process of "thresholding," or "quantizing," the scanner output in preparation for digital transmission is usually more involved. In a true digital system special circuits must be provided to "sample" the analog output at rapid—and discrete—intervals to determine whether a 0 or 1 (OFF or ON) pulse is to be generated at a given instant. There may even be provision for a *floating threshold*, so that whether the original image details are yellow-on-white, black-on-white, or any combination in between, they will still result in the generation of "mark" pulses, indicating the presence of details, but without degrading the system's ability to discriminate between signal and "noise." In a digital system in which separate binary codes will be generated for each of several shades of gray (as well as for black and white), an arrangement has to be provided whereby the sampled scanner output is *quantized* according to what pre-assigned thresholds its voltage level falls between at a given instant.

The circuitry to accomplish these conversions can be so complex that a detailed explanation would be somewhat beyond the scope of this book. For present purposes it is enough to say that as the emphasis shifts from analog to digital technology in communication electronics, the electronics of fax scanning tends to become more than just the conversion of light variations to an electrical analog.

One critical requirement for the electronics of scanning is that the d.c. power energizing the scan circuitry be properly stabilized. Slight fluctuations in voltage level, which would go unnoticed in an audio system, could easily result in false gray-level representations in a fax recording. The power to the scanner, including that used for illumination, must therefore be regulated within fairly tight tolerances.

RECORDING

As with scanning, most modern fax systems use electromechanical techniques for recording (visually reproducing) the received picture information. The prevalence of these techniques imposes certain speed limitations on the fax process, but the ceiling on speed is sufficiently high that for most purposes it is not an important factor. There are, in fact, cylinder recorders (and scanners) in operation capable of speeds as high as 3600 rpm.

Recorders, like scanners, exist in both the cylinder and flat-bed varieties, the former requiring that the recording medium be in the form of precut sheets to be wrapped around a cylinder, and the latter permitting continuous feeding of the recording medium from rolls.

In recent years a compromise technique has been employed in certain desktop transceivers. The recording paper is formed into a semicylinder for continuous passage *through* the terminal unit, while the recording head spins *within* the semicylinder. A similar arrangement for scanning was illustrated in Figure 3.5.

As applied to the actual recording process, electromechanical may be considered as denoting any system in which the recording transducer is in direct physical contact with the recording medium. Four basic processes fit this category, although within each there are mechanical variations. The oldest and traditionally most popular is *electrolytic*, which is actually an electrochemical process. The others are *electroresistive, electropercussive,* and *electrostatic*.

Of these processes the electrostatic is unique in that the image can be recorded either by a metal stylus or by an illuminated spot. The type of paper used will differ with the method, but image development (the process for which will be described later) is essentially the same for both methods.

The one established recording process that lies outside the electromechanical category (as defined above) is photographic recording, traditionally used in the facsimile reproduction of newspictures. In addition there has recently emerged a variation of photographic recording called the *dry silver* process, in which latent images produced by a light beam are developed by the application of heat rather than by chemical processing. It was developed by the 3M Company for use in photocopying machines and has been adapted to facsimile in machines produced by the Harris Corporation, notably the *Laserphoto* receivers currently being implemented in the AP newspicture network.

Two processes that do not exactly fit either the electromechanical or the photographic category are the Hellfax offset process and ink jet recording, which have in common the use of wet ink and ordinary paper. In the Hellfax process the transmitted "marks" are recorded in ink on a moving plastic ribbon and then transferred to plain paper a line at a time. In the ink jet process a combination of electromechanical and electromagnetic (or electrostatic) principles is utilized to modulate a fine spray of ink and direct it to the surface of plain paper.

The Hellfax process will be briefly described later. Ink jet recording is of interest mainly because of its potential for becoming a practical fax recording medium. Although introduced for that purpose some 35 years ago (see Chapter 1), its use in fax has been very limited. However, the advances made in the technique in recent years, notably by Prof. Helmuth Hertz[2] of the University of Lund in Sweden, have removed some of the previous drawbacks and have made it quite readily adaptable to use in an advanced fax system (which, in fact, had been the intention of the Magnavox Company when it was still in the fax business a few years ago). In recent years, variations of the process have found application as a substitute for impact printing in computer output devices and high-speed teleprinters.

There have been some other nonmechanical, nonphotographic processes developed, but not necessarily applied commercially as yet. An example is the experimental fax system developed at Bell Labs by H. A. Watson and others, in which a modulated laser beam burns holes in the thin metallic coating on a special 16mm film. No processing is required. The permanently recorded microimages are instantly visible by optical projection onto a screen.

By and large, all recording processes except conventional photographic fall under the general heading of *direct recording*, meaning that they require no additional processing (chemical or other). Electrostatic and dry silver recording rightfully belong somewhere between photographic and direct, inasmuch as they involve creation of a latent image, followed by one or two processing steps. But since they invariably include automatic processing, these methods may be loosely regarded as direct recording processes.

Electrolytic Recording

Electrolytic is the oldest and still one of the more popular fax recording processes. It makes use of the discoloration produced in a material saturated with a special electrolyte when an electric current passes through it. The color change is from natural to a darker shade, is permanent, and usually varies in relative darkness in proportion to the strength of the current.

Applying this effect to fax recording is merely a matter of placing the electrolyte-saturated material (here paper) between a fixed electrode and a moving stylus and passing the amplified fax signal current through it. As the stylus sweeps across the paper, the variations in current are recorded on the paper as a line of varying darkness. If each subsequent sweep of the stylus is displaced by the width of the recorded line, the variations in darkness will begin to form a pattern. With proper synchronism between the fax scanner and recorder, the pattern should form a facsimile of the scanned copy.

The first electrolyte used for fax recording—by inventor Alexander Bain in 1842 (see Chapter 1)—consisted of water, sulfuric acid, and a saturated solution of yellow prussiate of potash.[3] The marks made by the recording stylus on the electrolyte-impregnated paper were the result of decomposition of the iron stylus by electrolysis. Since Bain, a number of different electrolytes have been used successfully and patents have been secured by fax manufacturers on paper impregnated with the firms' own specially concocted solutions.

There is some diversity in marking characteristics among the various commercial electrolytic papers. The Alden Company's Alfax paper, for example, is instantly recognizable by the soft brown coloration of recorded details. Another brand produces marks that appear at times to contain a violet component. Others produce dark brown or black marks on a white background. Suppliers usually offer several paper types by name or number, each having different characteristics.

Some, but not all, electrolytic papers are *archival*, meaning that the images will not fade appreciably over a period of time, or that the paper will not deteriorate in storage. One popular type—*catechol*—is among the best from the standpoints of overall image quality and permanence, but has an unfortunate tendency to produce a permanent gray stain on other documents that come close to it.

The stylus/backplate technique of electrolytic recording lost favor some years ago to a more practical technique called helix-and-blade, which is used almost exclusively today in electrolytic recorders. The principle is illustrated in Fig. 3.12. The special drum containing the helix makes one complete revolution for each scan line. For the recording paper, the laterally moving pressure point between the rotating helix at the rear and the stationary blade in the front has the same effect that a laterally moving stylus in the front would have if it were pressing the paper against a stationary backplate. As the electrical potential between blade and helix varies during each rotation of the helix, marks of corresponding darkness are recorded along the length of the laterally moving pressure point. The stationary printing blade requires frequent replacement because of the eroding effect of electrolysis.

Although in its natural state electrolytic paper is moist—and it must remain so prior to and during recording—it usually emerges dry from the recorder because of a heating element just beyond the recording point. The change from moist to dry after recording may result in some shrinkage of the recorded image.

Electroresistive Recording

The electroresistive process, sometimes referred to (erroneously) as *electrothermal* and perhaps best known as the "burn-off" process, is similar to electrolytic in that the conductive recording paper is interposed between two electrodes, one of which is the recording stylus. It differs in that the marks made by the stylus on the paper are the result of the paper's white coating having been decomposed by the electric current passing through it. The arcing that occurs

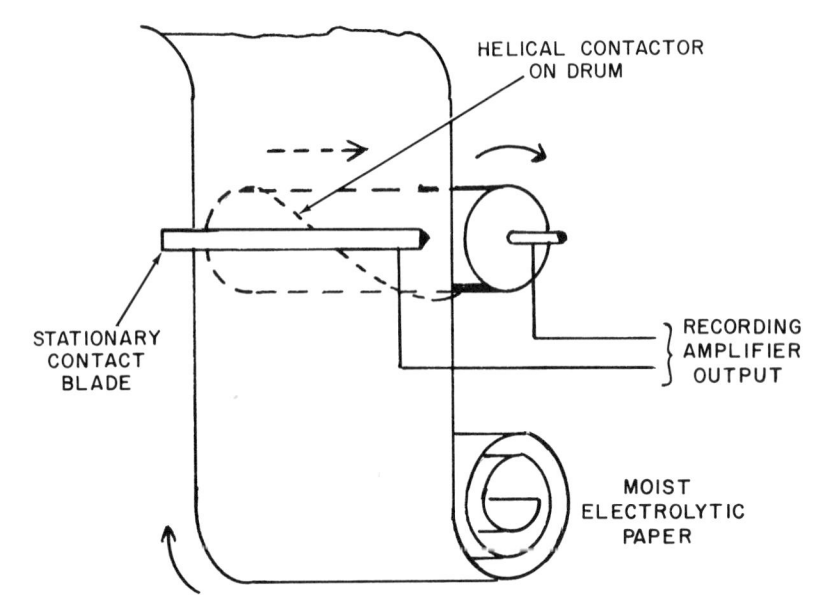

Fig. 3.12. Electrolytic recording by helix-and-blade technique. The drum containing the single-turn helical contactor makes one revolution per scan stroke. The junction of the helix and a stationary blade constitutes a laterally moving stylus (dotted arrow).

at the stylus may produce an occasional wisp of smoke and an accompanying odor, particularly when the subject copy contains large dark areas.

By far the best-known electroresistive paper for fax recording is *Teledeltos*,* developed by Western Union nearly 50 years ago and still in wide use in both fax and laboratory instrument recorders. The active ingredient of Teledeltos is a copper compound. Mixed with it is titanium oxide, which gives the paper its normal white coloration and which permenently discolors at the point where the current from the stylus passes through it.

The chemical make-up of the coatings on commercial electroresistive papers varies, and the variation usually has to do with the degree to which the equipment vendor is willing to sacrifice printing contrast for "smokelessness" or reduced odor.

Electroresistive paper is durable and generally unaffected by temperature changes, but it is usually slightly more expensive than electrolytic paper. The process has reasonable continuous tone capability, and the recorded images

---

*Teledeltos is a registered trademark of the Western Union Telegraph Co.

HOW FAX WORKS

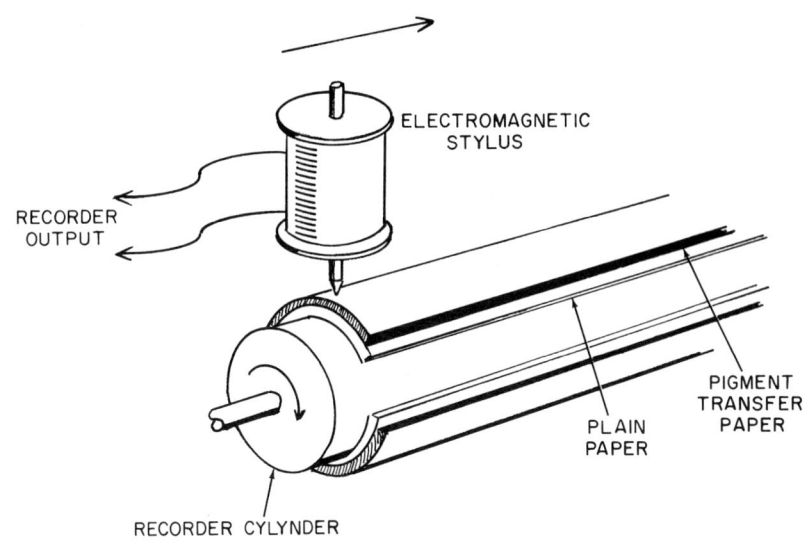

Fig. 3.13. Impression (or pigment transfer) recording utilizes a vibrating stylus in conjunction with ordinary carbon paper to produce fax recordings on plain paper.

are fairly "contrasty" and resistant to fading and background discoloration. They are, however, somewhat vulnerable to visible damage by scratching.

Electropercussive Recording

Electropercussive recording, also known variously as *impact, impression, pigment transfer*, or *pressure modulation* recording, has been in use on and off since the early 1930s. It has been popularized in recent years through its use in certain desk-top transceivers, notably Magnavox's *Magnafax*[*] line (now 3M) and the earlier Xerox *Telecopiers*.[**] It has also been used in some weather chart recorders.

The process is basically the same as that used in audio disk recording. The amplified signal is fed to a transducer in which a stylus is electromagnetically actuated in response to the signal current variations. In the facsimile application, visual recording is effected by interposing a sheet of carbon paper between the stylus and a sheet of plain paper. As the stylus vibrates during scanning, it produces a carbon impression on the paper, the darkness of the impression varying in proportion to the variations in strength of the picture signal (Fig. 3.13).

[*]*Magnafax* is a registered trademark of the 3M Co.
[**]*Telecopier* is a registered trademark of the Xerox Corp.

The technique produces good contrast recordings on nonchemical paper and permits recording in duplicate or triplicate through use of additional transfer sets. A possible disadvantage—depending on the mechanical design of the recorder—is that a malfunction resulting in obliteration of a recording or possibly no recording at all may go undiscovered until completion of transmission. The process is also somewhat speed-limited and acoustically noisy.

Electrostatic Recording

There are two basic electrostatic recording techniques. One is transfer xerography, which is somewhat analogous to offset printing and which permits recording on ordinary paper; and the other is a direct technique (also essentially a xerographic process), requiring specially coated recording paper. These two basic techniques are further divisible into illumination and metal-stylus imaging processes.

Transfer xerography of the illumination imaging variety is used in the Xerox Corporation's LDX^* (long-distance xerography) and Telecopier** 200 systems.

The LDX reproducer is a more or less conventional cathode ray tube on whose screen the received picture signal is reconstructed as a train of light variations. The image is projected onto a selenium drum, where it is converted to a latent image in the form of a varying electrostatic charge pattern. The drum is "dusted" with heat-sensitive powdered resin called *toner*, which clings by static electricity to the charged portions of the drum. The pattern formed by the toner is then transferred from the drum to paper and is fused to the paper by heat (Fig. 3.14).

Use of the transfer technique in the Telecopier 200 differs mainly in that the illumination source is an electromechanically deflected laser beam instead of a CRT.

In the direct technique, the reproducer may also be a CRT, or it may be a moving metal stylus or a linear array of stationary styli. If a CRT, it is likely to be one specially designed to permit direct "electronic" contact with the oxide-coated recording paper. It has no screen, but has metal pins imbedded in its faceplate, permitting the electron beam to extend through the glass to the outside for transfer of the signal energy directly to specially coated paper (Fig. 3.15). With the conventional tube, the scan spot may be either projected or "piped" (by fiber optics) from the phosphor screen to the paper (Fig. 3.16).

In the non-CRT process, "marking" voltage from the receiver output is applied directly to the stylus (or styli). In the case of a fixed-styli array, the moving stylus is simulated by a commutator arrangement through which each of

*LDX is a registered trademark of the Xerox Corp.
**Telecopier is a registered trademark of the Xerox Corp.

HOW FAX WORKS

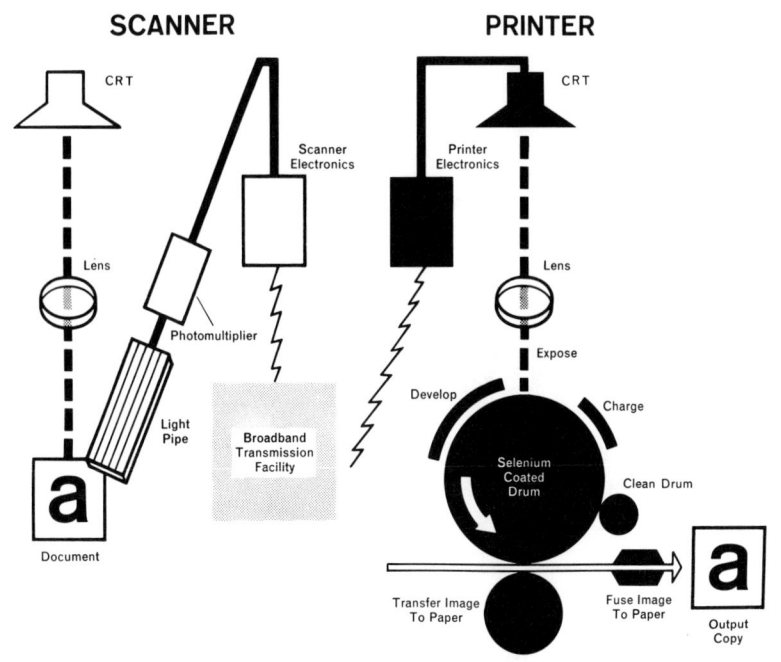

Fig. 3.14. Xerox's LDX system combines CRT scanning and reproducing with xerographic recording. (Courtesy Xerox Corp.)

the individual styli is sequentially connected to the receiver output. Multiple-styli electrostatic recorders in general (including the CRT type) are popularly referred to as "pin printers." All require use of specially coated paper to which the toner—either in powdered form or suspended in a liquid solvent—is applied directly and, as in transfer xerography, permanently fused to the parts where it clings.

A kind of paper called *dielectric* is sometimes used for direct electrostatic recording. Like electrolytic paper (but not to the same degree), it's effectiveness depends on its retention of a certain amount of moisture. Imaging problems can result when this type of paper is used in a very dry environment.

The direct electrostatic process offers exceptional gray scale capabilities and is, therefore, a popular alternative to the photographic and electrolytic processes for newspicture reception.

Hellfax Offset Recording

The wet-ink offset technique developed by Hell of Germany ranks as one of the more unusual fax recording processes currently in use in commercial equipment (Fig. 3.17).

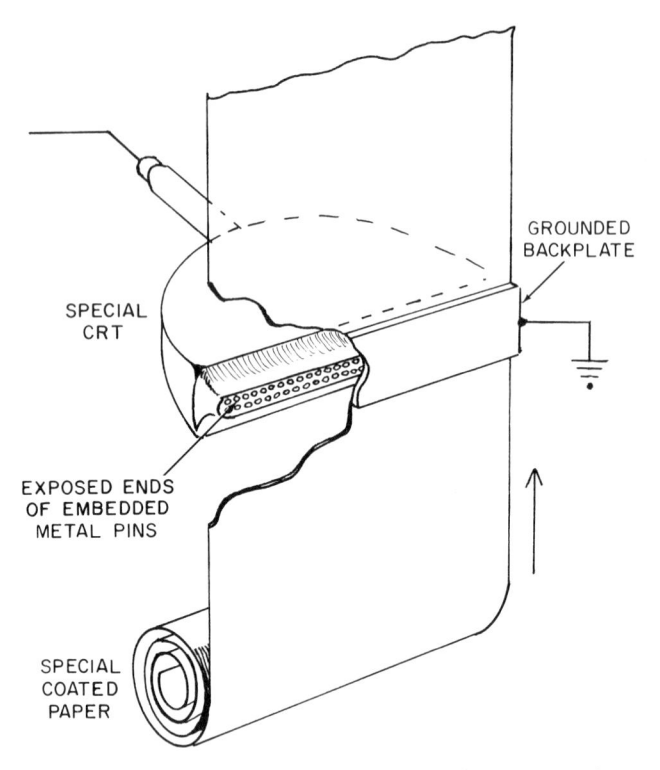

Fig. 3.15. "Videograph" recording technique uses a special conducting faceplate CRT to put a latent image directly on coated electrostatic paper. The image is then developed by application of electrostatic ink to the paper.

First, for the process to work properly, the received analog signal must be converted to a series of discrete, black marking pulses. The transducer (22), which responds directly to the processed signal, consists of a mechanically actuated inked roller, which reproduces the marking pulses visually, as they occur, on a moving belt of plastic ribbon (17). When one complete line has been thus reproduced, an electrically actuated printing bar (14) automatically transfers it from the ribbon to the paper on which the full recording is to be reproduced. Before the transfer belt completes a cycle, it comes into contact with a strip of consumable blotting paper (15), which removes any remaining ink and prepares that portion of the belt for reproducing a subsequent line.

The elaborateness of this process has to do with the machine's ability to record, flat-bed fashion, on paper fed from a roll. In Hellfax cylinder-type recorders utilizing the wet-ink process, the transducer contacts the paper directly. The ink used is a type that dries rapidly once it is applied to the paper.

HOW FAX WORKS

Electrothermal Recording

Although not among the processes listed at the outset of this section (because it is not among those that are widely applied), *electrothermal* recording nevertheless warrants a brief look as an emerging graphic recording technique. At the heart of the process are two elements: (1) a special, chemically treated paper capable of localized, permanent darkening in response to the application of heat beyond a certain threshold, and (2) a special resistive element capable of rapid temperature changes in response to variations in applied current. For graphic recording, the resistive element is in the form of a stylus, which contacts the

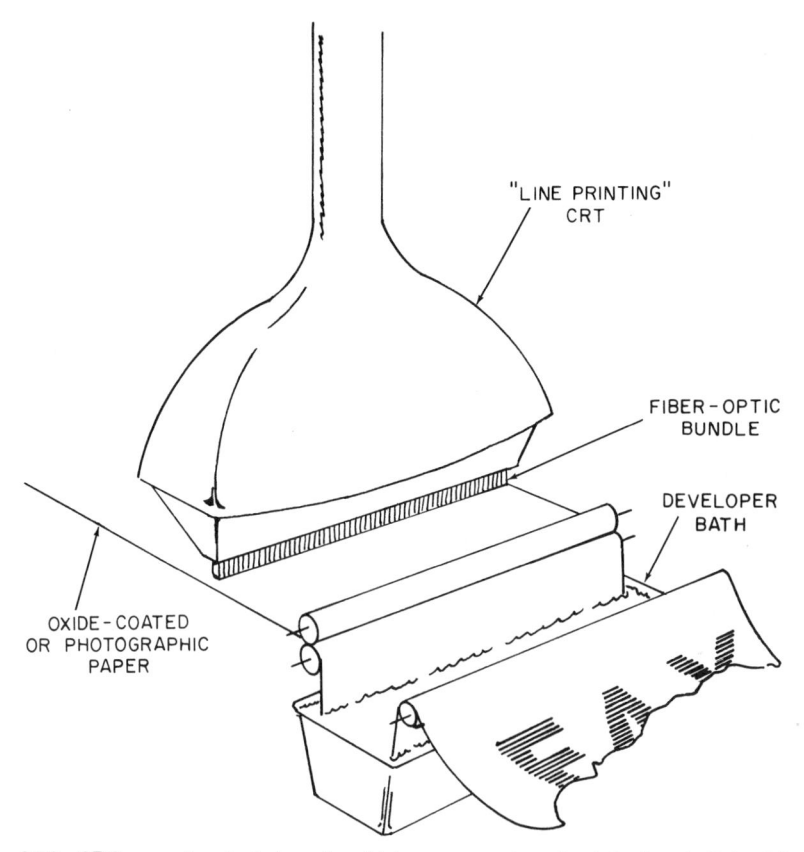

Fig. 3.16. CRT recording technique in which a scan spot on the tube face is "piped," via fiber optics, to the surface of specially coated paper within a lightproof enclosure. A latent image is formed on the paper and made visible by passing the paper through a liquid developer. The fiber-optic faceplate extension minimizes the effects of diffusion and reflections, and eliminates distortion due to curvature of the tube face.

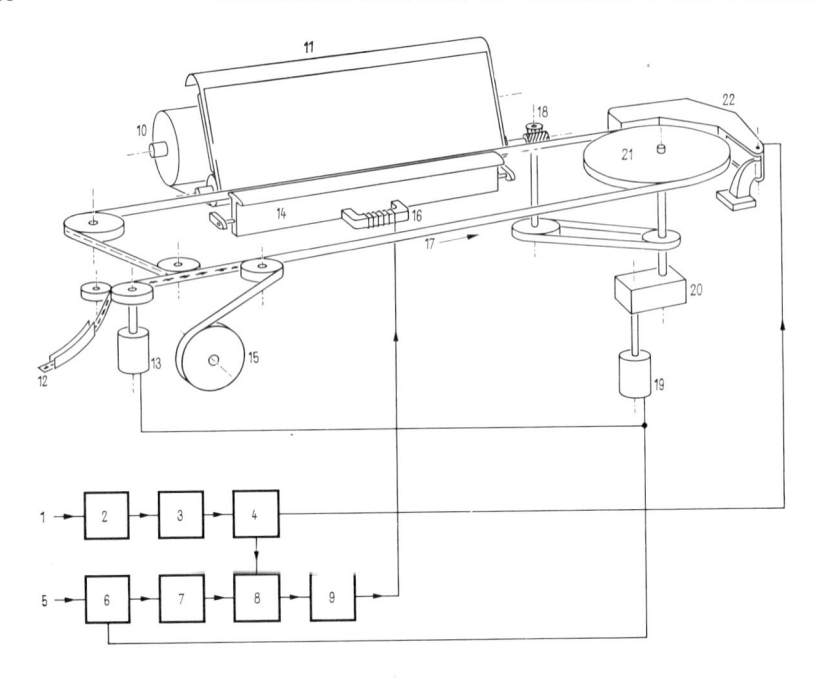

Schematic Representation of Operation

1 Signal input	9 Printing bar amplifier	17 Recording belt
2 Automatic gain control	10 Paper supply roll	18 Paper feed gearing
3 Demodulator	11 Paper guide	19 Main drive motor
4 Output stage	12 Erasing belt exit	20 Motor gear box
5 Power network	13 Erasing belt motor	21 Recording drum
6 Power supply	14 Printing bar	22 Recording system
7 Frequency divider	15 Erasing belt roll	
8 Automatic phasing circuit	16 Printing magnet	

Fig. 3.17. Hellfax wet-ink offset recording. Ink marks corresponding to black picture elements are printed by the transducer on a plastic belt and transferred to the plain recording paper by pressure. (Courtesy Dr. -Ing. Rudolf Hell of Kiel)

paper. The trick is to maintain the stylus just below the threshold temperature so that each pulse of signal current will raise the temperature sufficiently to produce a dark mark on the paper.

An advantage of this process is that both the electronics and the paper are quite inexpensive. A possible disadvantage is its questionable ability to maintain sharp detail at high speeds. At present it is being used effectively in character printers and in some handwriting receivers (see Chapter 8); Okifax of Japan is using it in a comparatively high-speed fax system.

Photographic Recording

The photographic process has been traditionally associated with the facsimile distribution of news photos, although it has been and still is used in other fax applications as well. It differs mechanically from direct recording in two ways: (1) recording must be done within a lightproof enclosure, and (2) the transducer does not directly contact the recording medium (photographic paper or film). It should be noted, however, that the transducer may contact the recording medium indirectly via a fiber optic "light pipe" (Fig. 3.16).

Traditionally, an added step has been required in the recording process, namely, darkroom development of the latent photographic image. In recent years, however, photographic recorders have been introduced in which processing is automatic within the recorder. The simplest way of achieving this has been through use of Polaroid film packs as the recording medium, but there are also more complex recorders in which the paper containing the latent image is automatically fed through a chemical bath and ejected from the machine as a processed photograph within seconds after recording is completed (Fig. 3.18).

The most recent innovation in photographic recording is "dry" processing, which eliminates the need for processing chemicals entirely. Representative of a trend in that direction are the Associated Press's new *Laserphoto* newspicture receivers and the related *Laserfax* line of receivers being offered commercially by Harris Corporation (manufacturer of the Laserphoto equipment). These machines use the 3M-developed dry silver paper previously described. The latent images are rapidly processed within the machine simply by passing the exposed paper through heated rollers. The result is a continuous-tone image practically indistinguishable in quality from a typical wet-process photographic print. Moreover, there are no chemical fumes and no waste disposal problems as there usually are with wet processing.

The transducer for photographic recording may be a conventional projection lamp combined with a light valve (e.g., a Kerr cell) or a special lamp capable of rapidly varying its brilliance in response to picture-signal fluctuations. Or it may be a modulated laser, one of the more recent innovations in fax recording technology. Of the three choices, the prevalent one at this writing is still the *crater*, or *glow modulator, tube,* a special lamp that provides a concentrated gas-discharge light source, whose intensity is variable directly by amplified signal currents. However, the laser's popularity as a photo-recording transducer is growing rapidly.

Fig. 3.18. A photographic recorder featuring automatic processing. There is a slight time delay between completion of reception and emergence of the finished recording from the processing section. (Courtesy Muirhead)

Figure 3.19 shows a typical photographic recording arrangement, and Figure 3.20 depicts two laser recording schemes.

AMPLIFICATION

Except for very high-speed systems, the frequencies involved in facsimile transmission are usually within the audio range. The basic difference between the output of a fax scanner and that of the optical sound head of a movie projector (the two devices have much in common mechanically) is that while the latter consists substantially of measurable audio frequencies, the former is usually made up of a more or less haphazard sequence of transitions. Unless the scanned subject copy has a reasonably symmetrical pattern of lines arranged obliquely or at right angles to the scan grain, the scanner output is in the audio range only in the sense of the number of transitions that normally occur within a second.

Another, more significant distinction is that while the audio range seldom extends below 20 Hz, the fax-scanner output frequencies can range all the way

HOW FAX WORKS

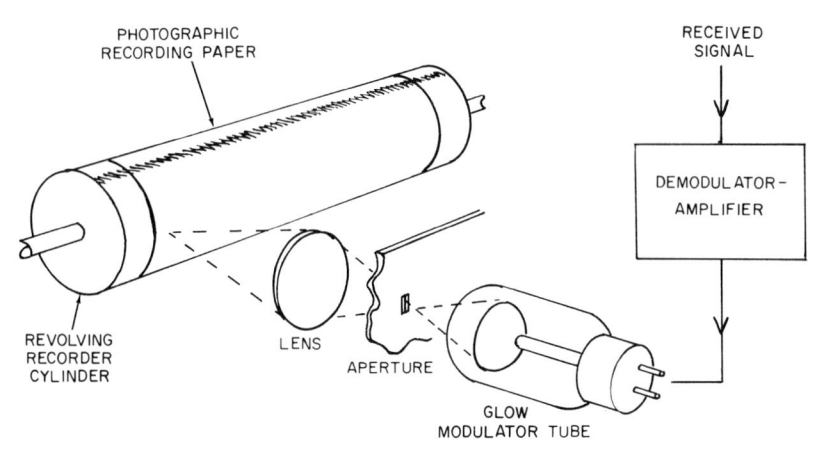

Fig. 3.19. Typical photographic recording scheme. The recovered baseband signal is amplified sufficiently to vary the illumination of the glow tube. The light variations, optically focused onto photographic paper, form a latent image, later made visible by conventional chemical processing. The cylinder must be in a lightproof enclosure during recording.

down to zero—a fact that presents a problem with respect to amplification and coupling. The essence of the problem is that although it is possible to amplify these *baseband* currents sufficiently to transmit them directly, over wires, the distance they can be sent depends on how far down the line the nearest trans-

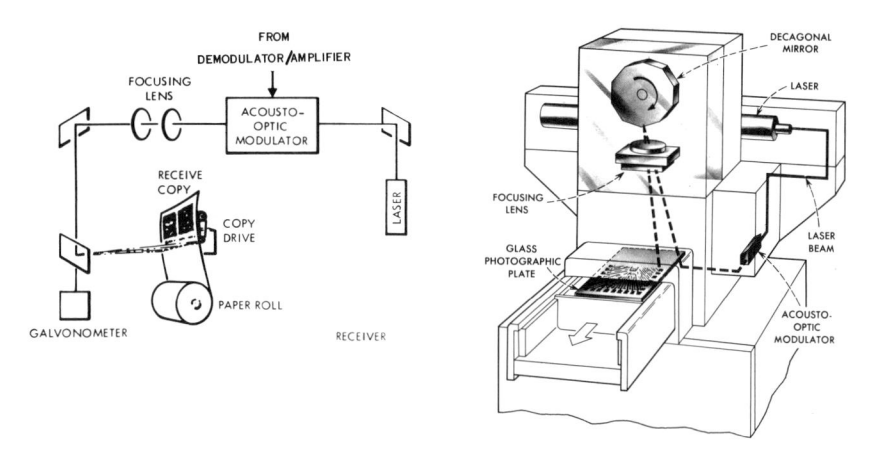

Fig. 3.20. Two typical laser recording schemes. The one on the left (picture courtesy Associated Press) utilizes a galvanometer-actuated oscillating mirror for lateral scanning, whereas the one on the right uses a continuously rotating faceted mirror for the same purpose. (Right-hand picture copyright 1970, Bell Telephone Laboratories, Inc.; reprinted by permission of the Editor, Bell Laboratories *Record*)

former or a.c. amplifier is. In a typical situation involving normal telephone lines, that distance is not very far.

The answer to the problem is modulation—the superimposing of the scanner output on an a.c. carrier signal of comparatively high frequency. This aspect of the facsimile process is explored further in Chapter 4.

Suffice it to say, for the moment, that amplification of the modulated fax signal involves use of conventional Class A amplifier circuits not unlike those found in audio systems, whereas the scanner output prior to modulation and the demodulated signal delivered to the recorder require direct coupling of amplifier stages (i.e., no capacitors or transformers in the signal path). Examples of the two types of amplifier circuits are shown in Figs. 3.21 and 3.22.

Automatic Background Control

There is one other aspect of fax system amplification that bears mentioning, and that is *automatic background control* (or *automatic gain control*). The

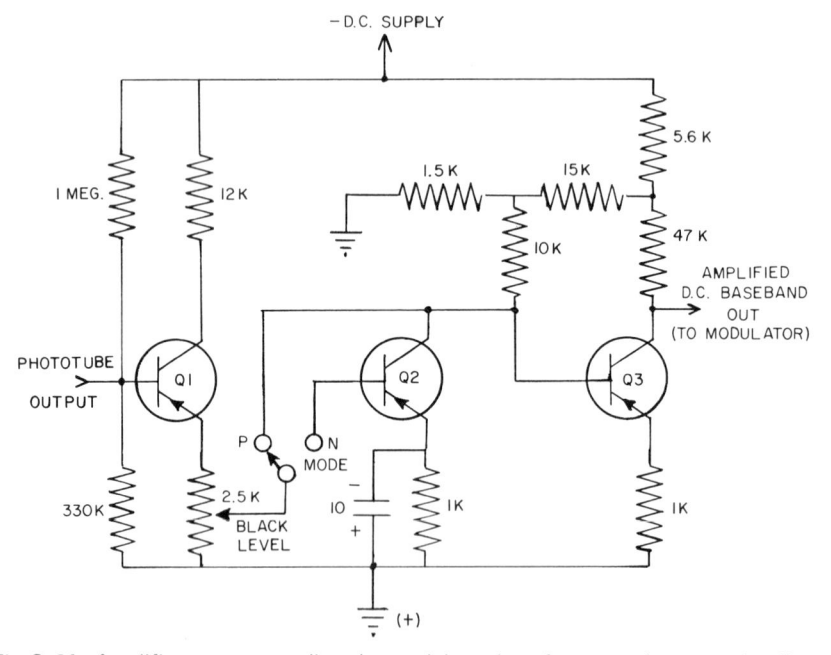

Fig. 3. 21. Amplifier stages preceding the modulator in a fax transmitter must be direct-coupled because of d.c. components in the scanner output. The same applies to stages following demodulation in the receiver. The *Mode* switch determines whether the signal will be transmitted with peaks representing black (positive) or white (negative) elements of the picture. (Courtesy Alden Research)

HOW FAX WORKS

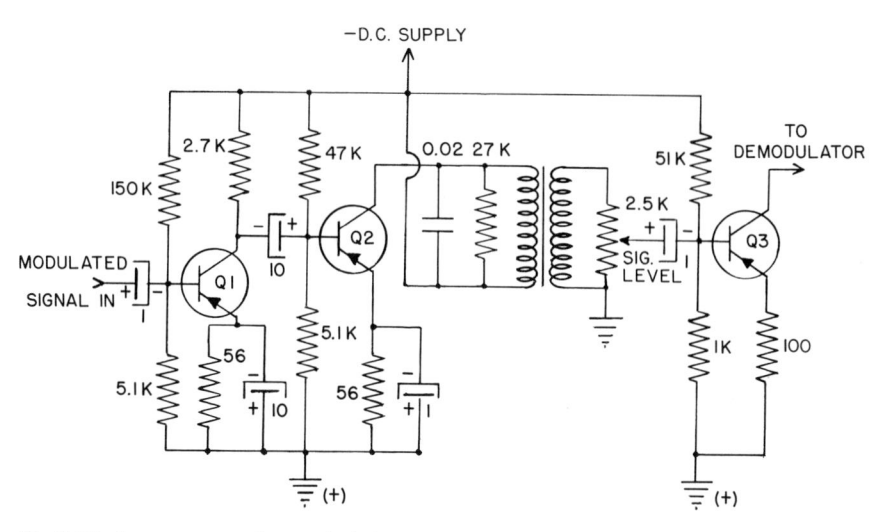

Fig. 3.22. Input stages of a typical fax receiver closely resemble a conventional audio amplifier. (Courtesy Alden Research)

object is to let the reflective properties of the subject copy determine the amplifier gain. One popular way of accomplishing this is to arrange the transmitter control circuitry so that the first reflectance value the scan spot encounters (usually the left-hand margin of the document) sets the amplifier gain for that entire scan stroke. If it is bright white, the gain will be reduced; if it is a darker shade, the gain will be increased.

In addition, to cover for possible changes in transmission circuit gain, most systems are arranged to transmit *white reference* pulses between scan strokes. When received, these serve as a check on the gain of the receiver amplifier, increasing it or decreasing it as required to compensate for transmission fluctuations.

PHASING AND SYNCHRONIZATION

It is one thing to get a picture signal from point A to point B, but it is another to reassemble the picture elements at point B in the same spatial order in which they were unraveled by the scanner at point A. To ensure the latter, two things are necessary: (1) at the commencement of transmission of image details, the positions of the scan and recording beams—or scan beam and recording stylus—must exactly coincide with respect to their relative positions within a stroke, and (2) this relationship must be maintained throughout transmission.

The first of these requirements is called *phasing*; the second, *synchronization*.

Phasing

Phasing can be accomplished in a number of ways, but they all employ the same general principle: retarding the mechanism at one end of a system (in standard practice the receiving end) until the start-of-stroke occurs at precisely the same instant at both—or all—terminals of the system. (With a proper accessory, a single transmitter can "broadcast" its output to a number of receivers simultaneously.) It will be recalled from Chapter 1 that in Bain's original facsimile system, which used stylus-equipped pendulums for scanning and recording, a leading pendulum was "grabbed" and locked by an electromagnet at the height of its stroke, to be released through a switching arrangement when the lagging pendulum caught up. This was the forerunner of the stop-start technique used in numerous facsimile systems right up to the present era.

The more popular method today is to retard, rather than completely halt, the recording mechanism until the start of a recording stroke coincides with that of a scan stroke. Depending on the design of the equipment, this may require several strokes prior to the actual transmission of image detail. In modern equipment the preliminary phasing step is completely automatic.

All conventional analog fax scanners are designed to transmit some form of phasing signal for a certain interval—usually several seconds—prior to the commencement of actual picture signal transmission. The phasing signal generally has a fixed amplitude and duration, occurring at scan-line intervals. It may vary in form from one system to another, but that shown in Fig. 3.23 is fairly typical.

Whatever the form, the object is to utilize the received pulse, in conjunction with a start-of-stroke pulse generated locally at the recorder, to free the temporarily retarded recording mechanism. Thus freed, the mechanism speeds up so that it is now in step with the scanner.

A common way of accomplishing this "catching up" is to temporarily alter, through a switching circuit, the electrical characteristics of the recorder drive motor (or of its power source) so that it runs at a slightly reduced speed for as long as the local and remote pulses remain out of phase. As soon as the pulses occur together, a relay closes and the motor is switched to normal operation.

Another way is to utilize the coincident occurrence of the two pulses to release an electromechanical, or magnetic, brake. Figure 3.24 shows one way that coincident pulses might be used to bring a receiver into phase. It is a typical pulse-comparing circuit arranged to operate a switch when the pulses coincide, and it may be applied to either of the phasing methods just described.

Synchronization (Analog Systems)

Synchronizing the scan and recording mechanisms of an analog fax system so that they remain in perfect step throughout transmission is usually achieved in

HOW FAX WORKS

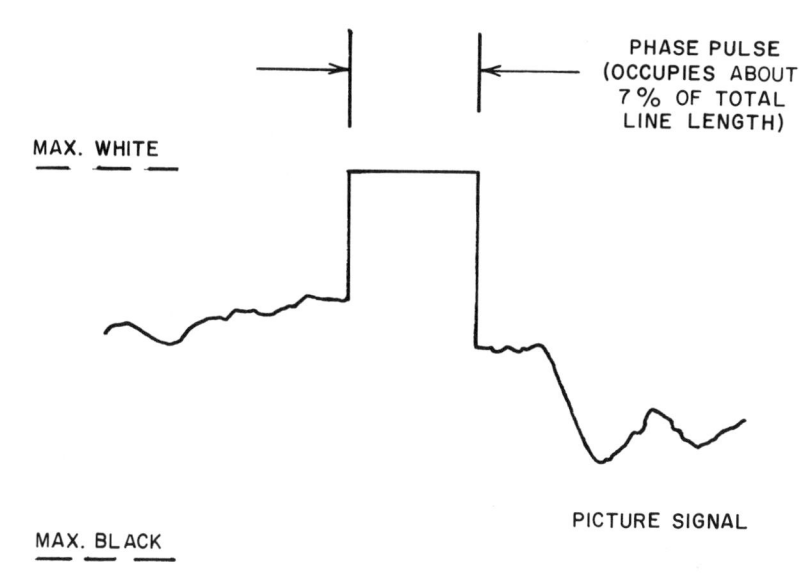

Fig. 3.23. Typical phasing signal is a full white (or full black, depending on polarity) pulse between scan strokes, occupying about seven percent of each scan cycle. During pretransmission phasing, the white pulse shown would be preceded and followed by a steady black signal.

one of three ways: (1) reliance on unification of commercial a.c. power sources serving system terminals; (2) use of matched precision power supplies; (3) transmission of a sync signal, or sync pulses, along with the picture signal.

The first two methods require the use of frequency-matched (normally 60 Hz) a.c. synchronous drive motors in scanners and recorders. Where the transmit and receive ends of the system operate from the same a.c. power grid, method (1) is applicable, and synchronization after initial phasing is automatic. This is by far the simplest method, and therefore a popular one in relatively inexpensive machines.

Where, however, the separate terminals must operate from isolated commercial power sources, it is necessary to resort to one of the other two methods. Method (2) requires that the scanner and recorder drive motors be operated from matched crystal-controlled a.c. power supplies, or *frequency standards*, as they are sometimes called. Accessory precision power supplies are available from some fax equipment vendors for more universal operation of terminals normally synchronized via the power grid.

For most purposes, the output frequencies of precision, or so-called *crystal sync*, power supplies must be matched within 0.001 percent (1 part in 100,000) to ensure adequate synchronization. Greater differences may cause a noticeable

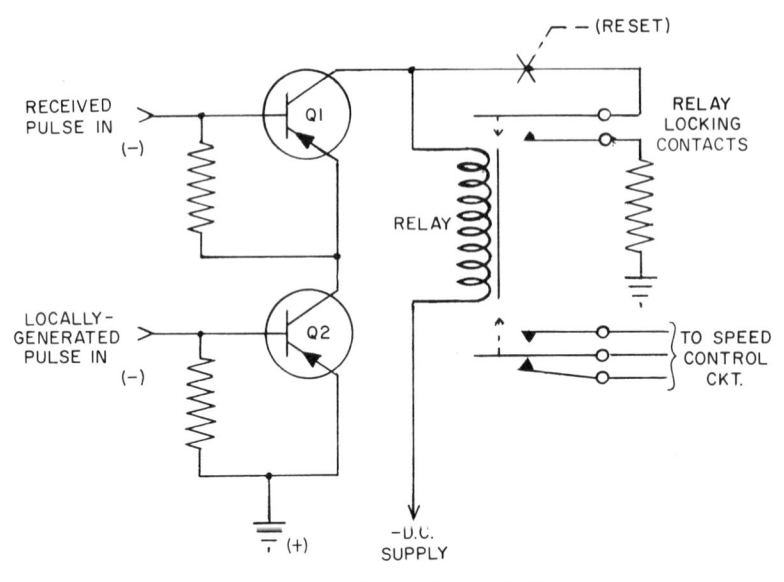

Fig. 3.24. Familiar "AND gate," or "coincidence detector," illustrates one approach to the automatic phasing of scanner and recorder mechanisms. With the system at rest, transistors Q1 and Q2 are cut off by lack of forward bias. Current cannot flow through the relay until coincident negative pulses at the inputs simultaneously unblock *both* links in the series path. The operated relay removes a temporary "drag" from the recorder mechanism.

skewing of recorded details. Elements normally at right angles to the scan axis appear to lean consistently one way or the other, the direction of skew depending on whether the recorder is running slower or faster than the scanner. The scheme of a typical crystal sync power supply is illustrated in Fig. 3.25.

Method (3), the "slaving" of the receiver to the transmitter by provision of a special sync signal, may be implemented in one of several ways. One is to transmit a continuous signal of constant frequency that is either below or above the highest picture-signal frequency transmitted. Amplified, the signal becomes the a.c. current on which the scanner and recorder drive motors operate. Where a high frequency is chosen, it will probably be a multiple of some standard motor operating frequency (e.g., 60 Hz), so that commercially available synchronous motors may be employed, the frequency being appropriately reduced through dividers.

This method requires effective filtration and separation of frequencies to ensure that the sync and picture signals do not interact. Design of the circuitry to accomplish this could be critical, depending on the amount of separation allowed by signal and channel bandwidths. Moreover, depending on the type of transmission system involved, the frequency could change sufficiently by the

HOW FAX WORKS

time it reaches the receiver to produce a noticeable skew in the copy. For the last reason alone, this method is seldom used in modern fax systems.

Another way of implementing method (3) is to send either a continuous frequency or short pulses to be utilized as a check on the speed of the recorder drive motor. This approach is similar to phasing, except that it is continuous throughout transmission of picture detail. It has the advantage of permitting use of d.c. or various other nonsynchronous drive motors, the one at the recorder being designed (or geared) to run slightly faster than the one at the scanner. "Stop-start" sync, rarely if ever used today in analog fax systems, is the most familiar example of this approach.

One other variation of method (3) is the use of between-the-line pulses to control the frequency of a motor drive oscillator at the recorder in much the same manner that transmitted sync pulses in a TV system control the sweep frequency of a local oscillator at the receiver.

There are several situations in which reliance on the commercial power grid for automatic synchronization is precluded. Among them are ship-to-shore and air-to-ground transmission, mobile operations, temporary military field installations, and most international facsimile operations. In each of these situations, some variation of method (2) or method (3) is required. Because of these limitations, and also because of the economies inherent in solid state circuit

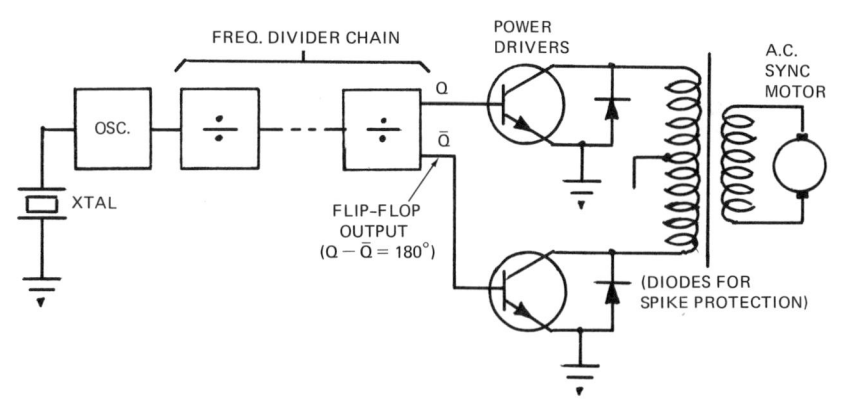

Fig. 3.25. A typical crystal sync circuit for precision operation of synchronous motors in fax terminals. The tightly regulated frequency generated by the crystal-controlled oscillator is in the radio frequency range (usually several hundred kHz) because of the more natural availability of quartz crystals at that resonance level. The oscillator output is reduced to the required motor frequency via a chain of perhaps a dozen dividers, each of which halves its input frequency. The power drivers boost the output of the final divider to motor operating level. (Courtesy Infolink)

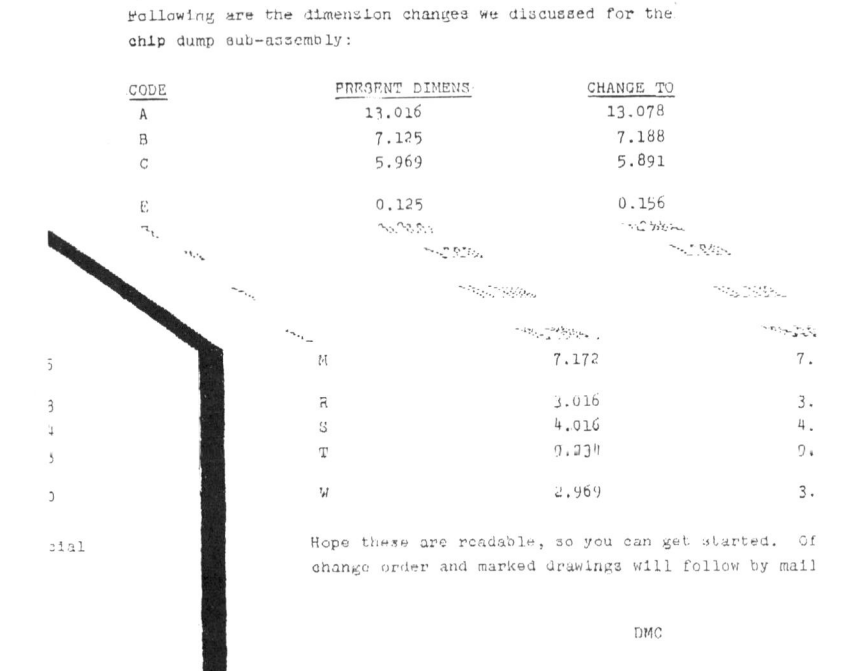

Fig. 3.26. Effects of synchronization and phasing malfunctions on a fax recording. Sync loss produces "tearing" similar to that resulting from loss of horizontal hold in a TV system. Recovery of synchronization without rephasing causes the recording stroke to start at the wrong point with respect to the paper. (simulated)

design, the general drift in recent years has been toward increased use of crystal sync.

Figure 3.26 illustrates the effects of phasing and synchronization malfunctions on a fax recording.

Digital Synchronization

In a digital fax system, there is a dual synchronization requirement. Not only do the scan and recording strokes have to start and end synchronously, but the sampling intervals *within* a stroke have to be synchronized as well.

The first requirement is usually met by some form of encoded line-by-line sync, and is made relatively simple by the fact that line advance in a digital fax system is usually on a *stepping* basis, rather than a continuous, fixed-speed movement as in an analog system. Thus, the start-of-line codes by which the "reading-out" of picture-element pulses for that line is triggered do not even have to occur at fixed intervals. This departure from the smoothly continuous, pendulumlike flow of data to which we are accustomed with the analog system satisfies one of the first requirements of a digital redundancy reduction—or "white space-skipping"—fax system. That requirement is the ability to temporarily store, in a "buffer," the line-by-line data coming into the receiver, and reading it out only when the terminal's own microprocessor circuitry has had a chance to digest and process it for recording. From this description, it may seem like a slow, hesitant process, but in fact it all happens very rapidly. (More on redundancy reduction in the next chapter.)

So much for the synchronization of line advance. As for the synchronous spacing of sampling intervals *within* a scan stroke, probably the best way to visualize that process is to examine a technique called *shaft angle encoding*. Application of the technique assumes a rotary scanning system, such as that depicted in Fig. 3.6, in which each full rotation of the rotary component (in this case a disc containing a spiral slit) constitutes the completion of one scan stroke. By simply adding some sort of commutator (or interrupter) to the drive shaft of the rotary component, such that with each full cycle the scanner output is switched on and off at many equally spaced intervals, we will have constructed a basic shaft angle encoder. The total number of on-off intervals within a stroke will be determined by the desired picture element resolution along the scan axis. It can range from a few hundred to several thousand. Popular numbers are 1024 and 1728, which equate to about 128 and 200 lines (elements) per inch, respectively. The principle of shaft angle encoding is illustrated in Fig. 3.27. The "commutator" consists of a perforated disc arranged to interrupt the light path between a light-emitting diode (LED) and a photodetector.

In practice there are matching commutators (encoders) in the send and receive terminals of the system. Assuming some sort of redundancy reduction system, the raw output of the scanner is converted to binary code "words" that in some way describe the content of each scan stroke. It is this string of codes that is transmitted to the receiver. At the receiver the code words for each separate line are reconverted to a string of black and white picture elements, which, on cue from the control circuitry, are sequentially released from the buffer to the recorder via the commutator.

Digital communication terminals, in general, are either *synchronous* or *asynchronous* in operation, although the distinction can be confusing.

Basically, in a synchronous system the terminals must be locked in phase with one another initially, and operated from precision oscillators sufficiently matched in frequency to ensure perfect synchronism for the duration of a fairly continuous flow of data. In other words, a synchronous digital data system is

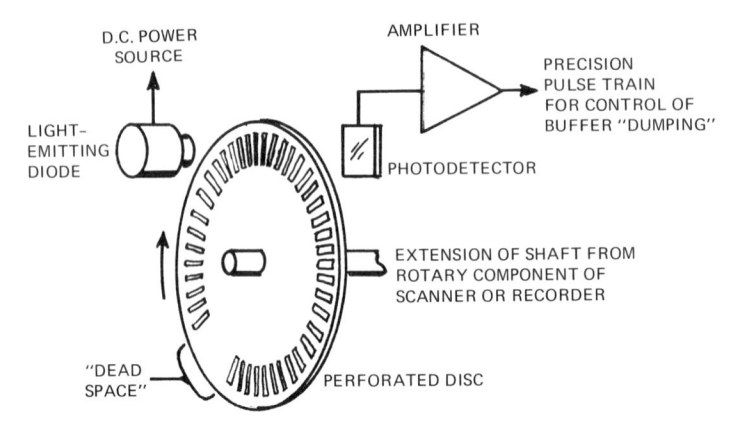

Fig. 3.27. "Shaft angle encoding" is a means of ensuring perfect synchronism of a scanner or recorder mechanism with the segmenting of a scan stroke into sampling intervals. A perforated disc affixed to the main rotating shaft of the fax terminal mechanism interrupts the illumination of a photodetector to produce the pulse train by which black-white samplings are "clocked" into or out of a buffer memory.

not unlike a typical analog fax system in terms of its basic synchronization requirements. (In fact, the maximum allowable difference in oscillator frequencies between terminals is about the same: approximately 0.001 percent.) However, most digital data systems have provision for resyncing the separate oscillators automatically at intervals of a few seconds or less.

An asynchronous digital system is distinguished by its need to remain synchronized only for the relatively short interval between the transmission of START and STOP codes—or, in a digital fax system, between occurrences of *line advance* codes. Such systems are therefore somewhat analogous to an earlier generation of analog fax systems using mechanical start-stop synchronization. A digital fax system of the sort described above would be considered an asynchronous system.

INDEX OF COOPERATION

What if the length of stroke at a recorder is less or greater than that at the scanner? Or what if the number of scan lines within a given linear dimension varies between the scanner and the recorder? The effect depends on the degree of disparity, but if the scan length varies without a proportional variation in scan-line density (scan lines per inch)—or vice versa—the result is a distorted recording.

Associated with every fax transmitter and receiver is an *index of cooperation*, which is the product of scan density and total stroke length.[4] If a transmitter

HOW FAX WORKS

and receiver have the same index, they are considered compatible, and the recording produced at the receiver will be a geometrically faithful reproduction of the subject copy transmitted. However, the reproduction will not necessarily be the same size as the original.

Consider, for example, a transmitter and a receiver each having an index of 816. This could mean that the scanner and the recorder produce the same number of strokes of the same length within a given linear dimension—e.g., 96 strokes per inch; 8.5 inches per stroke (96 × 8.5 = 816)—or it could mean that one produces fewer strokes of a greater length or more strokes of a shorter length than the other, the product in any case equaling 816. Regardless of the combination, the number of strokes necessary to result in a product of 816 will occur within the same length of time.

Assume that the scanner produces 96 $8\frac{1}{2}$-inch strokes in about half a minute, and the recorder, operating in synchronism with the scanner, produces 96 $4\frac{1}{4}$-inch strokes in the same amount of time. If both units have an index of 816, the

(a) TRANSMIT –

$8\frac{1}{2}$ INCHES

1 INCH

(ORIGINAL)

96 SCAN LINES (96 LPI)

$8\frac{1}{2}$ × 96 = 816

(b) RECEIVE –

$4\frac{1}{4}$ INCHES

$\frac{1}{2}$ INCH

(RECORDING)

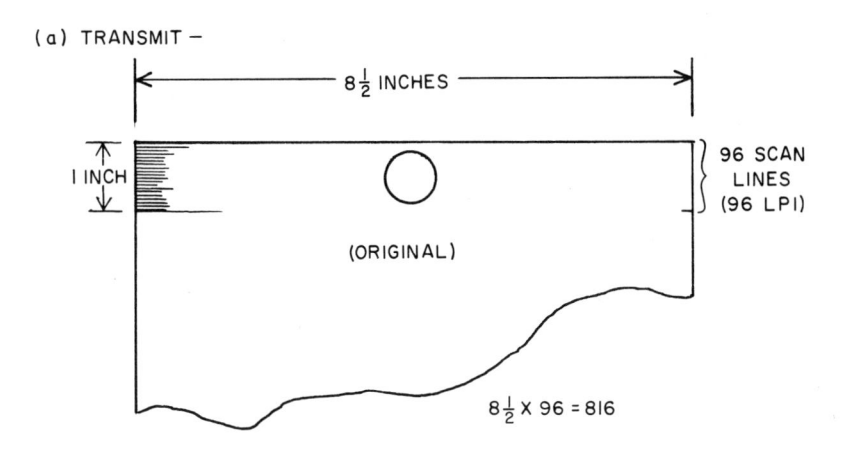

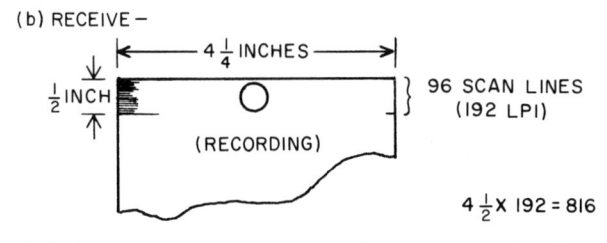

96 SCAN LINES (192 LPI)

$4\frac{1}{2}$ × 192 = 816

Fig. 3.28. Example of how matched indices of cooperation (816 in this case) will ensure against geometric distortion while permitting wide variations in relative size of original and recording. The scan *rate* (LPM) must be the same for scanner and recorder.

scanner's 96 strokes will cover only $\frac{1}{2}$-inch of space (192 strokes per inch). The result will be as illustrated in Fig. 3.28. While the recording is one fourth the size of the original, the vertical-to-horizontal relationship of image elements has been preserved.

Figure 3.29 illustrates one possible result of the linking of a recorder and a scanner with different indices of cooperation but equal scan rates. As noted, the

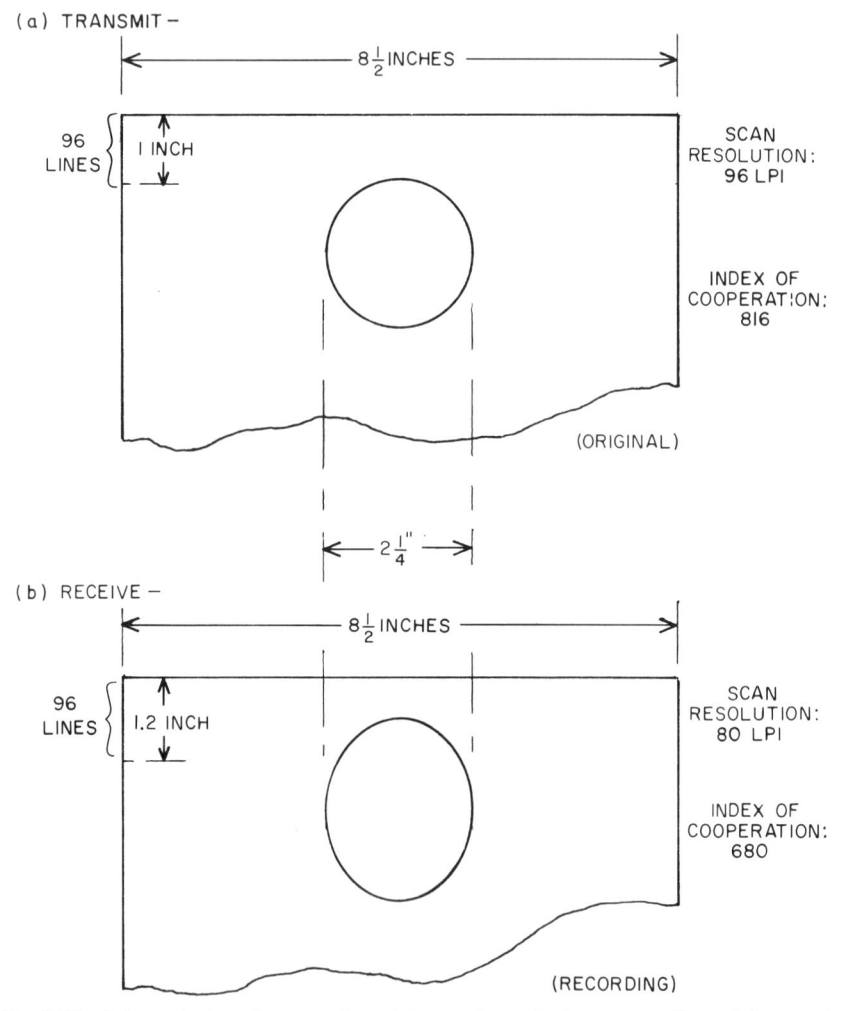

Fig. 3.29. A lower index of cooperation at the receive end of a system will result in stretching of image details in the direction normal to the scan axis.

HOW FAX WORKS

(a) TRANSMIT —

- 8½ INCHES
- 80 LINES / 1 INCH
- SCAN RESOLUTION: 80 LPI
- INDEX OF COOPERATION: 680
- (ORIGINAL)
- 2¼"

(b) RECEIVE —

- 8½ INCHES
- 80 LINES / 0.83 INCH
- SCAN RESOLUTION: 96 LPI
- INDEX OF COOPERATION: 816
- (RECORDING)

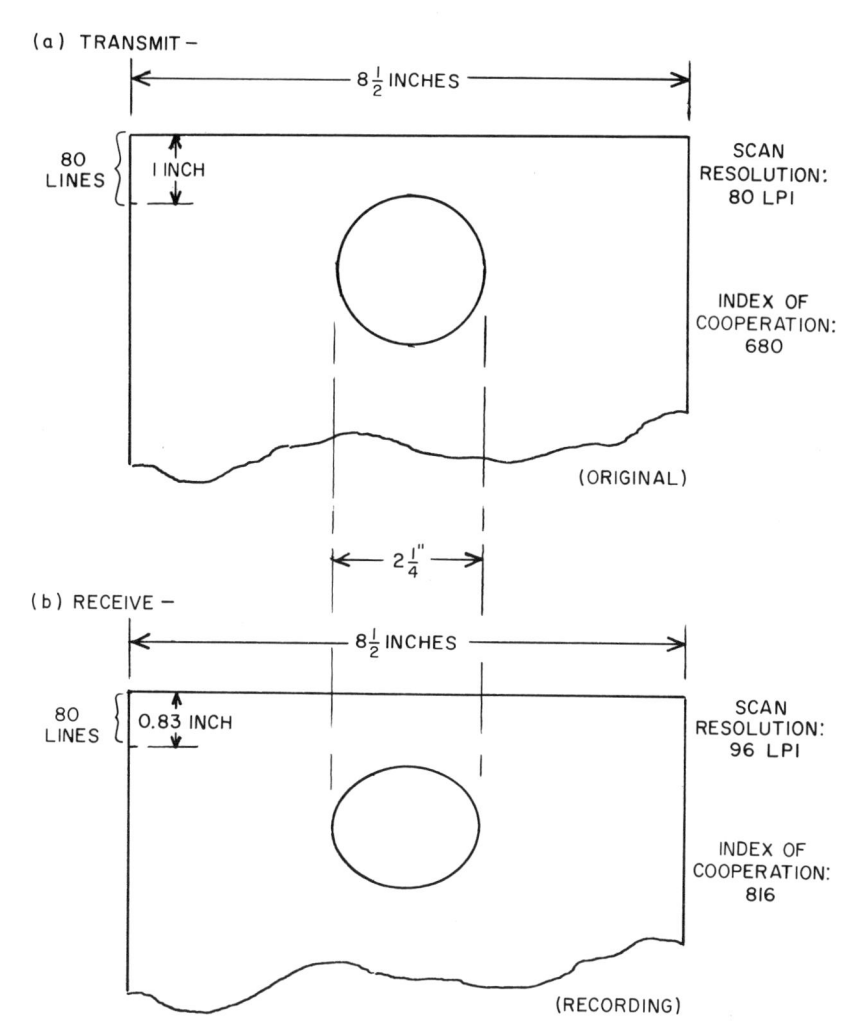

Fig. 3.30. A higher index of cooperation at the receive end of a system will result in compressing of image details in the direction normal to the scan axis.

scanner's index is 816, while the recorder's is 680. In this particular example, the 680 index is the product of 80 strokes per inch and 8.5 inches per stroke. But regardless of the combination of scan-line density and scan length, so long as their product is 680, the distortion pattern and the degree of distortion will be the same. Figure 3.30 illustrates the result if the indices of the scanner and the recorder are reversed.

The next chapter will discuss technical considerations and special problems involved in the actual transmission of a picture signal between the separated terminals of a fax system.

FOOTNOTES

1. IEEE No. 168, "Definitions of Terms For Facsimile."
2. Grandnephew of Heinrich Hertz, the German discoverer of radio waves.
3. *Automatic Printing and Telegraph Systems*, Part 1 (Western Union/American School of Correspondence, 1918), p. 9.
4. The international (CCITT) index of cooperation is based on cylinder diameter rather than stroke length, and is equal to 0.318 multiplied by the IEEE stroke length index.

4
TRANSMISSION

The output of a facsimile scanner is like that of a record player in that it is an electrical analog, varying in both amplitude and frequency, and that, typically, the frequencies it contains are within the audio range. Further, the amplitude transitions constituting the outputs of both devices may be either sinusoidal or essentially square and can therefore vary widely in harmonic content.

There are, however, two significant differences between the outputs of a fax scanner and a record player.

One, already briefly mentioned in the preceding chapter, is that the fax signal does not ordinarily contain "clean" frequencies such as those that would identify a flute or a bell in an audio signal. Having resulted from a succession of generally random light reflectance variations, the scanner output consists primarily of an irregular chain of transitions that might best be described as a kind of nondescript "hash" (Fig. 4.1).

The other basic difference is that the scanner output can vary in frequency from several thousand hertz all the way down to zero. It is this latter characteristic more than any other that necessitates altering the form of the fax signal before it can be sent out over a transmission circuit engineered to handle voice currents.

D.C. AND SUBAUDIO

To say that the fax scanner generates "zero frequencies" is to say that its output contains very slow variations that can approach a d.c. component. This fact presents a problem because virtually no communication transmission system in operation today will accept or pass subaudio frequencies. (Transmission has evolved into a highly sophisticated art since the day when the d.c. telegraph was the dominant means of communicating electrically.)

If the d.c. component of the scanner output represented only the white space of the copy being transmitted, there might not be a problem. Assuming an output polarity of maximum on black,[1] a steady white level would be analogous to

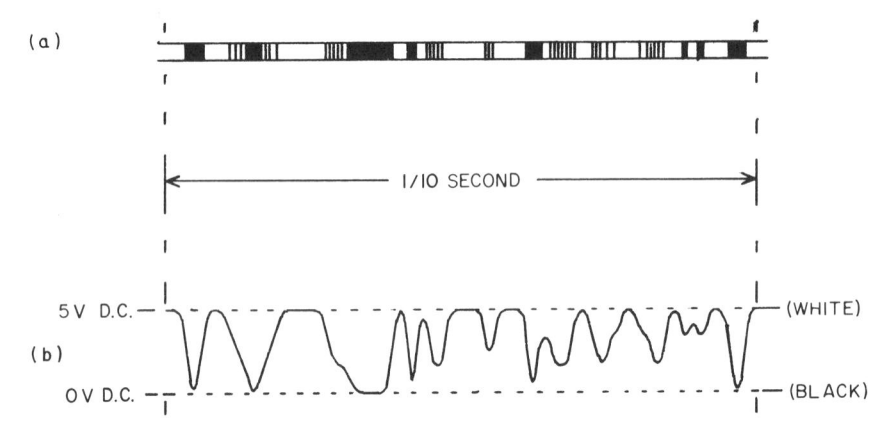

Fig. 4.1. Analogous fax baseband signal (b) resulting from random light variations within a scanned segment (a) of a subject copy.

a silent passage in an audio signal (e.g., a pause in a telephone conversation). The problem is that broad expanses of black—underscoring on a document, for example—and shades of gray, as in a photograph, are each represented by an analogous d.c. level. Moreover, there are digital (binary) fax systems in which the scanner output has been processed into "pulse trains"—series of square pulses having discrete d.c. levels. (Digital transmission will be discussed later in this chapter.)

However, even if the analog scanner output frequency did not go to zero and was not converted to d.c. pulses, but contained only very low frequencies (say in the neighborhood of 20 to 50 Hz), there would still be a problem in sending it over telephone circuits because of their poor low-frequency response. (This characteristic is, to some extent, deliberately built in to avoid a.c. power hum and to provide separation between channels.) The response curve for a single voice path in the telephone network is, in general, reasonably flat between 300 and 2500 Hz, but tends to fall off sharply beyond these points (Fig. 4.2). The channel is said to be *band-limited*.

In actuality, if a solid black, gray, or white area were to extend through several scan strokes, the resulting scanner output would not be a steady d.c. level—a steady zero—because of the regular occurrence of phasing, or "dead sector," pulses between strokes. Nevertheless, the resulting waveform would be such that it would present problems in transmission and, even if viewed as pulsating d.c., would represent a subaudio frequency in all but the highest-speed fax systems.

TRANSMISSION

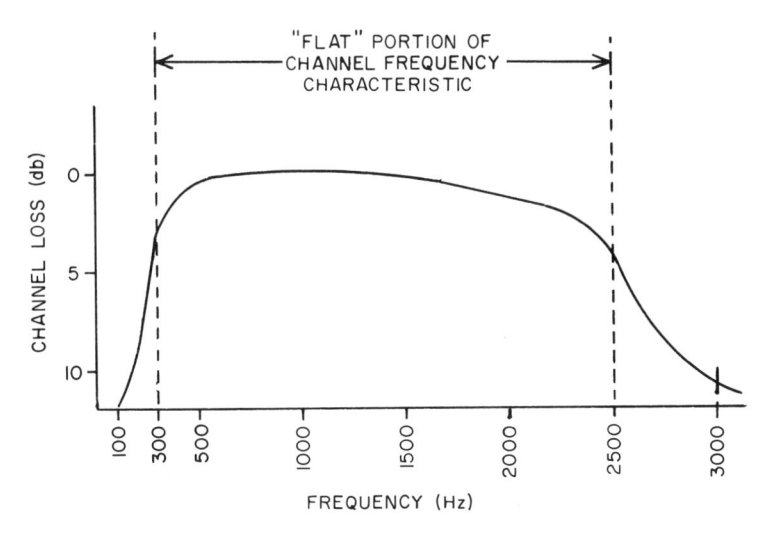

Fig. 4.2. Frequency characteristic of a typical telephone circuit.

MODULATION

The subaudio problem is solved in practice by impressing the scanner output on an audio frequency carrier. In that way the various d.c. levels are converted to audio rate a.c. of discrete frequencies or of corresponding amplitude levels of a fixed frequency, depending on the kind of modulation used. In addition, since the carrier frequency is customarily higher than the highest anticipated picture signal frequency, modulation has the effect of shifting the entire picture frequency range upward so that it is wholly within the flat portion of the transmission circuit's response curve.

Baseband, Harmonics, and Sidebands

In a conventional 96 LPI/180 LPM^2 message fax system, assuming equal vertical and horizontal resolution, the required picture signal frequency range—or *baseband*—is from 0 to about 1300 Hz. The 1300 Hz maximum assumes a 9-inch stroke interval, within each of which a total of 432 black-white-black transitions can be detected and reproduced. The 432 full-cycle transitions constitute a spatial frequency of 48 cycles per inch, or the equivalent of 96 contiguous black or white elements per inch. Since there are 96 scan lines per inch, this assures substantially equal vertical and horizontal resolution. At 180 strokes a minute, there are 3 strokes a second. Thus, $3 \times 9 \times 48 = 1296$ (about 1300) cycles per second (Hz).

The next consideration is how much additional channel capacity has to be allowed for harmonics and sidebands (the latter being the by-products of modulation). Just as a reproduced violin sound will lack fidelity if the audio system fails to reproduce the instrument's natural overtones or harmonics, a recorded image detail in a fax system will theoretically lack fidelity if harmonics are suppressed by too narrow a transmission channel.

Some purists have contended, in fact, that faithful reproduction of a modulating waveform requires a channel bandwidth sufficient to include the third harmonic of the highest fundamental frequency generated by the scanner. Moreover, it is a rule of thumb that the carrier frequency be at least double the highest baseband frequency to be transmitted. Any less of a gap between carrier and baseband will lead to an overlap of the latter and the lower sideband produced by modulation (Fig. 4.3), resulting in a distortion known variously as the *Kendall effect* and *aliasing*. It will usually manifest itself as a kind of fuzziness at the edges of recorded picture details.

In the case of the conventional 96/180 system, the third harmonic would be 3.9 kHz (3 × 1296 Hz), which would amount to a carrier frequency of 7.8 (2 × 3.9) kHz. Such a frequency far exceeds the capacity of a telephone voice circuit, even if allowance for sidebands is ignored.

The sidebands produced as a by-product of modulation consist of a range of frequencies below and above the carrier frequency, extending from the carrier *minus* the highest baseband frequency to the carrier *plus* the highest baseband frequency. (This pertains to *amplitude modulation*—AM—only. The sideband situation is somewhat different for frequency modulation—FM. The distinction

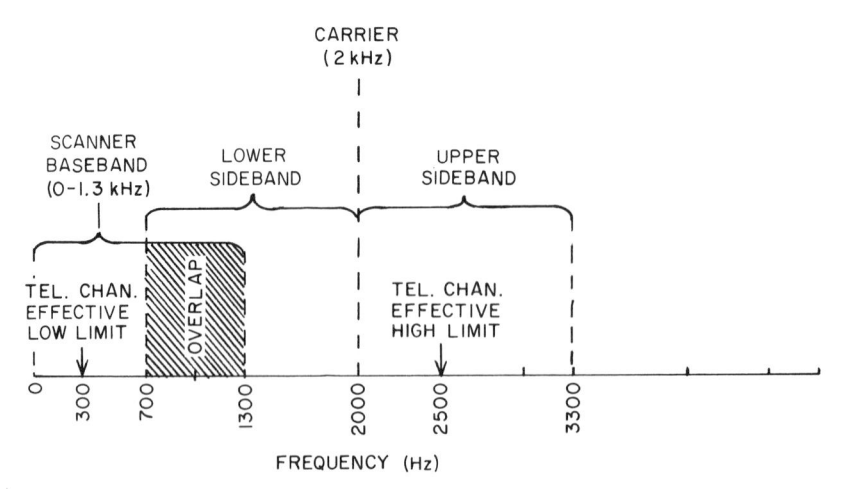

Fig. 4.3. Overlap resulting when the carrier frequency is less than twice the highest baseband frequency.

TRANSMISSION   **107**

will be discussed later in this chapter.) In the example just given, the lower sideband would range from 3.9 to 7.8 kHz, and the upper from 7.8 to 11.7 kHz. Theoretically, then, the transmission circuit for a conventional message facsimile system would have to be capable of handling frequencies as high as 11.7 kHz—which would, of course, completely rule out the use of conventional voice-grade telephone facilities.

So much for theory. In actual practice it has been found that harmonics can be pretty much ignored without risking any obvious degradation of recording fidelity in a fax system. Thus, from a practical standpoint, about the highest baseband frequency we need concern ourselves with in the 96/180 system is 1300 Hz. Experience has also shown that, Kendall effect notwithstanding, the carrier frequency can be substantially less than double the highest baseband frequency, and recording fidelity will still be adequate for most purposes.

But even when harmonics are ignored and no margin at all between the top baseband and carrier frequencies is allowed, the modulated carrier plus sidebands will just barely fit within the flat portion of the telephone network bandpass (Fig. 4.4). The bandspread of the fax transmitter has still somehow to be reduced.

Sideband Suppression

The one further step necessary to fit the fax signal into a voiceband channel is sideband suppression. In 1915, John R. Carson, an AT&T Company engineer,

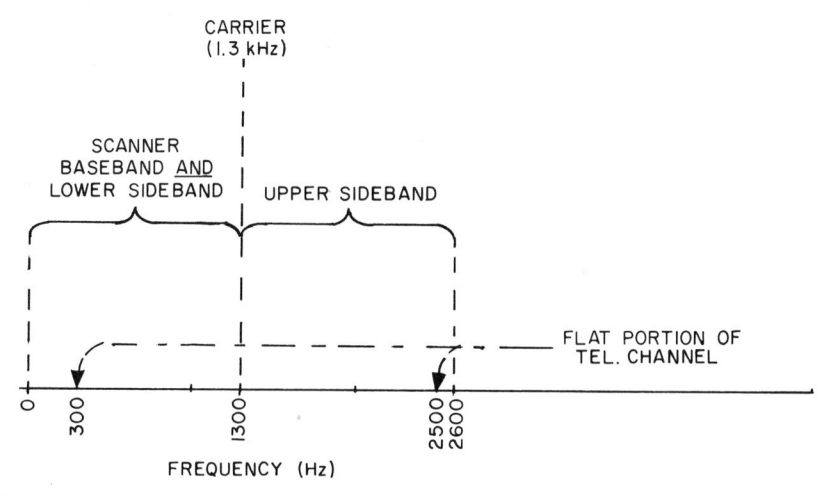

Fig. 4.4. Heavy channel space requirements make double-sideband AM a generally unsuitable fax transmission technique for the telephone network.

made some important observations regarding the role sidebands play in modulated carrier transmission. He established that identical information is conveyed by each of the two sidebands produced when a carrier is amplitude-modulated, and that each contains all of the intelligence present in the modulating signal. This being the case, it followed that one of the two sidebands could be suppressed with (theoretically) no effect on the intelligence being transmitted.

In fact, both an entire sideband (usually the upper) and the carrier itself can be suppressed by bandpass filtration without seriously affecting the fidelity of the received signal. The carrier frequency must, of course, be restored at the receiving end. However, while this *single-sideband technique* makes the most efficient use of a transmission channel, it is difficult in practice to suppress the upper sideband and carrier frequencies without also sacrificing a portion of the lower sideband. The portion lost will contain lower-frequency baseband components and will therefore result in total distortion of relatively broad picture elements in a fax recording.

A good compromise technique, and one that has been resorted to in a number of commercial fax systems, is *vestigal sideband transmission*. Here the entire lower sideband, the carrier, and a small portion of the upper sideband are transmitted. Using a carrier slightly higher in frequency than the top baseband frequency—e.g., 1800 Hz in the aforementioned 96/180 system—and retaining the portion of the upper sideband extending to, say, 2300 Hz, we have an audio frequency signal of a range that will fit most telephone circuits with room to spare (Fig. 4.5). A drawback of this technique is that positioning of the carrier and design of the bandpass filters may be critical.

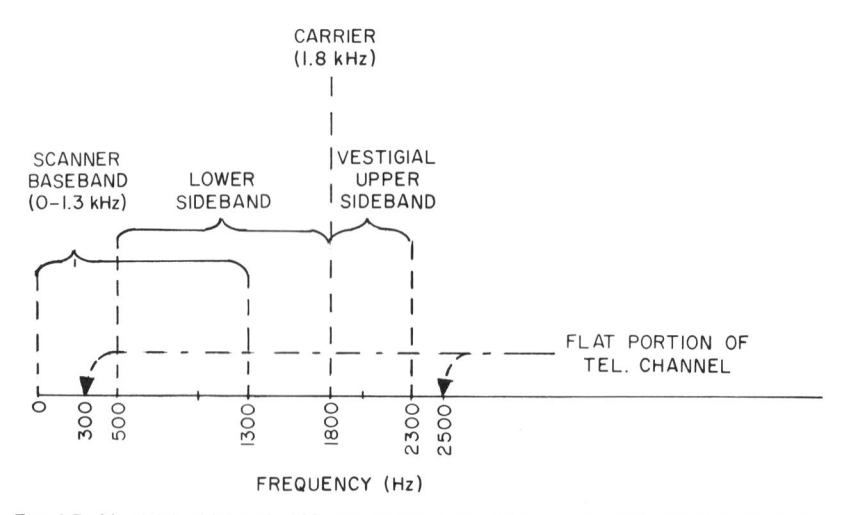

Fig. 4.5. Vestigial sideband AM transmission provides a practical compromise between single- and double-sideband techniques.

Limitations of the "Switched Network"

In the foregoing paragraphs occasional reference has been made to the "telephone network." By this is meant the switched, or dial, network to which a telephone subscriber has access only for the duration of a call. It is termed a switched network because the trunks connecting two telephone central offices are shared by all subscribers served by each office. When a party in one city phones someone in another, automatic switching equipment at the originating central office selects an interconnecting trunk that does not happen to be in use at the moment. A subsequent call placed between the same two points will very likely go over a different trunk.

Telephone lines can also be privately leased and specially conditioned for fax transmission. A disadvantage of the switched system for fax is that while slight differences in transmission quality between two trunks may not be detectable audibly, they may result in fax recordings of unpredictably different visual quality. Over the years, however, and particularly in the past decade, the transmission quality of switched trunks has steadily improved and has become more consistent.

So far, only amplitude modulation has been discussed as a fax transmission technique. There is also frequency modulation, which is said to offer certain technical advantages that make it the preferred modulation technique for systems designed to interface with the switched network. In recent years this matter has become the subject of much debate. Nevertheless by tradition systems intended for private line use have favored AM, whereas FM has been traditionally favored for systems intended primarily for use on the telephone dial network.

AM Versus FM

The fundamental distinction between AM and FM is that AM converts the varying modulating voltage to corresponding amplitude variations in a fixed-frequency carrier, whereas FM converts it to variations in carrier frequency. The distinction is illustrated in Fig. 4.6.

Ordinarily, the chief advantage of FM over AM is its relative immunity to the effects of electrical noise. Unfortunately, it is an advantage usually gained only at the expense of additional bandwidth—which the telephone network cannot afford.

It is nevertheless "noise" of sorts that makes FM somewhat preferable to AM where the switched network is concerned. We generally think of noise as the random, more or less constant, low-level voltage fluctuations that find their way into a transmisstion circuit and become an unwanted part of the transmitted signal. Amplifier-tube noise, power-line induction, and crosstalk are typical examples.

110 ELECTRONIC DELIVERY OF DOCUMENTS AND GRAPHICS

(a) BASEBAND D.C. (MODULATING VOLTAGE)

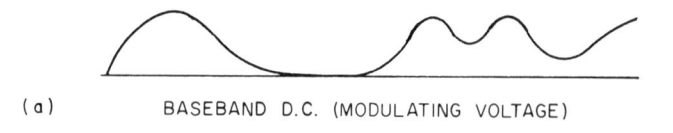

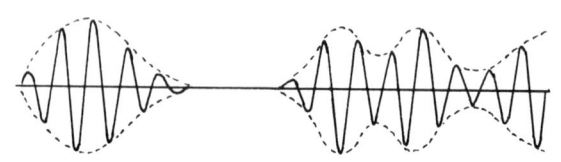

(b) AM: BASEBAND FLUCTUATIONS CONTROL CARRIER AMPLITUDE

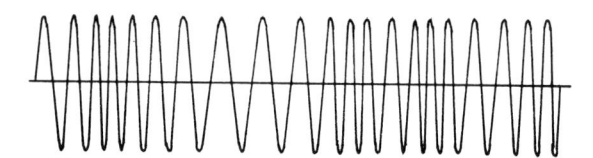

(c) FM: BASEBAND FLUCTUATIONS CONTROL CARRIER FREQUENCY

Fig. 4.6. Baseband d.c. from the scanner can modulate either the amplitude or the frequency of the carrier.

But there is also noise in the form of changes in signal level during transmission or from one trunk to another. At one time such fluctuations were particularly prevalent in the switched network, where they had a variety of causes, including the switching of batteries or other circuit components at any of the several exchanges a call may have had to pass through en route to its destination. Their occurrence could have an unpredictable and at times significant effect on the quality of fax transmission in an AM system. FM, with its relative insensitivity to amplitude fluctuations, is largely unaffected by them.

Another kind of amplitude fluctuation is the fading encountered in long-range radio reception, an atmospheric phenomenon that is unpredictable and generally unavoidable. Where fax signals are transmitted via long-range AM radio, the effects of fading—notably "rushes" of noise in the recording—might be disastrous, were it not for the expedient of employing an FM subcarrier (more on this in the discussion of radio facilities later in the chapter).

As for bandwidth, which FM normally needs plenty of, requirements are governed by two variables. It is an unfortunate peculiarity of FM that it produces not just two sidebands, as does AM, but a proliferation of them, each con-

TRANSMISSION 111

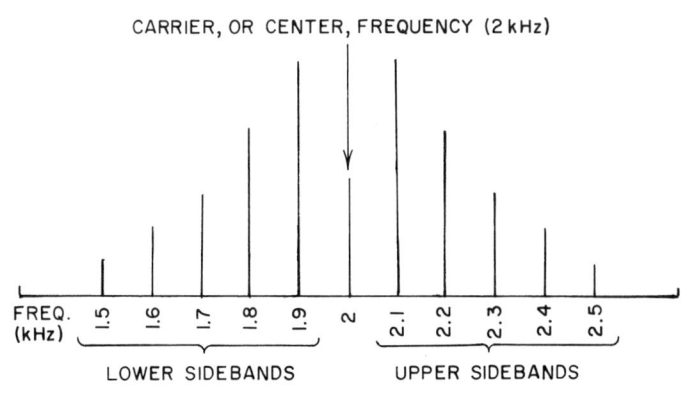

Fig. 4.7. Sideband proliferation is a characteristic of FM. Shown here is a partial array resulting from an instantaneous modulating frequency of 100 Hz.

taining a portion of the signal energy. Their spacing is in direct proportion to the instantaneous baseband frequency, and their number and relative significance in terms of energy content varies in proportion to the strength of the input signal and thus the degree of carrier swing (Fig. 4.7).

By constraining and balancing each of these variables (among other special design twists), it has been possible to perfect FM modulators that function quite effectively in "fax-by-phone" systems. A basic requirement is that the ratio of maximum carrier swing to maximum baseband frequency be as small as effective operation of the system will permit. A small ratio results in a low modulation index, which makes the FM signal practically indistinguishable from AM in terms of required bandwidth and resistance to noise other than changes in signal level.

Through skillful engineering (and some judicious compromising of quality), 96-LPI FM fax systems are now available that can transmit at the rate of more than 300 LPM over the switched network without apparently penalizing output legibility.

Modulation Schemes

What are the actual mechanics of modulation in a facsimile transmitter? There are probably more different schemes than there are makes of fax terminal equipment in use, but a quick review of some of the more venerable methods plus a brief analysis of a couple of modern circuits should suffice to give a general idea of what is involved.

During the reign of the vacuum tube, analog modulators in general were classified as either plate or grid, the grid being favored in facsimile because it

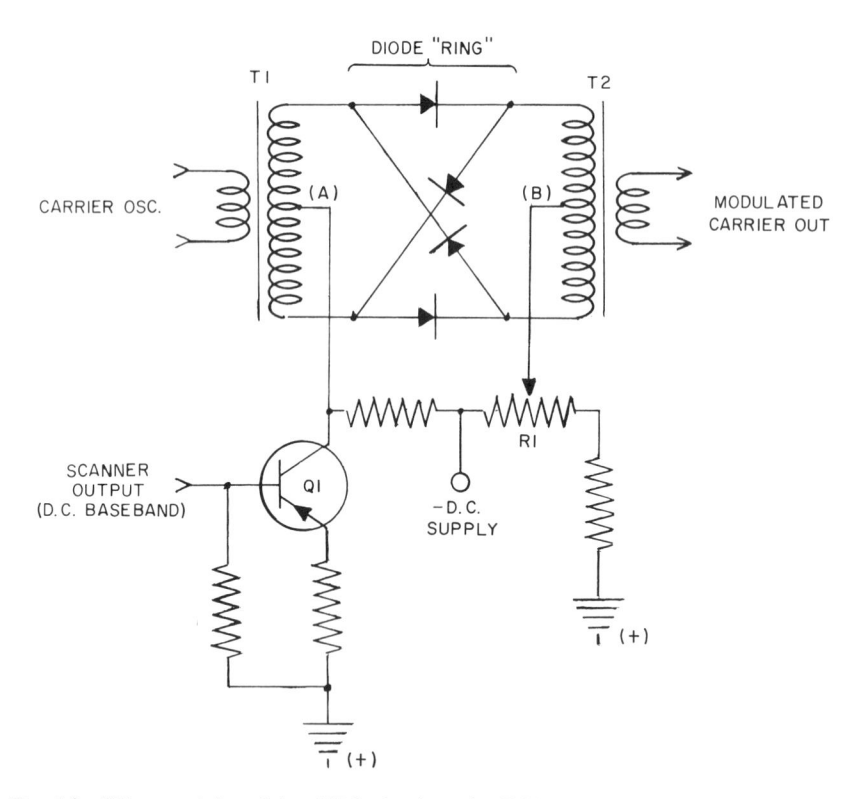

Fig. 4.8. "Ring modulator" for AM is simple and reliable. Balancing of d.c. bias between points (A) and (B) prevents flow of carrier from T1 to T2. D.C. signal fluctuations at the collector of Q1 produce a varying bias differential, controlling the flow of carrier in accordance with the baseband variations.

permitted use of a lower-level modulating voltage and thus minimized the required number of preamplifier stages. With vacuum tubes baseband amplification required relatively high d.c. supply voltages because the amplifier stages had to be directly coupled to preserve the d.c. component of the transducer output (see Chapter 3, *Amplification*). In some vacuum-tube fax transmitters, the transducer output was fed directly to the modulator, thereby eliminating preamplification entirely.

Another method employed in some older fax transmitters to eliminate the need for d.c. preamplification was to mechanically interrupt, or "chop," the scanner light beam at a carrier rate (approximately twice the highest baseband frequency to be transmitted). Modulation thus actually preceded the generation of the baseband signal, permitting the latter to be processed through a conventional capacitor- or transformer-coupled amplifier.

TRANSMISSION 113

Transistors have lent themselves much more readily to the design of direct-coupled amplifiers capable of operating at the low supply voltages characteristic of solid-state circuits, and have in general simplified the design of modulators.

A basic AM modulator circuit is shown in Fig. 4.8. It is a conventional *ring* modulator, so named for the arrangement of the four diodes in a kind of ring, or electrically circular configuration, which functions as a bridge. Adjustment of R1 balances the d.c. bias on the four diodes, in which state the diodes constitute a short circuit that blocks the carrier frequency from getting to T2 from T1. The biases become unbalanced with each change in voltage at the collector of Q1 because of the scanner output variations applied to the base. The bridge thus unbalances at the baseband rate and controls the flow of carrier current to T2 accordingly, resulting in an amplitude-modulated carrier at the output.

The FM modulator shown in Fig. 4.9 is essentially a free-running, symmetrical

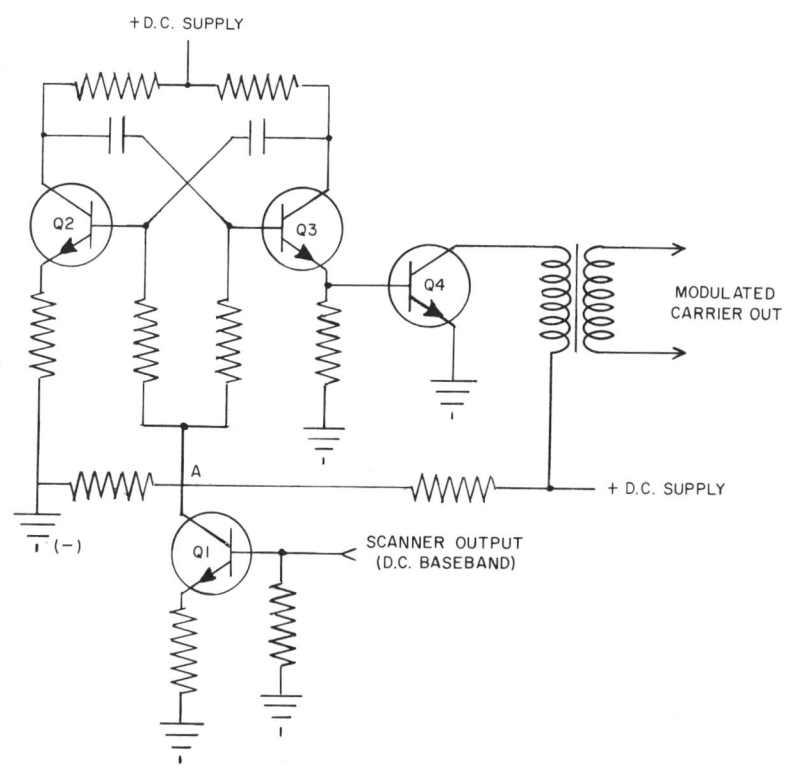

Fig. 4.9. FM modulator employing a free-running multivibrator (Q2 and Q3), the frequency output of which varies with supply voltage fluctuations. Baseband current flow through Q1 varies the supply voltage at point (A) at the baseband rate, thereby frequency-modulating the multivibrator output. Q4 isolates the multivibrator from the transmission line.

multivibrator (Q2 and Q3), the output frequency of which is controllable by varying the d.c. supply voltage. Modulation is accomplished by using the amplified baseband, d.c. to vary the supply voltage (via Q1). The multivibrator is isolated from the transmission line by Q4, preventing unwanted frequency shifts due to possible stray currents and line impedance changes.

Demodulation (Detection)

In a basic AM system, restoration—or detection—of the baseband signal at the receiver is principally a matter of rectification. As transmitted, the baseband is the *envelope* of the a.c. carrier and is electrically self-canceling. The first step in restoration, therefore, is to eliminate either the positive or the negative half-cycles of the carrier (Fig. 4.10).

In practice the retained carrier half-cycles also are smoothed (by filtering) so that only the broader d.c. fluctuations represented by the envelope remain. The latter are, in effect, the restored scanner output, which is amplified and fed to the recorder.

Whereas AM signals are detected by rectification, FM detection is basically a filtering action, the object being to attenuate the carrier's constantly shifting frequency on a sloping curve so that the higher frequencies (for example) will be passed more freely than the lower ones. Thus processed, the carrier now varies in strength with frequency and can be treated as an AM signal. Rectified

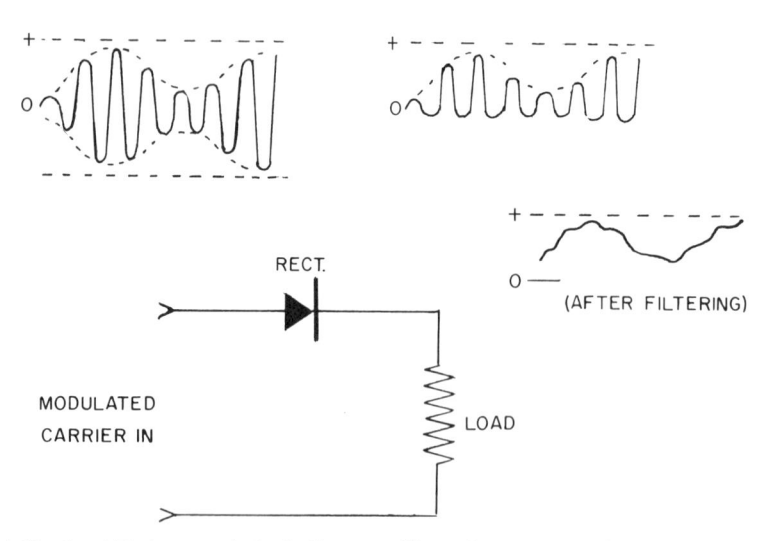

Fig. 4.10. An AM detector is basically a rectifier, whose purpose is to convert the a.c. carrier to pulsating d.c. Subsequent filtering effectively eliminates carrier pulses, leaving only the restored d.c. baseband variations contained in the carrier envelope.

TRANSMISSION

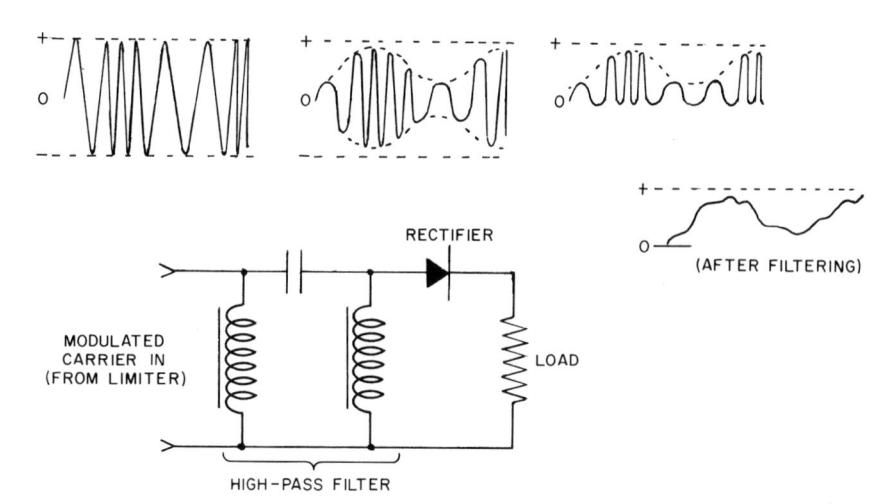

Fig. 4.11. FM detection is a combination of "slope" filtering and rectifying. The circuit shown illustrates the basic principle. The modulated carrier emerges from the high-pass filter with its amplitude attenuated at lower frequencies. It thus acquires an equivalent AM envelope, enabling recovery of the baseband by conventional rectifying and filtering.

and filtered, it becomes a faithful reproduction of the scanner output. The general principle is illustrated in Fig. 4.11.

Immediately before detection, the received FM carrier passes through a *limiter*, which clips its amplitude peaks, eliminating unwanted level variations that may have been picked up in transmission. As pointed out earlier, it is this aspect of the FM technique that makes it attractive for fax transmission via the public telephone network, where level variations may be a problem. In principle the limiter is really nothing more than a conventional amplifier stage that is overdriven by an excessive input-signal level.

TRANSMISSION IMPAIRMENTS

Some of the sources of impairment of a fax recording have already been broadly discussed, particularly as they relate to the use of the public switched telephone network as the transmission medium. It might be well at this point to briefly review some of those already mentioned and to explore others that are likely to be encountered regardless of the transmission medium employed.

External Interference

Lightning, power-line induction, and "crosstalk" are among the principal external interferences that can blemish a fax recording. Lightning and other "arcing" interferences are usually momentary, but their spasmodic occurrence could obliterate essential details of the transmitted copy. However, as pointed out in Chapter 2, one of the virtues of the fax process is its relative immunity to occasional impulse "hits." The extent of the effect on the received copy will generally depend on the type of system (whether analog or digital data compression), the relative coarseness of the scan resolution, and the nature of the transmitted copy.

AM radio systems and open-wire telephone lines (the latter are fast approaching extinction) are most vulnerable to atmospheric disturbances, but all telephone connections are subject to impulse noise from arcing interferences.

Single-frequency interference such as power-line induction, unlike interference from spasmodic arcing, is usually continuous and symmetrical, and therefore will generally result in a kind of herringbone, or diagonal-stripe, pattern over the entire fax recording (Fig. 4.12).

Crosstalk is a problem almost as old as the telephone. Despite elaborate measures taken to minimize it, it persists as an elusive malady in most transmission systems having multiple parallel paths. It occurs when the transmission level in a wire circuit or carrier channel exceeds prescribed limits and causes transmission paths to overlap.[3] Its effect on a fax recording is somewhat like that of single-frequency interference (Fig. 4.12), but without the clean symmetry.

There is actually no fixed threshold of signal strength beyond which crosstalk will occur in all cases. The maximum tolerable signal current can range from 1/20 to 1 milliwatt. The coupling circuitry generally required for connection of privately operated terminals (including fax transmitters) to the telephone network is designed to hold the maximum signal energy within prescribed limits for a given installation.

Internal Noise

Various "noises" can originate *within* a transmission path as well. Typical examples are contact noise in a switching system, tube and transistor noise within terminal amplifiers and repeater (relay) stations, and intermodulation noise caused by nonlinear circuit elements.

Some, if not all, of these noises are likely to be present in varying degrees in almost any transmission system. Naturally, the more complex the system, or the longer the transmission link, the greater the variety of noises and the greater their potential effect.

For the purposes of this book, the main point to be made about these unwanted, but usually unavoidable, noises—and this applies in general to external interference as well—is that in a typical system they can usually be ignored if their level remains at least 30 decibels (db) below the maximum picture-signal

TRANSMISSION 117

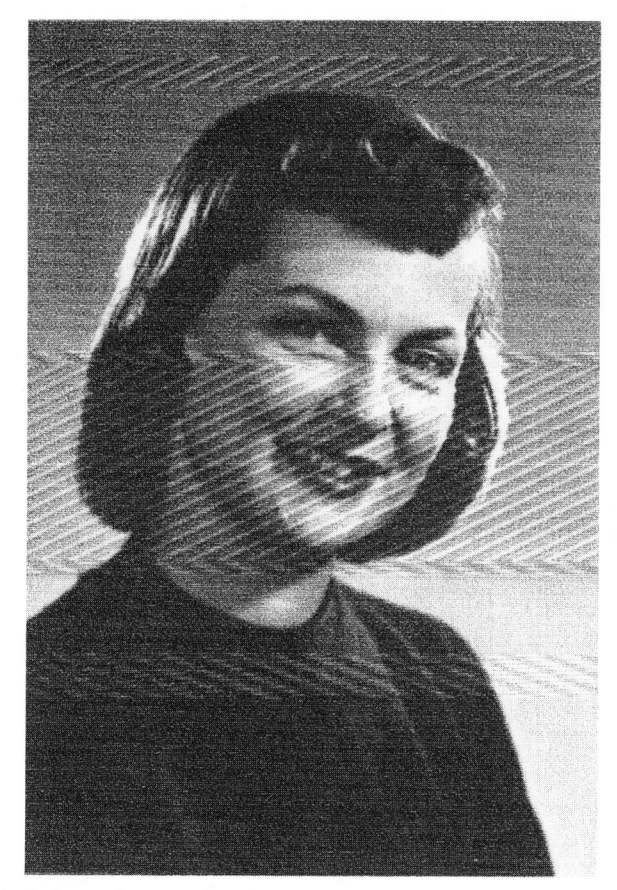

Fig. 4.12. Single-frequency a.c. interference produces a herringbone pattern in a fax recording.

level. If they are of a single frequency, such as the "singing" that sometimes results from feedback in two-wire telephone circuits, the signal/noise separation may have to be slightly greater.

In general the effect of these "noise" impairments, even at higher levels, may be merely aesthetic. However, as previously indicated, this will depend somewhat on the nature of the system—particularly the extent to which it employs sophisticated digital techniques. For example, random (broadband or *gaussian*) noise sufficient to produce a slight "spatter" effect in analog copy could disable a 9600-bit-per-second digital system operating over a voiceband analog circuit.

Level Fluctuations

Fluctuations in transmission level can occur for a variety of reasons, some of them associated with intrinsic characteristics of long telephone circuits, others with natural phenomena that are beyond human control (specifically the fading encountered in long-range radio reception), and still others that are purely accidental. The last category consists of intermittent troubles that can occur anywhere within the system, including the fax terminal equipment. These need not be discussed, since they are not inherent conditions.

An inherent (though not necessarily critical) source of level fluctuations in the telephone system is the in-service substitution of equipment—batteries, amplifiers, etc.—at telephone exchanges. The longer the circuit and the more exchanges it passes through, the greater the chance that such fluctuations will

Ed:
Following are the dimension changes we discussed for the chip dump sub-assembly:

CODE	PRESENT DIMENS	CHANGE TO
A	13.016	13.078
B	7.125	7.188
C	5.969	5.891
E	0.125	0.156
F	0.812	0.844
G	1.250	1.281
J	11.188	11.172
L	15.344	15.328
M	7.172	7.125
R	3.016	3.078
S	4.016	4.094
T	9.234	9.203
W	2.969	3.000

Hope these are readable, so you can get started. Official change order and marked drawings will follow by mail

DMC

Fig. 4.13. Effect of a sudden change in transmission amplitude level in an AM fax system. The weak signal portion at the top also illustrates the effect of random noise. (Level change is simulated.)

TRANSMISSION

occur. These abrupt changes in transmission net loss are small enough to go unnoticed in aural communication, but sufficient to visibly blemish a fax recording (Fig. 4.13). The effect is most likely to be discernible in a system having broad tonal capability, e.g., a newspicture system. Less than one decibel of amplitude change may conceivably have a visible effect on a fax recording, while it takes at least three decibels of change to be detectable in speech transmission.

As already mentioned in the discussion of modulation, FM is practically impervious to transmission level changes and has therefore become popular for systems utilizing the switched network, as well as for long-range, broadcast-type radio systems. In leased-wire systems, where the problem is less likely to be encountered, AM is generally preferred.

Echo

A telephone user will occasionally have the annoying experience during a long distance call of hearing an echo of his own voice whenever he speaks. The effect is the result of conditions similar to these that produce acoustic echoes in large empty halls: Each time a sound wave encounters a physical obstacle, a portion of it bounces back toward the source. The greater the distance of the obstacle from the source, the longer the delay between original sound and echo, and the more detrimental the effect on acoustical communication.

Likewise, in a long telephone circuit linking several separate facilities, any significant mismatch in impedance at a circuit juncture can cause a portion of the signal current to bounce back toward the source. Since a two-way communication path is in essence a loop, the echo can travel completely around the circuit to be picked up at the receiving end as well as the originating terminal.

In practice long circuits include special design features intended to prevent echo. One prevention is a *balancing network*, whose function it is to optimize impedance matching at circuit junctures. Another is the insertion of *fixed loss* in the trunks interconnecting the customers' switching offices, and a third one is a *voice-controlled echo suppressor*, through which additional attenuation is introduced into the return path of a two-way circuit during transmission and automatically switched to the transmit path during reception.

The first of these, the balancing network, is only partially effective because, through switching, it has to match a number of different circuits, all varying slightly in impedance. Hence, the impedance match it provides can only be a compromise. The second method can be applied only to relatively short-distance trunks, and the third method may pose a potential problem in fax transmission because of the possibility of suppression elements being switched into the transmission path during a low-level interval in the transmitted signal.

In leased-line systems it is possible to tailor the balancing networks at circuit

Fig. 4.14. "Ghosts" caused by echo in a long-distance telephone circuit. This example was artificially produced with an AM system, and is somewhat more extreme than the typical echo condition. Echo in an FM system would more likely be manifested as a train of "ripples" extending out from the edges of tonal transitions in the picture. (The echo that produced this distortion, incidentally, was not readily discernible to the ear.)

junctures to exactly match the impedances involved and thus improve the networks' effectiveness in preventing echo. This is easily achieved, since no switching of lines is involved in these systems.

As for problems with echo suppressors, these are being minimized in two ways. One is the designing of terminals to provide a steady signal throughout transmission (or a special signal by which the suppressor can be disabled for the duration of a call), and the other is an ongoing program by the common carriers to improve overall circuit quality and thus minimize the need for echo suppressors. (In the past several years a substantial number of the devices have already been eliminated from Bell System trunks.)

Figure 4.14 illustrates the effect of echo on a fax recording.

Delay Distortion

Somewhat akin to echo is envelope delay distortion (EDD), one of the more chronic of transmission impairments and one that can visibly affect the quality of fax copy. It can in fact have a devastating effect on the operation of a digital fax system. It is a distortion of signal frequencies near the low and high cutoff points of the channel's response curve, resulting from a retarding effect that can slow the fringe frequencies of a signal and thereby deform the signal envelope. The free-space velocity of electrical signals is 186,000 miles a second, the same as that of light. However, telephone circuits contain a variety of inductive and capacitive characteristics that tend to decelerate signal currents.

In voice transmission the effect of excessive low-frequency delay is somewhat like that of shouting into a well, while an excess at the high end will cause sharp sounds to be accompanied by a "pinging" that sounds like the plucking of a high-pitched string. The visual effect on an analog fax recording is similar to that of echo, except that it is more a *smear* than a *ghost* and may not manifest itself uniformly throughout a recording (Fig. 4.15). Depending on how bad the distortion is, the effect on the digital fax system may be no output at all. The receiver may automatically reject the transmitted data because of a tendency of the received pulses to smear together. The higher the bit rate, the greater the impact of envelope delay distortion.

Experience has shown that for effective fax transmission within the voice band, delay must be held uniform (±300 microseconds) for all frequencies within the full frequency range of the transmitted signal. This can be accomplished by the common carrier by installation of devices called delay equalizers at intervals along the length of a circuit.

It is a relatively simple matter to equalize (or "condition") leased lines to ensure good fax transmission. However, in the switched network, circuits are equalized on an "as needed" basis, and there can be no assurance that equalization will in all instances be adequate to produce quality fax recordings.

Digital fax systems and some of the newer analog systems are likely to have limited equalizing capabilities of their own. However, this may be confined to merely compensating for the *attenuation* characteristics of the circuit at various frequencies. Some digital systems circumvent the problem by providing the ability to switch to a lower bit rate when the EDD or other circuit characteristics are such as to impede or preclude transmission at the higher rate. This expedient results, of course, in a reduction of transmission speed.

Other Impairments

Another transmission impairment to which digital systems in particular are sensitive is phase *jitter*, the accidental shifting of transmitted frequencies at

Fig. 4.15. Envelope delay in the transmission circuit was the probable cause of the distortion noticeable along the right periphery of the girl's head in this fax recording.

some fixed rate. It is caused by the appearance of a comparatively low-frequency tone at some vulnerable point in the transmission chain—60 Hz a.c. hum in telephone system multiplexing equipment, for example, resulting in a shifting of signal frequencies at a 60 Hz rate.

Both the rate and degree of the jitter will determine the extent of its effect on received signals, and as with EDD and various other transmission impairments, the effect of a given level of it is likely to be more significant at higher bit rates (4800 and up). Analog systems are affected by phase jitter only when it is pronounced, and then it usually manifests itself as a herringbone pattern in the copy, like that produced by any form of single-frequency interference. (See Fig. 4.12.)

Harmonic distortion is an impairment that can cause detection errors in digital transmission schemes using multiple frequencies to transmit several bits simultaneously. Harmonics present at sufficiently high levels can confuse the precision detection circuitry and cause damaging errors in the received data. The 2nd and 3rd harmonics are usually the chief culprits, the higher ones generally having been filtered out. Special conditioning of leased voiceband lines is sometimes required to prevent the problem.

"Hits" are sporadic and momentary changes in amplitude or phase that can cause transmission errors in a digital system. In systems having an error correction capability, the effect of hits can be minimized, but a high incidence can drastically reduce the system's throughput speed. Systems that can detect, but not correct, such errors may simply cease transmission if too great a number of hits of sufficient intensity occur within a given time span.

Other transmission impairments already briefly discussed are those resulting from the Kendall effect ("aliasing") and from synchronization and phasing errors and the mismatching of indices of cooperation (Chapter 3). All that need be said further about these is that all are usually avoidable through proper engineering of the system terminals.

OTHER FACILITIES

So far the emphasis of this chapter has been on voice-grade fax transmission. There are, of course, broader bandwidth circuits available from common carriers like the telephone companies and Western Union for use with higher-speed fax systems. And not to be overlooked is the use of radio where wire service is impracticable.

The heading "Other Facilities" may be somewhat misleading in one respect, because conventional phone lines frequently form the basis for broader bandwidth wire links. But it is justified because our concern here is with how existing facilities are utilized and how they are augmented by special amplifying and equalizing devices.

There are a few commercial analog facsimile systems available (usually of an older generation) that have bandwidth requirements lying somewhere between voice-grade and broadband—e.g., 5-15 kHz. Transmission service to satisfy these odd bandwidths is not a standard common-carrier offering—at least not for fax transmission. For that reason, channel requirements beyond the limits of voice-grade or *Telephoto* facilities will usually require custom engineering by the local telephone company or the leasing of standard broadband channels with bandwidths that may greatly exceed requirements.

In spite of this bandwidth gap, the variety of available transmission services and facilities is impressive, as will be seen in the paragraphs that follow.

Telephoto

Still leasable from common carriers (but on the brink of being phased out) are so-called *Telephoto* circuits, known also as "Type 4002," or (to an older generation) "Schedule II," facilities. They are specially conditioned voice-grade wire links intended primarily for the transmission of photographs. Their practical maximum bandpass is about 800 to 2800 Hz, which is generally adequate for distortion-free transmission of newspictures (about 120 scan lines to the inch at 90 lines a minute, using AM double-sideband transmission) or for "message" systems of higher-than-average speed (e.g., 96 LPI at 360 LPM).

As might be expected, Telephoto facilities are somewhat more expensive to lease than conventional voice-grade circuits. Depending on where the termination points are located, the difference can range from 25 to as much as 53 percent more per month for a 100-mile private link. (As of mid-1977, a 100-mile Series 3002 voiceband data circuit cost a minimum of about $225 a month, including termination costs. The Telephoto circuit for the same distance was priced at $476.) While the additional cost is usually justified for transmission of photographs, where quality requirements are critical, some of the benefits of the special engineering of the Telephoto facility are lost on general-purpose systems that merely take advantage of the slightly increased bandwidth to effect a modest gain in speed or resolution.

Current plans call for an eventual phasing out of the present Telephoto service (which has existed virtually unchanged since the late 1920s) in favor of a relatively wideband digital service. The replacement service is expected to result in faster, more efficient, and more versatile newspicture operations.

Western Union offers an approximate equivalent of Telephoto circuits (equivalent mainly from the bandwidth standpoint) on a switched, "dial-up" basis, permitting the customer to use the facilities only as needed rather than lease them on a full-time basis. In most cases, the dial-up mode will afford savings to the customer compared with leased-line service. The economic crossover point will be determined principally by the average copies-per-month volumes invloved. It will vary with mileage and with the customer's objective in going to a higher bandwidth in the first place—i.e., whether he uses the extra bandwith to gain *speed* or *quality*.

One disadvantage of the dial-up mode is that the customer does not have the

TRANSMISSION

same assurance he has with a dedicated facility that results will be consistent from one transmission to the next. This drawback is minimized in Western Union's *Broadband Exchange* service (the switched service referred to above) by the fact that all intercity trunks are specially conditioned and there are no alternate routings of dialed calls. At this writing,[4] the Broadband Exchange service costs the customer a flat rate of $30 a month, plus a per-call charge of 20 to 75 cents a minute, depending on mileage. (The range is roughly 0 to 3000 miles).

"Grouped" Channels

If it is absolutely necessary, a local phone company or other communication common carrier will specially engineer a transmission circuit for a customer when no standard service meets his particular specifications. Such special engineering, however, is expensive and can usually be done only within a restricted geographic area. The usual alternative where a broadband facility is required is to lease channel space on the basis of the number of voice channels being displaced. Twelve full, 4-kHz voice channels are considered a *group*; 60 voice channels, a *supergroup*; and 240 voice channels, a *mastergroup*.

In terms of interstate tariff a 48-kHz group (or "8000 type" channel) is the next step after Telephoto in ascending bandwidth for common-carrier analog wire links.

For an added charge a group can also be subdivided into two *half groups*, each with a usable bandwidth of about 20 kHz, or a maximum of six voice channels. Supergroups and mastergroups ("500 type" channels) can be more finely subdivided. But the smallest grouping a customer can lease is still 12 voice channels (a full 48-kHz group), which can be used as a single broadband facility or, with proper terminal equipment, subdivided as required.

At present a 48-kHz group leased between two points 100 miles apart costs the customer a basic "line charge" of $1620 a month. A 240-kHz supergroup leases for a flat $32.50 a mile per month, and a 960-kHz mastergroup leases for a flat $92.05 a mile per month. For each of these services there is an additional charge of about $450 per month for each termination, plus installation and possible other charges that will vary according to the extent to which the full channel is subdivided. (See Chapter 6.)

Xerox's high-speed, high-resolution LDX (long-distance xerography) system is one of the few available commercial fax systems that requires the full group or supergroup channel space. Though no longer manufactured, the LDX terminals are, at this writing, still available from Xerox. Using the full bandwidth of a 240-kHz analog channel, the system can transmit a full letter-size page in about eight seconds at better-than-average resolution (135 scan lines per inch). Other "copy-a-minute" analog message fax systems (96 LPI/960 LPM, for example) may require the leasing of a 48-kHz group, even though only a portion of the full bandwidth of the channel may actually be used.

Digital Circuits

All of the transmission facilities and services discussed so far are analog by nature. That is to say, every aspect of their design is aimed at optimizing the passage of a.c. waveforms within a given frequency range. This means extensive use of inductive and capacitive elements to ensure the desired bandpass characteristics and to isolate the a.c. signal currents from the d.c. required to energize telephone instruments. It also means the use of amplifiers to compensate for losses in signal current over relatively long hauls.

Unfortunately, despite all of the ingenious innovations to counteract it, the various noises and distortions picked up by the signal over a long-haul circuit get amplified too (as was illustrated in the preceding section on transmission impairments). The result is that the received a.c. signal never looks exactly the same as it did when it left the transmitter.

One way around this problem is to convert the analog currents to pulse codes that describe the fluctuations in sufficient detail to permit their reasonably faithful reconstruction at the receiver. That way the d.c. pulses—all of which look exactly alike—need merely be *regenerated* periodically as they become weakened and distorted.[5] In other words brand new pulses take the place of the original ones in relay race fashion. It is a principle that harks back to one of the earliest innovations in electrical communication: the d.c. telegraph repeater.

The advent of solid-state electronics in the 1950s made possible the design of relatively simple, compact high-speed pulse-generating and -regenerating devices, and implementation of pulse-code techniques in the voice communication realm was soon to follow (notably in the Bell System's pioneering T1 and T2 digital carrier systems). With the concurrent growth in data communication, it was inevitable that these strictly digital facilities would be made available for that more natural purpose as well.

While considerable use is still being made of existing analog facilities for digital data communication, the trend is definitely toward a gradual phasing out—or revamping—of these older facilities, so that virtually all electronic communication will eventually be on a digital basis. Illustrative of the trend toward provision of direct digital facilities without need for *modems* (see the next section) per se is the Bell System's Dataphone Digital Service (DDS), currently avaiable within and between selected cities and gradually being expanded. Eventually it is expected to be available on a switched basis.

Typically digital service is available at selectable bit rates (bits per second), the standard ones being 2400, 4800, 7200, and 9600. Bit rates currently go as high as 56 kilobits, and Bell plans to expand that to 1.54 megabits in the near future.

TRANSMISSION

What all of this means for digital fax systems is that with such facilities available the expensive modems now required for use of these systems on voiceband analog circuits can conceivably be replaced by much simpler devices, thereby significantly reducing terminal costs. As for transmission costs, they naturally rise with the data rate, but on a curve that makes the higher bit rates a better bargain than the lower ones—provided the service is efficiently utilized. For example, the present basic line charge for a 100-mile, 4800-bit circuit in the Bell System's DDS offering is $175 per month, whereas for a 56-kilobit line—nearly 12 times the data rate—the charge for the same mileage is only five times as high: $876 per month. (Chapter 6 will go into more detail on the various costs associated with these and other transmission services.)

Besides those *facilities* that have been engineered specifically for digital transmission, there are special fax *services* available that use digital techniques to provide subscribers with the ability to send messages on a store-and-forward basis and to interconnect normally incompatible terminals. One of the first such services is *Graphnet*, which uses a digital principle called *packet switching* to achieve optimum transmission efficiency. (See Trends: Fax Networks, Chapter 2.) Similar fax-oriented services have recently been announced by ITT and Southern Pacific Communications (SPC).

Graphnet Incorporated is a "value-added" carrier in the sense that it uses its own speical equipment in conjunction with conventional common carrier phone circuits to provide unique services to subscribers.

Radio Facilities (Terrestrial and Satellite)

Theoretically the factors affecting required bandwidth for distortion-free transmission by wire facilities should apply to radio as well—and for the most part they do. Radio transmission by nature must rely on the generation of alternating currents, which act as carriers of the transmitted information. In that respect it is not fundamentally different from voiceband wire circuits when used for the transmission of fax signals. It does, however, allow for broader bandwidths and is therefore not normally as susceptible to edge-of-band distortions. Also, not being as band-limited, it offers the potential for higher transmission speed without the need for some of the critical engineering requirements that characterize relatively long-haul, wideband wire circuits. As a rule, an additional modulation step is required for radio transmission of fax signals. For convenience (and to some extent for technical reasons) the modulated carrier output of the fax transmitter is usually applied as a *subcarrier*, which means that it becomes the wavering audio "tone" by which the radio frequency carrier is in turn modulated.

Another difference between wire and radio—a negative one for radio—is the latter's potentially greater susceptibility to external interference. AM systems in particular are affected by lightning and other atmospheric and cosmic disturbances, while radio systems in general—with the possible exception of microwave—are vulnerable to harmonic and crosstalk interference from one another. Fading, caused indirectly by atmospheric temperature inversions and by movement of ionized layers in the upper atmosphere, is a common problem in shortwave and microwave systems. Such systems are also subject to high power losses between stations and at very high frequenices can even be affected by rain.

Still another difference is that, theoretically at least, radio signals are somewhat more accessible to interception. With the possible exception of microwave, radio is by nature a broadcast medium—i.e., the signals are more or less scattered and therefore receivable by anyone within range who has a compatible instrument.

Obviously radio's chief advantage is that it can reach places that cannot be reached by wire. Morever, it is possible to utilize existing broadcast facilities on a multiplex basis for subsidiary transmissions that are normally only accessible to paid subscribers through a selective reception scheme. The FCC's *Subsidiary Communications Authorization* (SCA) policy, for example, permits the broadcasting of data and facsimile messages via a subcarrier within a commercial FM channel. At least two vendors (Fax Net, Incorporated, of Chatham, N.J. and Digital Broadcasting Corporation of McLean, Virginia) are currently operating such services on a subscription basis via commercial FM outlets in various cities around the country. At present these services are confined to the communication of alphanumeric data via teleprinter terminals, but are readily adaptable to fax and slow-scan TV[6] as well.

In modern common-carrier communications there is no longer a meaningful distinction between wire and radio facilities, particularly from the customer's point of view. Within the telephone network, for example, long-distance calls (including fax and data signals) may on one occasion go by wire, on another by radio—and the customer will not know the difference. In fact nearly half of all long-distance carrier telephone nets existing today are in the form of microwave radio relays. Figure 4.16 shows a familiar common-carrier microwave-relay tower.

With proper licensing, sufficient capital for the initial equipment investment, and a willingness to sustain continuing maintenance costs, a business concern can install its own microwave network between separated branches. However, maintenance alone can be a major problem. Even if the need for repairs is infrequent, specially trained technicians have to be on call in the event of a system breakdown, as well as for periodic tests and adjustments. This is one of the reasons—possibly the prinicipal one—why to date relatively few private businesses have attempted to operate their own communications networks.

In recent years a number of specialized common carriers have gone into

TRANSMISSION

Fig. 4.16. A Bell System microwave relay tower typical of those that dot the American landscape. This one is part of the pioneering TD-2 radio system implemented in 1951. The system is currently capable of carrying 1500 telephone circuits per channel, or more than 16,000 per typical route. The towers are usually spaced about 25 miles apart. (Courtesy Bell Telephone Labs)

business in competition with the established telephone and telegraph carriers on high-density intercity routes. As might be expected, these newer carriers tend to be radio oriented (microwave, specifically). One of the first to establish a nationwide network is MCI Telecommunications Corporation, which at this writing connects several major cities coast to coast, including New York, Washington, D.C., Chicago, St. Louis, Houston, Dallas and Los Angeles. As with the established carriers, the nature of the services provided by the specialized carriers and the rates charged are controlled by government regulatory bodies (notably the FCC).

One of the more venerable commercial applications of radio to facsimile transmission is the so-called *Radiophoto* service for international dispatching of news photos. Introduced by RCA in the 1920s, Radiophoto was later taken over by the major wire services (AP and UPI) themselves, which today lease the radio facilities from common carriers on a full-time basis. It operates in the 4-27 MHz HF band, and its operation is governed by FCC regulations. Picture signals are transmitted via an FM subcarrier to minimize the effects of fading.

RCA Global Communications (or RCA Glōbcom, as it is more commonly known) is one of a handful of international common carriers known as "record" carriers because of their traditionally exclusive license to handle other-than-voice communications (i.e., fax and data) on overseas routes. Others are Western Union International, ITT WorldCom, Cable & Wireless, and TRT Telecommunications. The term "traditionally exclusive" is used above because at this writing that situation appears almost certain to change. On January 8, 1976, the FCC adopted an order authorizing AT&T to file application for authority to provide International Dataphone Service. What this means to fax users is that if and when the pursuant red tape (notably challenges filed by the existing record carriers) is untangled sufficiently to permit implementation of the new service, worldwide fax communication via the public telephone dial network will become legally possible for the first time.

Meanwhile, RCA Glōbcom is (at this writing) drawing up plans to provide a new high-speed fax service via its worldwide facilities. The terminal machines will employ an advanced digital data compression scheme (see below, *Redundancy Reduction*) to achieve an average speed of less than half a minute per typewritten page. Initially the machines are to be operated centrally out of Glōbcom offices, but later may be installed on customers' premises as well.

Another area in which radiofacsimile has proved particularly useful is the gathering and distribution of weather data. Since 1930, weather maps have been transmitted to ships at sea from land-based radio stations. Present-day equipment operates in the 1.5-25 MHz HF band and uses the popular *frequency-shift* method of two-level transmission (1500 Hz for black and 2300 Hz for white). Conventional fax gear is used for scanning and recording.

Of more recent vintage is the Automatic Picture Transmission (APT) system by which television views of segments of the earth's cloud cover are transmitted at regular intervals from orbiting satellites several hundred miles high and reconstructed by facsimile at ground stations all around the globe. The slow read-out signals from the space-borne vidicon storage tubes amplitude-modulate an audio subcarrier, which is then transmitted by compact, 5-watt FM transmitters operating in one of two frequency ranges: 4-30 MHz (HF) or 135-139 MHz (VHF). Transmission is at the rate of 240 lines a minute, with a full 800-line picture being

TRANSMISSION

sent in a little more than three minutes. (Newspicture and weather data dissemination are among the fax applications that have been discussed in a more general way in Chapter 2.)

In regard to earth-orbiting satellites, one of the more recent advances in long-distance radio communication—and one that has tremendous future potential—is the use of *geostationary* satellites as multi-channel relay stations. By *geostationary* is meant the location of the satellite in an equatorial orbit about 22,000 miles above the earth, at which height its motion is synchronized with the earth's rotation and is therefore stationary with respect to the earth. The first such satellite to go into operation commercially as a common carrier was the *Early Bird*, launched by the Communications Satellite Corporation (Comsat) in 1965. It accommodated 240 separate telephone channels.

Today the *Intelsat* global communications consortium has a satellite network serving nearly 150 earth stations in over 80 countries, and domestic systems are in operation in Canada and the United States. Collectively, these global and domestic satellites ("Domsats") provide sufficient channel space for well over 100,000 simultaneous telephone messages. As noted in Chapter 2, Dow-Jones & Company leases a wideband channel on Western Union's Westar I for fax-dispatching the daily *Wall Street Journal* from Massachusetts to Florida via the paper's own earth stations. It has been estimated that the satellite operation is saving Dow-Jones approximately $900,000 a year as compared with the cost of terrestrial microwave between the same two points.

At this writing, a single voiceband domestic satellite channel costs from $500 to $1100 a month, depending on point-to-point terrestrial mileage, plus termination and local access charges of $40 to $60 a month. There are quantity discounts for multiple channels. Wideband rates run between $5000 and $9000 per month for a single 48-kHz channel. At least four established common carriers are already operating domestic geostationary satellite systems (Western Union, RCA, AT&T, and GT&E), and a joint-venture group called Satellite Business Systems (SBS) has won FCC approval to join those four. SBS is a consortium made up of IBM, Comsat General, and Aetna Casualty & Surety. It plans to have its competing system operational by mid-1979.

COUPLERS AND MODEMS

To prevent excessive signal levels that may present hazards for maintenance personnel and cause trouble for other customers, connection of "unregistered" communication terminals to a common-carrier network has to be made through an appropriate interface. Moreover, where the terminal to be connected is anything other than a telephone instrument—e.g., a fax or data terminal—an appropriate *modem* (modulator-demodulator) has to be provided to ensure that the transmitted signal is compatible with the transmission circuit. The modem may be integral to the main terminal or separate, and it may be furnished by either the common carrier or the customer. But in any case it must meet FCC requirements under the recently enacted *registration* program, or must be coupled to the line via a carrier-provided *Data Access Arrangement*. The one exception to this rule—ostensibly, at least—is *indirect* acoustic/inductive coupling via the handset of a standard, carrier-provided telephone instrument.

"Indirect" Couplers

While the rules governing "interconnection" specifically forbid direct electrical connection of unregistered customer-provided terminals to the switched network, *indirect* coupling by acoustical or inductive means is permitted, provided certain requirements are met regarding the level and frequencies of the transmitted signal. Partly for this reason but mainly as a matter of convenience, couplers that permit use of a standard telephone set as a "mechanical" interface between the terminal unit and the dial network are widely used with conventional analog fax transceivers. In fact in many cases they are integral to the terminal machine.

Use of the indirect couplers—or "phone couplers," as they are sometimes called—requires that the sender place a call through a standard telephone set to the receiving party, whom he knows to possess a fax transceiver compatible with his own. Having established voice contact (which may be optional, depending on the kinds of automatic features the terminal units have), both parties place their telephone handsets in the coupler units, which are physically designed to accept all standard handsets. One type of separate coupler unit is designed to clamp to the telephone set in such a way that the handset may be left in the coupler without affecting normal use of the phone. A built-in solenoid automatically raises the phone's "hook switch" upon receipt of ringing tone, thereby permitting the fax machine to automatically "answer" a call (Fig. 4.17).

Electrically the indirect coupler is so designed that in the SEND mode a built-in reproducer induces its output into the telephone handset, either acoustically—by "speaking" into the mouthpiece of the handset—or inductively, by tranformer action. In the RECEIVE mode the coupler functions as an acoustic or inductive pickup, relaying the incoming audio signal from the handset to the fax transceiver for recording.

A disadvantage of the indirect coupler is its susceptibility to external acoustic and/or inductive interference and to impairments introduced by the telephone set. Moreover, its method of transferring signals is not the most efficient, and can result in losses both in quality and strength of the processed signals. However, in a well-designed coupler, the effects of these shortcomings can be minimized.

TRANSMISSION

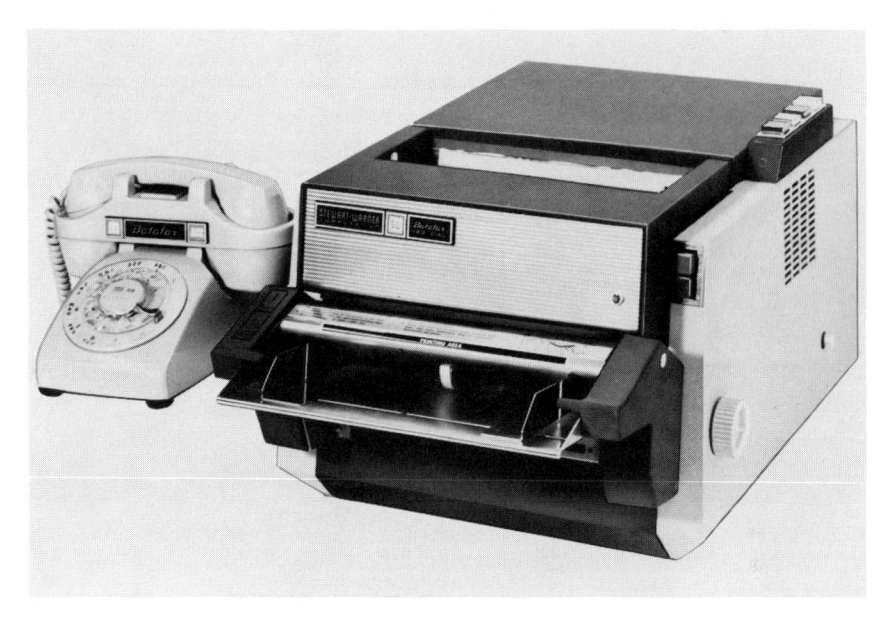

Fig. 4.17. This accessory acoustic/inductive coupler (below) mechanically attaches to any existing phone, as shown, without affecting normal use of the phone. It permits unattended fax reception by automatically responding to ringing current on an incoming call and physically raising the telephone handset off-hook. (Courtesy Stewart-Warner Datafax)

Network Protection Requirements

Until 1976, direct "hard-wire" hook-up of *all* customer-provided terminals—fax machines, digital modems, etc.—to the telephone network required a protective interface called a *data access arrangement* (DAA), furnished by the local telephone company for a nominal monthly charge. Under recently revised FCC rulings, the arrangement is now required only for those terminals that have not been suitably "registered" with the FCC. (At this writing, the exact status and ultimate scope of that ruling are still somewhat vague.)

Fig. 4.18. The Bell System's 602 Data Set was the pioneering signal converter that officially opened the public telephone network to facsimile use in the early 1960s. (Courtesy Bell Telephone Labs)

In its simplest form, the DAA consists of the addition of a protective coupling circuit to a standard telephone set, along with a manual key through which the instrument can be switched from its normal operating mode (voice) to a data input mode, in which it functions as a passive interface between the fax equipment and the line. More elaborate couplers are also available to provide for automatic origination and answering of calls, as was described earlier.

Inherent in the FCC's registration/certification concept is the assumption that the protective circuitry formerly embodied in the DAA can be integrated into the terminal unit, or, alternatively, that the fax terminal vendor will be able to provide a properly registered separate coupler.

602 Data Set

The Bell System's 602 Data Set (Fig. 4.18) was the pioneering analog modem that officially opened the public telephone network to facsimile use in the early 1960s. As a separate device, it is now mainly of historical interest. However, inasmuch as the modem circuitry now commonly embodied within the typical commerical FM/analog fax terminal is essentially the same as that housed separately in the 602, a brief description of the latter will serve as a useful introduction to the basic concept of modems.

TRANSMISSION

The 602 is basically a *signal converter* intended to provide a standard means for transmitting low-speed, or "message," facsimile signals over regular phone lines.

As a modulator in the SEND mode, the unit converts the d.c. variations of the scanner output to a linear carrier frequency swing of 1500 Hz, corresponding to 0 volts of scanner output, to 2450 Hz, corresponding to the upper limit of +7 volts. In the RECEIVE mode, the FM carrier is *clipped* by a limiter to remove amplitude variations due to noise acquired in transmission, *rectified* to convert it from an a.c. signal to a series of positive pulses, and *reconverted* to the original baseband signal by an "averaging" technique. This last step is accomplished by the combination of a *monopulser*, which emphasizes the alternate "bunching" and separating of the positive pulse train by adjusting all pulses to uniform width, and a *low-pass filter*, which filters out the actual separate pulses and passes only the lower-frequency variations, thus recovering the original baseband frequencies. Figure 4.19 illustrates the sequence of signal-processing steps involved in modulation and demodulation.

303 Data Set

For group and supergroup service, the Bell System's 303 Data Set (Fig. 4.20) is the standard interface between terminals and transmission facilities. It is a versatile piece of equipment, considerably different in design from the 602 Data Set. For one thing it is designed to operate in a *digital* rather than an analog mode.

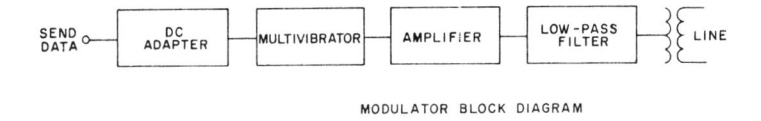

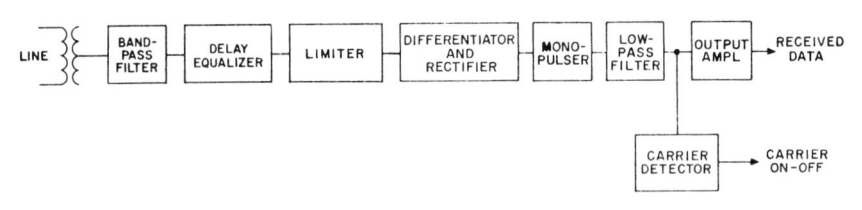

MODULATOR BLOCK DIAGRAM

DEMODULATOR BLOCK DIAGRAM

Fig. 4.19. Functional scheme of the 602 Data Set. Each set has both transmit and receive capabilities. (Courtesy Bell Telephone Labs)

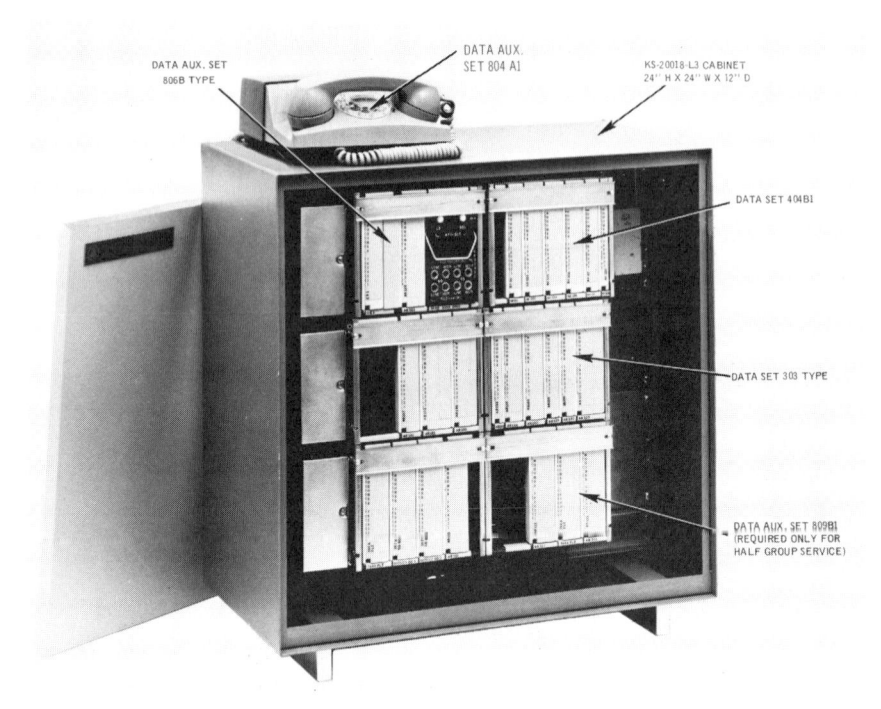

Fig. 4.20. The Bell System's 303-type wideband data station permits displacement of up to 60 voice channels for high-speed fax transmission. The station is of modular design, and consists of a composite of individual terminal devices. (Courtesy AT&T Co.)

In the use of this data set with analog facsimile equipment, digital operation means that the analog output of the fax scanner is converted to a sequence of ON and OFF pulses, each pulse being determined by whether a given point along the signal envelope is interpreted as black or white. In other words, regardless of what gray tones may exist on the original, the scanner output is transmitted as a binary, or two-level (black and white), signal.

Part of the 303's versatility is that it can be operated in either a synchronous or asynchronous mode. In the synchronous mode, precision oscillators, or "clocks," at each terminal are synchronized with each other so that the rapid switching between ON and OFF conditions at the transmitter can be accurately followed and regenerated at the receiver and at repeater stations along the way. This is roughly analogous to the use of standard-frequency power supplies in fax systems when the terminal mechanisms are served by different commercial power grids. It is also somewhat analogous to the use of sync pulses in TV systems, except that in digital transmission the sync pulses and the data being transmitted are more directly interrelated.

TRANSMISSION

The synchronous mode is intended primarily for the linking of business machines—e.g., a tape reader at a computer center with a high-speed printer at a distant office. As an interface for fax gear, the 303 is normally used in asynchronous mode because it is more convenient to resynchronize a fax system on a line-by-line basis.

In the SEND mode, the input to the 303 is the d.c. scanner output, taken directly from the scanner or recovered at the fax transmitter output by an accessory converter. An *amplitude-quantizing* circuit in the 303 provides a threshold above which all d.c. signal levels represent an ON, or black, condition, and below which all represent an OFF, or white, condition. The resulting stream of ON and OFF pulses goes out on the line to the nearest exchange, where it is then modified to conform to the requirements of the broadband carrier facility.

In the RECEIVE mode, the 303 samples and interprets the demodulated pulses and feeds them to the fax recorder, which translates them to a two-level graphic reproduction of the original.

Two other features that contribute to the 303's versatility are (1) its applicability to subdivision of the full broadband channel into narrower bands and (2) its provision for simultaneous voice or code coordination over a separate channel included for that purpose in the group service package. The 303 Data Set, incidentally, is generally included as part of the monthly station charge for this class of broadband service (basically $460 per month per main station, as of mid-1977).

Voiceband Digital Modems

The subject of digital modems has so far been touched upon only with regard to the effects of transmission impairments. The purpose of these devices is essentially the same as that of the analog modems previously described, namely, to convert the output of the fax scanner to a form suitable for transmission via voiceband circuits and to reconvert the transmitted voice frequencies to d.c. pulses for recording. Like its analog counterpart, the digital modem may either be integral to the fax terminal or separate from it.

Typically the digital modem uses some form of frequency-shift keying (FSK) to accomplish its purpose. The principle is virtually identical to that employed in the 602 analog Data Set described above, except that in the basic digital modem there is one fixed frequency for the black condition and another for the white, with no variation in between.

In practice, of course, there is more to it than the mere shifting of frequencies between two states. For one thing the black and white (or *zero* and *one*) states represented by the transmitted frequencies must somehow be "clocked"—i.e., precisely timed and phased so that there is always the same, fixed number of pulses per scan stroke and so that those in each stroke fall properly in line with the corresponding points in the preceding stroke. Inasmuch as a digital fax system may not operate in the steady, repetitive manner of an analog system—a digital data compression system, for example, may have to pause occasionally to permit a "buffer" to catch up with the processing of the received information— there are some special synchronization considerations. However, these were covered in Chapter 3 and need not be repeated here.

In addition some measure of error control is required. Since digital communication is generally in the form of multibit code "words" (a string of zeros and ones constituting a given instruction or picture element), an impairment-produced error in just one digital pulse can have a far more devastating effect than that of the same impairment on a single picture element in an analog system.

Beyond these special requirements, there are also several variations of the basic signalling scheme. For example, amplitude or phase modulation may be used instead of FSK. Then there are some exotic variations like differential phase-shift keying (DPSK) and quadrature AM (QUAM). Either of the latter may be used to permit transmission at greater bit rates than the channel would ordinarily allow. This is accomplished by *multi-level signalling*.

The multi-level transmission scheme is perhaps best explained by considering a system in which the channel bandwidth is divided into eight discrete frequencies, as illustrated in Figure 4.21. The modem is designed to recognize each frequency as representing a preassigned 3-bit "word," and also to recognize that each pair of successive bursts therefore constitutes one of a possible 64 6-bit words. What this means is that compared with a basic FSK system in which there are only two frequencies—one representing a zero bit and the other a one bit— three times as much information can be transmitted in the same amount of time via the same channel.

It follows that if the two frequencies for each 6-bit word could somehow be transmitted simultaneously (multiplexed) and detected in the intended sequence, then the system could operate *six* times as fast (7200 bits per second). Modems with these and even higher speed capabilities are in fact commercially available. At present the highest bit rate possible with commerical modems on voiceband lines is 9600, and that speed can generally be achieved only on very "clean" lines (virtually no impairments).

The chief limiting factor of multi-level schemes in general is that the greater the number of levels, the more critical becomes the task of discriminating between one level and the next at the receiver. The task is made even more difficult when the individual frequencies are somehow altered or delayed as a result of transmission circuit characteristics.

The more complex and faster the modem, the more likely it is to be syn-

TRANSMISSION

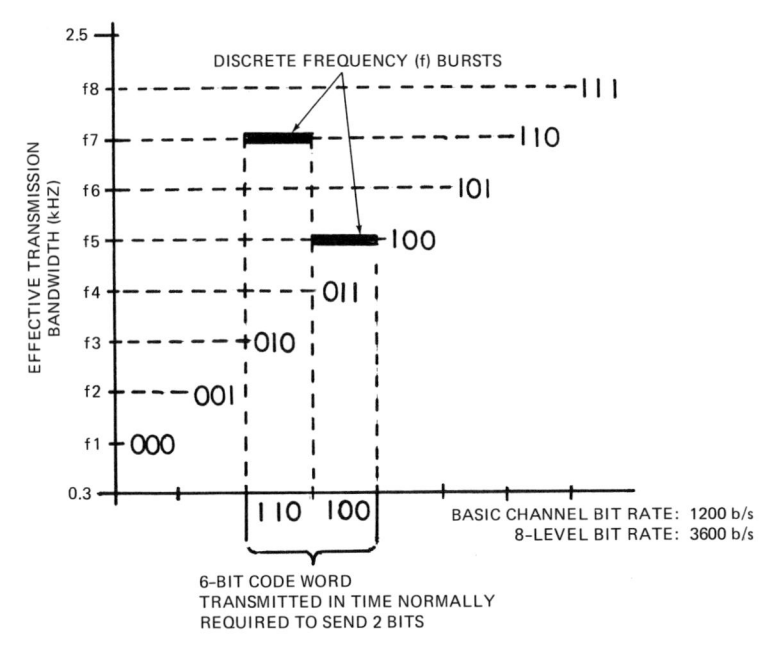

Fig. 4.21. The normal maximum digital bit rate of a transmission channel can be at least tripled by using a given frequency to represent a 3-bit code "word" (or "half-word"). Thus, the sequential transmission of only two frequencies can convey a full 6-bit word. With the necessary separation circuitry at each end of the system, the two frequencies can be sent simultaneously to achieve a 6X increase in bit rate on the same channel.

chronous as opposed to asynchronous. (The distinction was explored in Chapter 3.) Digital modems are available from the common carriers and from a number of private vendors—notably Codex, ICC/Milgo, Intertel, Rixon, and Rockwell International—or, as previously mentioned, may be an integral part of the digital fax terminal. Figure 4.22 shows a typical separate modem.

REDUNDANCY REDUCTION

Sideband suppression (discussed earlier in this chapter) was the first successful technique for reducing bandwidth in modulated carrier transmission systems. Transmission having always been a comparatively costly proposition, it is natural that engineers have persisted through the years in seeking new ways to conserve bandwidth in communication facilities.

Until the advent of data transmission in the 1950s, emphasis had been on *bandwidth compression*, the actual reduction of the total range of frequencies required for effective transmission, or the more efficient use of available band-

Fig. 4.22. A separate digital modem provided by the common carrier. This one, slightly larger than a typical attaché case, permits transmission of digital signals at the rate of 9600 bits per second via a voice-grade circuit. As a general rule, the higher the bit rate, the larger the modem. (Courtesy Bell Telephone Labs)

width. With the ensuing spectacular growth of data communications, the emphasis has shifted to *data compression*.

It is sometimes difficult to determine where one leaves off and the other begins, although bandwidth compression is usually thought of as being confined to analog systems, and data compression to digital systems. Methods used in voiceband digital modems to make optimal use of an analog channel for data transfer—e.g., the multilevel FSK scheme just described—may be regarded as a form of bandwidth compression, whereas a process by which the analog output of a fax scanner is reformatted into a kind of digitally encoded shorthand is normally considered a case of data compression. It is pretty much up to the individual to decide which of the two terms (if either) is appropriate in a given situation. The main point is that whichever name we choose, the basic objective is the same: to reduce the redundancy of the transmitted message. Both schemes, therefore, fall under the general heading of *redundancy reduction*.

It would take nearly an entire book to discuss the whole gamut of compression techniques that have been developed to date. What follows is a condensed examination of some of the more noteworthy ones, past and present.

The Vocoder

After sideband suppression the next significant step in bandwidth compression technology was the development of the *Vocoder* at Bell Labs in the late 1930s.

TRANSMISSION

It is not applicable to facsimile, but is described here to illustrate a basic redundancy reduction concept.

Basically the original Vocoder was a device for analyzing and artifically reproducing speech. It was developed for that purpose by Homer Dudley of Bell Labs in 1936, and its potentialities as a bandwidth compression system were not fully appreciated until sometime thereafter. As a compression device, instead of transmitting the actual pitches representing a series of spoken words, it transmits an electric current whose variations represent the *transitions* from one pitch to another. Since in normal speech there are fewer than ten such transitions per second, it is easy to see how this technique can conserve on bandwidth.

Inasmuch as the actual pitches are not transmitted, they must be re-created artificially by oscillators in the receiver. The transmitted signal, then, merely determines what pitches are to be reproduced and in what sequence. In practice there is more to it than that, but for the purposes of this chapter the foregoing will suffice as a basic description of the Vocoder.

The Vocoder has had a few isolated successful applications, but facsimile has not been one of them. As noted earlier in this chapter, the baseband frequencies from a fax scanner, while usually in the audio range, tend to be of a random, basically impure nature. The Vocoder is limited to reproducing pure, or nearly pure, oscillator-produced frequencies common in speech or music. Figure 4.23 illustrates the Vocoder principle.

Variable-Velocity Scanning

Various other analog-type compression systems have been developed and demonstrated, but for one reason or another have not found application in fax transmission. An exception is variable-velocity scanning, which may be applied to analog and digital systems alike—and may even combine both modes in a single system.

The object is to vary the scan rate in accordance with the density—or degree of congestion—of tonal transitions the scan beam encounters within a stroke. Typically this is accomplished by momentarily storing the scanner output in a buffer (a recycling magnetic recorder) while its content is analyzed by a built-in computer.

The computer's function is twofold: (1) to determine the pattern of speed variations at which the temporarily stored signal is to be "read out" onto the transmission line, based on variations in density of tonal transitions and (2) to provide, along with the processed picture signal, the codes necessary to tell the recorder at the other end when and by how much it must change speed to keep in step with the variations in read-out speed.

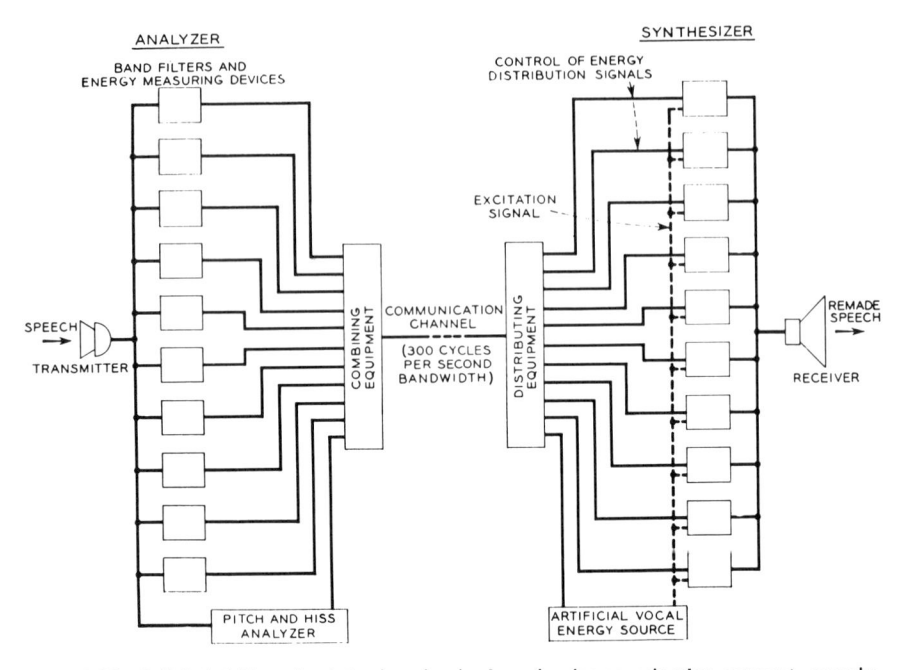

142   ELECTRONIC DELIVERY OF DOCUMENTS AND GRAPHICS

Fig. 4.23. Bell Labs' Vocoder introduced a basic redundancy reduction concept, namely, the transmission of data representing the *transitions* from one state to another in the original signal, rather than the signal itself. The original signal is artificially reconstructed at the receiver. (Courtesy Bell Telephone Labs)

Variable-velocity scanning is one of the simpler compression techniques—at least in principle. It is also a proven technique, inasmuch as variations of it have been successfully applied in commercial equipment. For typical text material, it can afford compression ratios up to about 6:1, which means that a typical business letter can be sent in about a minute via the DDD.

A pioneer in the commercial use of this technique in a compression fax system was Comfax Communications Industries, whose patented system was used in the *FAX I* transceiver, manufactured and (until recently) marketed by Electronic Associates Incorporated, a firm specializing in analog and hybrid computers. It is a dual-speed technique in which a CRT scan spot normally sweeps the copy at the higher of two rates, but by interaction of the sweep circuit with the system's control logic is caused to switch to the lower rate whenever image details are encountered. Obviously the variable-velocity commands controlling the scanner must also be transmitted to the receiver, along with the image detail and synchronizing information. The same basic scheme (see Fig. 4.24) is used in Comfax's yet-to-be-marketed *ComputerPix* microfilm-input fax (microfacsimile) system.

TRANSMISSION

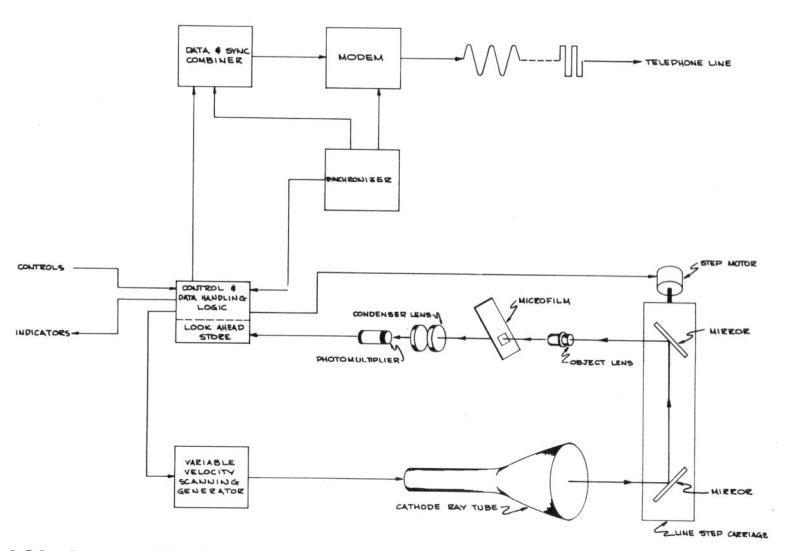

Fig. 4.24. ComputerPix data compression scheme. The scan beam in effect "looks ahead" to determine the density of detail to be encountered, and the speed of the system varies accordingly. (Courtesy Comfax Communications)

In the Comfax system, as in most data compression fax systems, the image details are initially *thresholded* into black or white elements, with the in-between gray tones being divided into one category or the other depending on where the particular shade falls relative to the threshold.

Another variable-velocity scanning system recently developed by a new firm, Faxon Communications Corporation of Pasadena, California, is able to preserve any gray tones in the copy, while only partially sacrificing its average 6:1 speed advantage. Like the Comfax system, the Faxon system employs CRT technology for both scanning and recording. Also like the Comfax system, it is a true white-space-skipping system, varying its scan speed in accordance with the relative informational value of the image details encountered. But unlike the Comfax system, it is (in its present configuration) strictly analog and does not require a digital-to-analog modem for use on conventional voiceband circuits.

Commercial fax terminals employing these redundancy reduction schemes—or for that matter any of the schemes described in this section—naturally bear a higher price tag than conventional analog terminals. But despite their cost, they offer the potential for significant net savings where the volume of use, need for fast service, or economies realized through shorter line time or use of narrower band circuits over long distances are important factors.

Run-Length Coding

Of the various scan compression techniques that are purely digital, run-length coding (RLC) was among the earliest to be developed and is perhaps the best known. It views all image elements as either black or white, and periodically samples (at a very rapid rate) the content of a scan stroke to determine which of the two conditions obtains at that instant. However, instead of the resulting serial bit stream being transmitted bit for bit, binary codes are generated to signify *transitions* from one state to another and the *locations* of those transitions. Typically, since the end of one state will always be followed by the beginning of the other, only one of the two states need be encoded as to occurrence and length. Assuming black is the choice, the system will probably be arranged to generate a START code when a white-to-black transition is encountered, and an END code when the next black-to-white transition is encountered. It may further be arranged to assume a white condition at the beginning of each scan stroke, thus making white the normal, or "resting," condition of the system. Obviously the "address" at which each transition occurs within a stroke must somehow also be reflected in the transmitted codes, and there must also be codes signifying the start and/or end of each stroke.

At both the sending and receiving ends of a typical system, the data must pass through a buffer stage to absorb the slight time lapses that ensue between detection of an image element and encoding at the sending terminal and subsequent interpretation and recording at the receiving terminal. With the received codes having been translated back into a chain of black and white elements momentarily held in storage, reconstruction of these elements in their original spatial relationship is a simple matter of their being synchronously "clocked out" of storage, on cue, and fed to the recorder. A typical process for accomplishing this was described in the section on synchronization in Chapter 3.

As an alternative to the transmitter buffer, variable-velocity scanning may be used to vary the scan speed in accordance with the code generator's ability to keep pace.

Depending on the coding scheme used and the ratio of white space to image detail on the subject copy, run-length coding is capable of a wide range of compression ratios. It can in fact range down to a point where it becomes totally impractical, as when the average subject copy contains an abnormally high degree of graphic detail—a page of a telephone directory, for example. The average number of transitions per scan stroke could conceivably reach the point where the descriptive codes would require greater transmission time, or bandwidth, than the actual picture elements. With a typical coding scheme, this situation is likely to obtain for consecutive runs of fewer than three sampling intervals in length.[7]

TRANSMISSION

DFC-10 TRANSMITTER BLOCK DIAGRAM

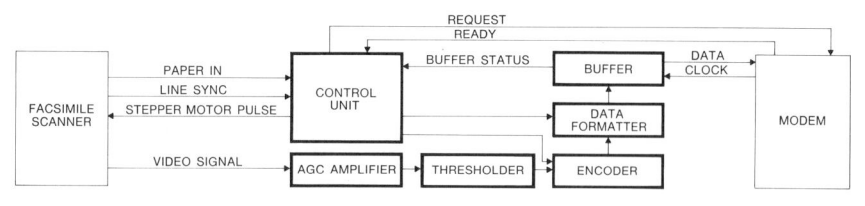

DFC-10 RECEIVER BLOCK DIAGRAM

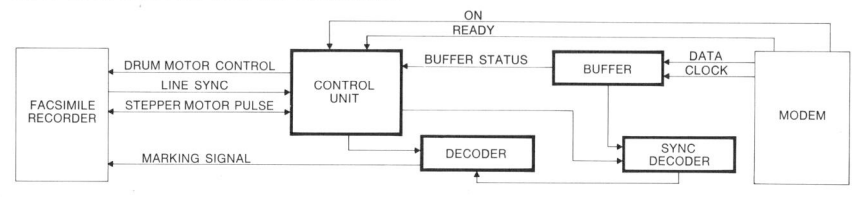

Fig. 4.25. Basic Dacom approach to data compression. Heavily outlined blocks are basic components of the DFC-10 compressor and reconstructor. In addition to the Dacom unit, a modem is required at each terminal as an interface between the DFC-10 and the transmission facility. (Courtesy Dacom, Inc.)

Most modern RLC systems use *adaptive encoding* schemes in which the code-word length is varied according to the nature and amount of information contained in a given scan stroke. The preprogramming required to accomplish this reflects the results of studies of the relative probability of occurrence of a given condition and the optimum code length required to define that condition.[8]

Examples of commercial systems employing RLC techniques are the Dacom 400 series transceivers, the Dacom-developed Rapifax 100 system, the Dacom and Litton-Datalog newspaper-page transmission systems, and the recently introduced 3m Express 9600 transceiver. Figure 4.25 depicts the basic workings of the Dacom compressor and reconstructor.

Run-length coding is sometimes also referred to as *differential-coordinate* coding.

Multi-Line Coding

RLC principles may also be applied on a multi-line basis with the possibility of saving some transmission time. In a typical multi-line system the information content of two adjacent scan strokes is simultaneously encoded. At this writing, at least three Japanese companies, Fujitsu, Matsushita, and Nippon Electric, have such systems in prototype being readied for commercial marketing. (The recently announced *Panafax* digital system, scheduled for commercial availability in 1978, will use dual-line coding.)

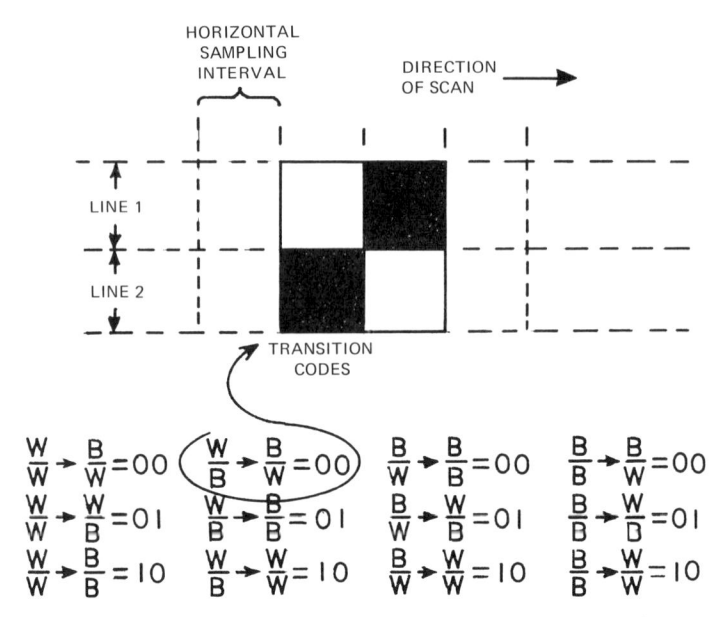

Fig. 4.26. Dual-line run-length coding. Transitions occurring in two adjacent scan lines are encoded as a unit. The particular transition illustrated is $\frac{W}{B} \to \frac{B}{W}$. A total of 12 such transitions is possible, as shown. However, the run-length code structure can be arranged so that a given 2-bit code can be applied to each of four different transitions, thus minimizing code-word length. Compared with single-line RLC, the system can yield a net gain in transmission efficiency.

The principle of multi-line coding is illustrated in Fig. 4.26. Assuming a dual-scan spot, the variety of transitions that can be encountered jumps from two (B-W and W-B) to a total of 12. However, only the four basic vertical combinations need be encoded, namely, $\frac{W}{W}, \frac{B}{W}, \frac{W}{B}, \frac{B}{B}$. Experimental results indicate that, with more or less conventional coding, such a system can yield a discernible gain in transmission efficiency. One reason is the probability that a given pattern of transitions will remain unchanged through two or more scan strokes, in which case a dual stroke will generate no more data than a single one would have.

2-Dimensional Compression

In conventional run-length coding, the compression of data is "one-dimensional"—i.e., no attempt is made to compress the data on the basis of transitions occurring *from one scan stroke to the next*. Considerable experimental work has been done on such *2-dimensional*, or *"vertical,"* compression schemes, with results ranging from about 1.5 to perhaps as much as 3 times the compression ratio normally expected of a conventional RLC system for the same input material.

The idea is to exploit the high probability that a transition occurring at a certain point in one scan stroke will be repeated at approximately the same point in the succeeding stroke. That can be accomplished by encoding all transitions in the first stroke in the normal manner, but thereafter encoding only the *differences* in position of those elements forming essentially vertical details in the copy (i.e., image lines that lie at approximately right angles to the scan axis). The difference codes can be appreciably shorter than the full positional codes.

Among the earliest such systems to be demonstrated experimentally has been the "blob-encoding" technique, which has been under development at Bell Labs by Amalie J. Frank and others since 1970. The essence of the technique (also known as "patch-encoding" or "feature-encoding") is to collectively encode entire areas of like tonality. If, for example, a playing card were being scanned, the symbol identifying the suit (heart, spade, etc.) would be encoded as an *entity* rather then as a number of separate run-lengths, each with its own code.

The principal developmental thrust for this system at Bell Labs has been in the direction of computer-controlled photocomposition of *Yellow Pages* advertisements. However, it has also been applied experimentally to facsimile communication. As with any redundancy reduction scheme, the degree of data compression obtainable will vary with the nature of the graphics involved and the desired output resolution. It has been demonstrated, however, that at reasonably high resolutions (about 200 LPI), and with a typical business letter page as the input, ratios of 15:1 are easily obtainable with this system. Interestingly the compression efficiency tends to improve in direct proportion to an increase in scan resolution—i.e., at 400 LPI, a compression ratio of about 30:1 is obtainable for the same input material.

Bear in mind that we are speaking here of *compression efficiency* and not necessarily transmission *time* compression, as is usually the case when we compare digital data compression fax systems with conventional analog systems. Obviously at higher scan resolutions it is going to take longer to physically push a piece of paper through a scanner or recorder. Thus, there are certain tradeoffs to be considered in this regard.

In a 1971 laboratory experiment, Singer Simulation Products of Sunnyvale, California, succeeded in scanning and recording a typical engineering drawing at a time compression of 41:1 as compared with element-by-element reconstruction. This compared with 19:1 for the same drawing on a one-dimensional basis, using an optimum code. At a scan resolution of about 200 elements per linear inch, the entire drawing consisted of 78 million black and white elements. The 2-dimensional RLC system reduced these to a total of 604 black/white transitions, which in turn were encoded into 1.9 million bits.

In October, 1975, the experimental capabilities of a Japanese prototype 2-dimensional RLC fax system were reported at a technical symposium in Geneva. Compression ratios ranging from 1.6 to 2.2 times as large as that for a conventional one-dimensional system had been achieved with the prototype equipment. The system is similar in several respects to the Bell Labs blob-encoding process.

A novel compromise between one- and 2-dimensional compression exists in the form of an "error-masking" technique that has been effectively demonstrated by Fujitsu Limited of Japan and by Hasler Research Laboratories of Berne, Switzerland. The process typically requires dual buffers in each terminal, one of which is unloading the line just processed while the other is processing the next line. For each line, the functions are automatically switched, and the switching is so arranged that if a line contains an error, it is simply skipped and the previous line automatically repeated. In that way, a statistically appropriate "filler" is substituted for what might otherwise have been a conspicuous gap or totally distorted line.

Here, as in other redundancy reduction techniques (blob-encoding, for example), the effectiveness of the process tends to improve with increased resolution: the higher the system's scan resolution, the less the effect of a given error and therefore the easier it is to achieve greater time compression simply by allowing the system to tolerate a greater number of errors.

Table 4.A compares the speed capabilities of the present generation of data compression fax systems with projected speed capabilities of those of the next generation. The projections assume use of 2-dimensional compression techniques.

TABLE 4.A. Typical Present and Projected Speeds for Data Compression Fax Systems at Various Digital Bit Rates.

Bit Rate	Present-Generation System (Approx. Avg. 6:1 Compression)	Next-Generation System (Est. Avg. 12:1 Compression)
*4,800	60	30.0
*7,200	45	22.5
*9,600	30	15.0
19,200	15	7.5
40,800	7	3.5
56,000	5	2.5

Time in Seconds to Send an "Average," Letter-Size Text Page at a Medium-To-High Resolution (Approx. 20,000 Picture Elements per Square Inch)

Note: Some of the data shown are actual (based on tests with existing commercial or experimental terminals), and some are merely theoretical.

*Present scope of voiceband circuits (6/77). Transmission times shown assume an ideal circuit. No allowance is made for the probability of error-caused delays at the higher bit rates.

TRANSMISSION 149

Multi-Level and Bipolar Techniques

A comparatively simple and inexpensive way of achieving an assured 2:1 bandwidth compression in a "thresholded" (black-white, or binary) fax system, regardless of density of image detail, is by the addition of a second white level or a second black level, depending on signal polarity. The object is to convert each white-black or black-white swing of the signal to a two-step *unidirectional* transition, thereby halving the number of zero-crossings.

The scheme depicted in Fig. 4.27 will suffice to illustrate the basic principle, although there are variations in the way it may be applied. For simplicity the figure shows the principle applied to a three-level d.c. representation of the scanner baseband output. Zero-crossings are therefore represented by the dipping of the signal to zero level. Halving the number of zero-crossings means halving the transmitted frequencies, which permits the choice of either utilizing a narrower bandwidth transmission facility or transmitting twice the data over a given facility.

The first commercial fax device to use the three-level principle was the *Pacfax* modem introduced in the 1960s by John H. Clark's HELAC Electronics Company. The Pacfax system was adopted by the U.S. National Weather Service to double the speed of map transmissions. The circuitry has since been embodied in the Scanatron fax line, marketed until recently by Victor Graphic Systems and now by a new company, Infolink.

Although the above description is confined to a thresholded scanner output, in which all gray tones between black and white are eliminated, essentially the

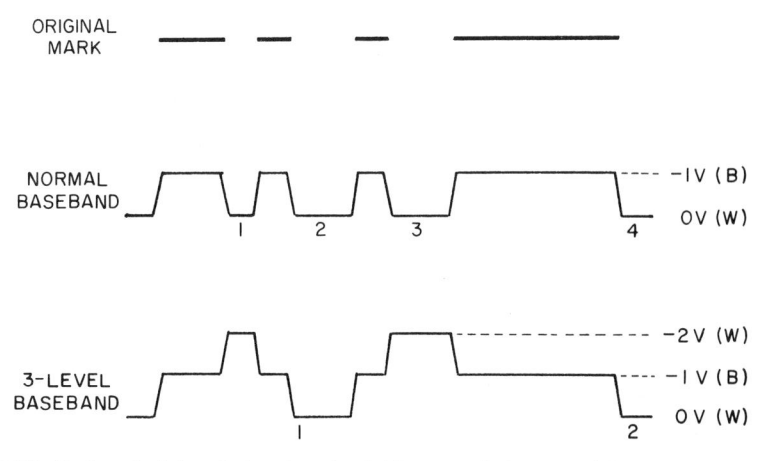

Fig. 4.27. Basic principle of the three-level binary technique used in some commercial analog fax systems. Reshaping the waveform in the manner shown has the effect of halving the frequency of the transmitted picture signal.

same technique can be applied with the gray scale retained. A triggering signal to alternate the waveform polarity must still be generated from the baseband peaks (or pronounced "inflections"), but the *slope* of the baseband pulses and the lesser ripples between peaks need not necessarily be altered. Such a technique is applied in the Graphic Sciences *dex** line of dual- and multi-speed analog fax terminals to achieve a 2:1 speed-up in transmission time over voiceband circuits without sacrificing gray scale. A variation of it is also the basis for the recently finalized international CCITT "Group 2" recommendation for a universal dual-speed message fax standard. (See Chapter 7.)

Although it includes certain basic differences covered by individual patents, the 3-level approach to analog compression has come to be referred to generically (and perhaps erroneously to some extent) as *duobinary,* or *spectral, compression*.

Another multi-level scheme is described in a Xerox patent (#3,555,428, D. A. Perreault, 1971). It takes advantage of the fact that the higher of the two frequencies in an FSK pulse transmission system can theoretically accommodate more pulses per second than can the lower frequency (assuming one pulse per a.c. cycle). Therefore, rather than let the lower frequency determine the fixed pulse length for the system, means are provided to vary the pulse length with the frequency. Since the black elements on a page of copy normally represent only 5 to 10 percent of the total elemental structure of the copy, the initial step in designing a fax system using this scheme would be to assign the lower frequency to the black condition.

Implementation of a practical system along these lines would, of course, also require some means of varying the spacing of the pulses in transmission. This suggests a digital system with variable "clock" rates automatically switched according to the state being depicted. It also suggests the need either for buffering at each terminal or for some means of variable-velocity scanning.

Other Schemes

There are probably as many different compression schemes as there are communications engineers who have at one time or another said to themselves "There ought to be a better way." For the purposes of this chapter, it will suffice to briefly survey some additional ones that have been developed and applied either experimentally or commercially.

In a system called RACE (Radiation Adaptive Compression Equipment), developed in the late 1960s by Radiation Incorporated of Melbourne, Florida (now the Electronic Systems Division of Harris Corporation), compression ratios in the neighborhood of 5:1 were achieved by removal of redundant samplings in an analog-to-digital conversion process. Although developed primarily to im-

---

*dex is a registered trademark of Graphic Sciences, Inc.

TRANSMISSION 151

prove the quality of military digital voice communications, the process is also applicable to fax.

Figure 4.28 illustrates the RACE principle. Only those data points are retained that are necessary to reconstruct a straight line-equivalent of the original waveform. By itself the system is capable of reducing the digital bit rate from about 48 kilobits per second to 9.6 kilobits. The terminals are designed to interface with commercial modems that enable transmission via voiceband analog circuits.

By today's standards, and in view of the various advanced technologies that have been laboratory-tested, a 5:1 compression ratio is relatively slow. At the other end of the existing spectrum is a technique know loosely as *vector encoding*, or *vectorization*, which offers the potential for transmission speed compressions of well over 100:1, but which so far has been applied only to computer graphic storage and editing activities (by Bell Labs and others). The trick is to abandon the traditional line-by-line *raster* scan approach and instead have each image detail sought out and *followed* by the scan spot, which in its travels generates codes that will subsequently instruct the recording spot when to shift direction and at which angle, when to turn on and off, etc. In a highly sophisticated system arcs as well as vectors can be encoded and reproduced.

As a process vectorization represents an overlap of digital fax and optical character recognition (OCR) technologies. As might be expected, it requires considerable logic and buffer space and comparatively elaborate circuitry and

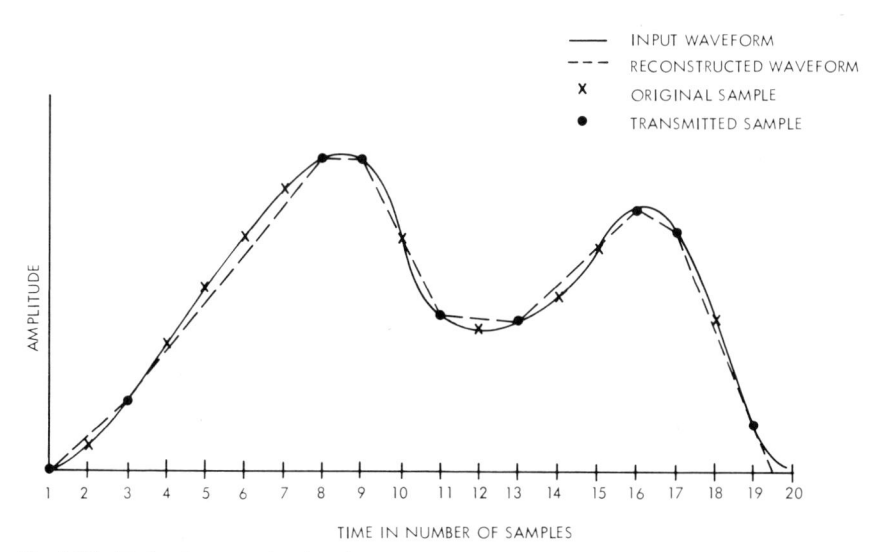

Fig. 4.28. Redundancy reduction by removal of redundant samplings in an analog-to-digital conversion process. (Courtesy Harris/Radiation Div.)

hardware. In short, while it offers a potential for significant data compaction, vectorization is a decidedly expensive proposition.

Dr. N. S. Szabo of Singer Simulation Products has done considerable experimental work with the process, part of which was the basis for his doctoral thesis, "Digital Representation for storage and Editing of Line Drawings" (University of California–Berkely, 1976). In a typical experiment he succeeded in reducing an engineering drawing containing a total of 78 million bits to a mere 2.5 million bits of reconstruction coding: a compression ratio of approximately 32:1. While not quite as impressive as that obtained in an earlier experiment with 2-dimensional RLC (41:1), this result is by no means optimum, and therefore has considerable potential for improvement as the technique becomes further developed.

Although they are perhaps outside the context of redundance reduction per se, we might mention techniques developed (by Litton-Datalog in this country and by Fujitsu in Japan, for example) to vary scan resolution automatically in accordance with the density of detail encountered in scanning. The Litton-Datalog scheme is applied to the transmission of newspaper page proofs, in which a higher resolution is required for screened halftone pictures than for text. (See Chapter 5.) Obviously, where there are both fine and coarse details within the scanner's sweep, the fine details will be favored and a higher scan resolution automatically selected for the portion of the document where that condition exists.

Techniques have also been developed to automatically switch a digital transmitter to a higher transmission bit rate in the event that a particularly "clean" (impairment-free) circuit is encountered on the dial network. However, the automatic circuit condition determination will usually be made at the outset of transmission, and therefore the selected bit rate will generally apply for the duration of the message, even though the circuit condition may change within that period.

TRANSMISSION SEQUENCE

Until quite recently (before fax became more automated and in general more sophisticated), the standard procedure for sending a document consisted of a few words of voice coordination by the terminal operators, followed by a simple manual action, like pressing a button, at each terminal. A fallacy of that system is that if either operator does something wrong—like selecting the wrong speed or forgetting to work the exclusion key on the phone—the terminals may go into action anyway, and keep going for the normal transmission duration. The result is delivery of a blank or distorted page as well as several minutes of wasted line time.

Today most systems are designed to follow a predetermined signalling sequence

TRANSMISSION 153

known variously as the *control procedure* or *transmission sequence*—or, particularly in data-processing circles, the *protocol*. Assuming a switched facility such as the telephone dial network, the first step in the sequence is always *channel establishment*, and the last step *channel release*. In between, besides the actual *message transmission*, are *pre-message signalling*, *phasing*, and *post-message signalling*, the first two of which (and sometimes the third as well) are often referred to as the terminal *"handshake."* The basic sequence is illustrated in Fig. 4.29.

In a private leased-line system, in which there is normally no dial-up or automatic ringing as in the public switched network, the first step in the sequence—channel establishment—may prove a problem in some cases. Ordinarily the common carrier furnishing the service expects the customer to provide an appropriate signalling scheme through which one terminal machine can "awaken" another. It is also expected, however, that the start-up signalling scheme will be compatible with line requirements. As a rule either the customer or terminal vendor must work this out with the carrier to arrive at some means suitable to both.

While the separate components and the order of their transmission do not vary significantly from one system to the next, the subroutines and the nature of the signals may vary appreciably. For example, the pre-message signalling component alone may consist of from two to perhaps as many as 10 or 12 back-and-forth signalling steps. Normally the first step following interconnection of the terminals is for the SEND machine to emit an identifying signal, which in a dual- or multi-speed system might serve to identify the speed setting of the transmit-

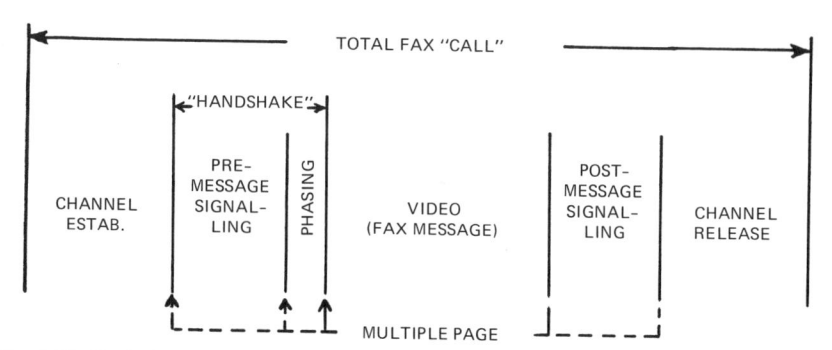

Fig. 4.29. Basic fax transmission sequence. The pre-message signalling will usually consist of an acknowledgment by the called station that a connection has been established, followed by a calling station identifying signal (possibly containing information on speed, line condition, whether the transmission is to be single- or multi-page, etc.), possibly followed by an acknowledgment from the called station that it is ready to receive. The post-message signalling may be merely a channel release signal from the transmitter upon completion of the message.

ter. This is normally followed by an *acknowledgment* signal from the receiver, the receipt of which assures the transmitter that there is a workable interconnecting link. In a relatively simple system this latter step will mark the end of the pre-message signalling component (the "handshake"), and will be followed by the phasing step and then the actual message transmission. In a more complex system a whole sequence of "set-up" signals may follow, conveying such information as the scan resolution and bit rate at which the copy will be sent, whether there are to be two or more pages sent consecutively, etc.

As for the individual signals, Xerox uses the following: an interrupted 1100-Hz tone for transmitter identification; a 1500-Hz tone for receiver acknowledgment (amplitude-modulated at a 2-Hz rate if the receiver is set for the higher of two selectable speeds); an alternating 2425-Hz and 1500-Hz (or 1250-Hz, depending on the selected speed) tone to signal the start of the phasing step, followed by 15 seconds of 1500-Hz phasing pulses at scan stroke intervals; an FM signal continously variable from 1500 Hz to 2450 Hz for the actual message transmission; and, finally, an 1100-Hz post-message STOP tone to signal the end of transmission.

In the GSI *dex** analog systems, the transmitter identifies itself by 2-second bursts of either the 2048-Hz carrier frequency alone or the same frequency in combination with a 1500-Hz tone (depending on the speed setting). The receiver responds with like tones, but of a steady nature. Phasing is accomplished by automatic drum start upon the next receipt of carrier by the receiver, the carrier this time being amplitude-modulated for message transmission. The end-of-message is signalled by a 1500-Hz STOP tone from the transmitter.

These are merely examples. At the moment, efforts are underway both domestically and internationally to arrive at some universally acceptable standards on transmission sequence.

TRANSMISSION OPTIONS

So far this chapter has dealt mainly with the bare essentials of fax transmission. Following are brief descriptions of some of the more interesting transmission-related options available to users of modern fax systems.

Encryption

Compared with voice communication, fax may be regarded as a relatively secure transmission medium. After all a compatible machine many times more costly than a telephone set is required to reproduce an intercepted fax message. Nevertheless that requirement may be no obstacle for someone to whom the intercepted information has great value. Consequently an *encryption* (or message-

---

**dex* is a registered trademark of Graphic Sciences, Inc.

TRANSMISSION

scrambling) capability has been made available to discourage unwanted interception.

Figure 4.30 illustrates the basic encryption principle. The scanner baseband must first be converted to a series of on-off (0 and 1) pulses in accordance with a clocked sampling of the scanner output. Each of these pulses is then combined with a synchronously clocked series of quasi-random pulses know as the *key*, the combination resulting in a third pulse stream called the *cipher*. It is the cipher stream that is transmitted. Applying the identical key at the receiver reconverts the received cipher to the original baseband pulses.

By some means of coordination between operators at the various stations, the same key is set—and changed periodically—at each terminal. This is usually accomplished by the manual setting of thumbwheel switches arranged to yield

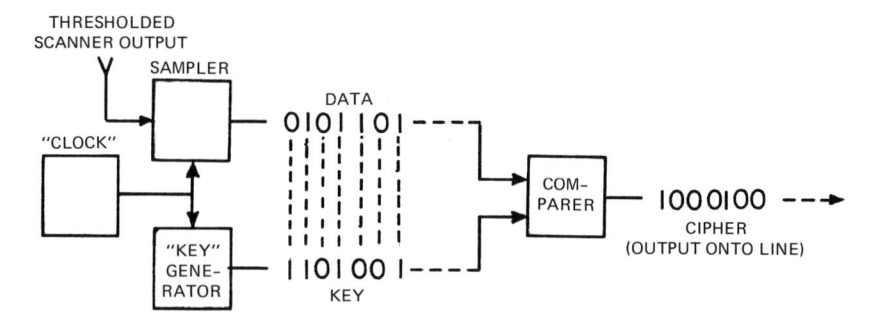

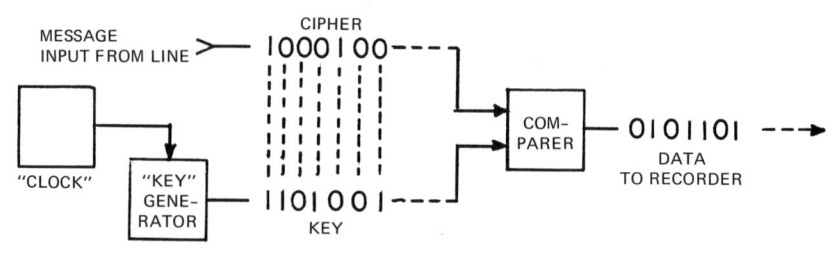

Fig. 4.30. Basic encryption principle used in the "scrambling" of transmitted fax signals. The synchronous binary representation of the scanner output is compared with a "key," a pseudo-random bit stream controlled by the same clock. The two streams are combined in such a way that, where the simultaneous data and key bits differ, a 1 results, and where they are the same, a 0 results. The resulting difference signal, or "cipher," is transmitted as the scrambled fax signal. At the receiver, the incoming cipher is compared with a key identical to that at the transmitter (and in precise step with it), and the resulting difference signal is the recovered scanner data. In the absence of the identical key, the cipher bears no relationship whatever to the original data signal.

156   ELECTRONIC DELIVERY OF DOCUMENTS AND GRAPHICS

several million different combinations. If any thumbwheel is off by even one step, the recorded copy from the terminal will remain scrambled. Figure 4.31 shows a comparison of "clear" (nonencrypted) copy, scrambled copy, and deciphered copy.

The encryption/deciphering capability has been built into some fax systems—especially military systems—but is also separately available in a selection of commercial units with which the fax terminals can interface. The encryption devices often cost more than the terminals with which they interface. The popular Datotek DF-300, for example, designed for use with analog fax ter-

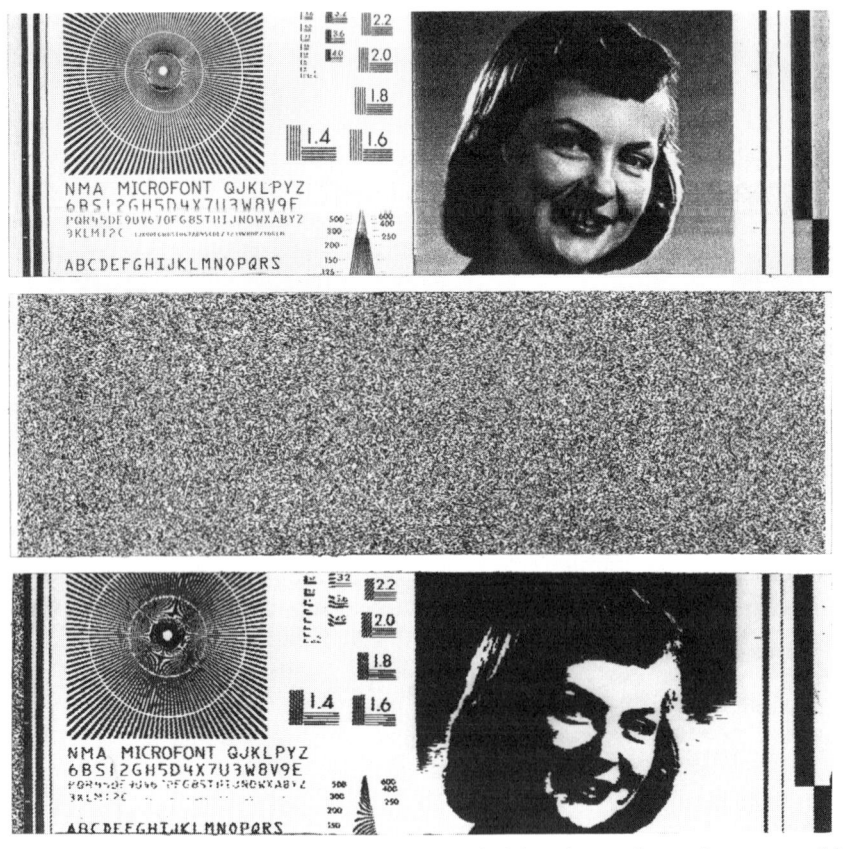

Fig. 4.31. Comparison of "clear," scrambled, and deciphered copy from a fax system with an encryption capability. The clear copy (top) was produced with the encryption circuitry completely switched out, and is therefore a normal analog transmission. The scrambed copy (center) was produced from an encrypted, but undeciphered transmission, and the deciphered copy (bottom) is the result of the signal's having gone through the full encryption/deciphering process. Note the conversion from a continuous tonal (linear) to a 2-level (binary) output mode. (Courtesy Datotek)

minals, currently markets for more than $4000 (or about $250 a month) per terminal.

Broadcasting

In modern fax parlance *broadcasting* does not necessarily imply a radio service. It simply refers to the selective interconnection of one transmitter with multiple receivers, either simultaneously or auto-sequentially, and usually by wire links. Separate broadcast modules for use with leased lines have been commercially available for some time from certain fax terminal vendors.

Recently, similar devices have become available for use with dial lines as well. For example, Telephonic Equipment Corporation of Irvine, California, markets the *Datapatcher*, a 9-channel module purchaseable for under $500. Such devices have also become available as options with some commercial fax terminals (Rapifax and GSI *dex*, for example). Obviously, their operation requires as many separate phone lines as there are terminals to be reached simultaneously.

The alternative method for broadcasting via the DDD would be a sequential operation preferably involving a preprogrammed auto-dialing capability. This would, of course, also require automatic recycling of the page or pages being sent. At this writing, such systems are largely still in the talking stages.

Two notable examples of simultaneous fax broadcasting systems that have been in operation for some years are the newspicture operations of the major wire services (AP and UPI), and the nationwide distribution of weather charts by the National Weather Service.

Polling

Polling, or *automatic transmission*, is sequential broadcasting in reverse. It is an arrangement by which a single receive station may sequentially (and selectively) contact any number of send stations, the latter being arranged to automatically answer an incoming call and respond to the secure polling signal by transmitting the preloaded copy. If a particular send station has nothing to transmit, it will signal the receiver accordingly and will be quickly disconnected so that the poller can move on to the next station.

Polling is often used by large vendors of commercial products to funnel orders from branch sales offices to headquarters. It can be a manual or automated operation, the latter being facilitated by an auto-dialing capability. It is available as an option from several commercial fax terminal vendors.

Automatic Sequential SEND/RECEIVE

Through use of an auto-dialing capability such as that recently introduced by Xerox (and even more recently by the 3M Company), variable combinations of automatic sequential transmission, automatic polling, and automatically delayed transmission can be achieved. In both the Xerox and 3M systems the numbers to be dialed sequentially are entered as pencil marks on preprinted forms functioning as coded control cards. These cards are then placed in the fax terminal's automatic stack feeder just in front of the related documents, and after a few seconds of scanning of the mark codes, connection is made and communication with the dialed terminal commences. Other auto-dialing systems have been developed that permit storage of phone numbers in memory, to be called forth automatically and in proper sequence.

To take advantage of lower late-hour rates, or to allow for the crossing of time zones, the starting time for the automatic operation can be preset by manual control. In the Xerox system, and in other systems currently under development (mid-1977), a record is automatically kept of any uncompleted transmissions or receptions.

Store-and-Forward

This appears to be one of the more promising fax transmission options. It is the ability to input documents for delayed automatic transmission, but, unlike the operation just described, requires that both the dialing instructions and the video signals representing the related documents be temporarily stored via magnetic recording. Besides taking advantage (as does any auto-dialing/auto-feed system) of the fact that many modern fax terminal machines are capable of completely unattended operation, store-and-forward also permits inputting of documents at speeds exceeding that at which transmission will subsequently take place, and it permits *recycling* of the input. The recycling lets it bypass a called number that is busy and return to it later. This cannot be done with a simple auto-dialing system, which has to physically "dump" the documents intended for that number before proceeding to the next call.

The combining of all of these various capabilities in a single terminal machine is certainly feasible and offers the possibility of a most interesting, fully automated document distribution system.

FOOTNOTES

1. Depending on the manner in which it is accessed and the type of photoelectric transducer used, the output of a scanner may represent white portions of the copy as zero and black as maximum signal, or vice versa.
2. Ninety-six lines (or strokes) per inch and 180 lines per minute, in accordance with the Electronic Industries Association message facsimile standard RS-328.
3. The reference to *carrier* here concerns the multiplexing systems by which multiple signals are sent simultaneously over a single wideband facility. Most long-distance telephone calls today are transmitted via carrier systems.

TRANSMISSION

4. References to the time of writing, including such phrases as "at present" or "to date," or the word "now," may be interpreted as meaning mid-1977, unless otherwise noted.
5. The term "d.c. pulses" is used here mainly for illustrative purposes. In actual practice, a pulse will more likely be represented by a short burst of some discrete frequency.
6. See Chapter 8.
7. See Wyle, et al., "Reduced-Time Facsimile Transmission by Digital Coding," IRE *Transactions on Communications Systems*, September 1961, pp. 215–222.
8. The concept of *statistical*, or variable-length, coding was first advanced by C. E. Shannon and R. M. Fano in the late 1940s, and was subsequently modified by D. A. Huffman (1952) and others. Actually, Morse had applied similar principles in the 1800s in working out the simple telegraph code that bears his name.

5
QUALITY

Before the planning of a facsimile system (or for that matter any electronic graphic scanning system) can properly proceed, at least two of three basic variables have to be tied down within reasonable limits. The three are *cost*, *speed*, and *output quality*. Given any two, the third can usually be derived without much difficulty.

There may, of course, be additional factors to complicate the task, such as size or portability of terminal equipment, availability of facilities to meet transmission needs, and the psychological impact of the physical and copy qualities of the system output. In the past, many a fax and video system has failed because it simply did not live up to the promise inherent in the system concept, and a high percentage of those failures were tied to output quality.

Of the three basic variables output quality is perhaps the most involved—primarily because it cannot be computed or predicted with mathematical precision, as can the other two. This chapter will therefore confine itself to a discussion of various factors governing legibility and overall quality of a scan-system recording. The interrelationship of quality with other variables, particularly cost and speed, will be explored in Chapter 6.

Some of the factors influencing quality in fax systems have already been discussed in preceding chapters. Chapter 3 described distortions produced as a result of phasing and synchronization errors and of mismatched indices of cooperation, while Chapter 4 covered the detrimental effects of various phenomena inherent in the transmission process.

In this chapter, the emphasis will be on conditions and principles that are fundamental to the scanning and recording processes—which takes us beyond straight mechanics into the realm of individual judgment.

RESOLUTION

Before we can legitimately assess the output quality of a graphics communication system, we need to know something about resolution. As applied to

QUALITY

optics, resolution may be defined as *the degree to which adjacent elements of an image are distinguishable as being separate.* As applied to scan systems (fax and TV), the term is used loosely to specify the *scan density*, i.e., the number of scan lines within a given linear dimension. In TV it is the number of lines constituting an entire image—e.g., 525 for American commercial TV—whereas in fax it is the number of lines per inch—e.g., 96 in conventional message fax. (The fax-equivalent resolution of a properly adjusted picture on a 21-inch, diagonally measured TV screen would be about 35 lines per inch.)

Actually, in scan systems, there are two kinds of resolution—or, more correctly, the resolution of the system is governed by two sets of variables, one affecting the width and the other the height of the reproduced image. While resolution *across* the scan "grain" (i.e., at right angles or oblique to the scan axis) is controlled essentially by the number of scan lines within a given linear dimension, resolution *along* a scan line is governed by the speed with which the system as a whole can react to tonal transitions.

Lines Versus "Line Pairs"

Careless use of the word "lines" in a discussion of resolution can lead to serious misunderstanding. A *scan line* is one kind of line, while the line that is a component of a resolution test pattern may be quite another. The latter may consist of a *line and space* of equal width, while scan lines are (theoretically) contiguous. The significance of this difference is that it takes two scan lines to resolve one line and space—or *line pair*, as it is sometimes called—of test pattern resolution (Fig. 5.1).

Electrically a visual line-space pair represents one cycle of a theoretically

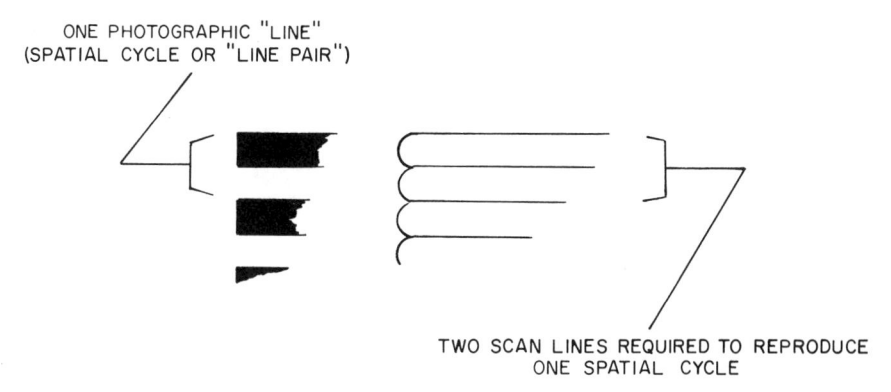

Fig. 5.1. It takes at least two scan lines to resolve one "line pair," or spatial cycle, of a resolution test pattern when the transitions occur along an axis normal to the scan axis.

square waveform (why not actually square will be explained later in this chapter). However, even where there are no electronics involved, a visual line-space pair is in itself a *cycle of spatial frequency*. Throughout the remainder of this chapter the word *cycle* will therefore be used to refer both to a visual cycle and to its electrical counterpart. The term *hertz* will, of course, apply as a unit of measure when we speak of electrical cycles as occurring within a particular time frame (specifically, per second).

The IEEE Facsimile Test Chart (Fig. 5.2) contains a variety of test patterns for measuring the resolution of a system. At the top of the chart are vertical bar patterns ranging from 10 to 96 contiguous lines (5 to 48 cycles) per inch. A bit farther down are clusters of triple-line groups known as repeating tri-bar patterns. These permit resolution readings ranging from 61 to 406 contiguous lines (30.5 to 203 cycles) in 12 steps.

The radial line pattern, in the form of a square at the left of the chart below the bar patterns, permits resolution readings at any angle, ranging continuously from 50 contiguous lines per inch (outer circle) to 200 (innermost circle). The wedge, or fan, pattern diagonally below it is calibrated in the number of contiguous lines (black *and* white) per inch, and is the most effective pattern on the chart for determining at a glance the spatial frequency of the system along the scan axis.

The variety of typefaces in the lower left quadrant of the chart are self-explanatory (more on character size later). The six rectangular gray-scale patterns just below the left edge of the portrait photo consist of two halftone dot patterns—65 and 120—each reproduced in 10, 50, and 90 percent tints. These are useful in checking for the probability of encountering moiré patterns in the transmission of halftone material.

The National Bureau of Standards (NBS) microcopy resolution test target appears in two locations on the chart: between the radial-line pattern and the portrait photo, and in the lower right-hand corner. It contains 23 patterns, each consisting of two sets of lines and spaces. The five lines and four spaces constituting each set are of equal width, and the two sets within a pattern are arranged at right angles to each other. The number associated with each pattern is the quantity of spatial cycles (lines and adjoining spaces) per millimeter.

This is actually a condensation of the standard NBS target, which consists of 26 patterns, ranging from 1 to 18 cycles per millimeter. However, in all other respects it is unaltered. It is a particularly valuable set of patterns inasmuch as it is standard for microphotography and therefore provides the basis for comparisons of microfilm and facsimile quality standards. Such comparisons will become increasingly important as the growing use of microfilm—both for document storage and as a computer output medium—continues to impact on the input requirements for modern fax systems.

QUALITY 163

Fig. 5.2. IEEE Facsimile Test Chart, 1975 Edition. (*Note*: The chart as reproduced here cannot be used for actual tests.) (Courtesy IEEE)

The angular positioning of the corner NBS target permits measurement of resolution along two axes oblique to the scan axis (by about 45 degrees).

All that need be said further about the various resolution patterns constituting the IEEE Chart is that depending on whether scanning is vertical or horizontal, the patterns may be measuring either the effective system bandwidth (indirectly, of course) or the effect on resolution of the segmenting of the image by the scanning process. The radial-line and NBS patterns measure both, regardless of scan direction.

Lines Versus Elements

To avoid confusion it is a good practice to think of resolution in terms of elements rather than lines. Since there is ideally a 1:1 relationship between scan lines and image elements (in the across-the-grain direction), the distinction between vertical and horizontal resolution effectively disappears when each is expressed in elements. (Image elements are often referred to in technical documentation as *picture elements*—or by one of two accepted abbreviations thereof, *pixels* and *pels*.)

The same 1:1 relationship exists between image elements and *bits* (binary digits) in a two-level ("mark and space") digital system. Thus we can talk about the reproduced subject copy as consisting of a total of so many thousands of elements, or *kilobits*, thereby facilitating the computation of transmission time and required channel capacity.

For example, let's say we are sending an $8\frac{1}{2} \times 11$-inch document over a system having a scan resolution of 100 lines per inch. Assuming that resolution is to be equal horizontally and vertically, we will need a transmission facility capable of accommodating 850 × 1100 (a total of 935,000) elements of picture information within a given amount of time. If we are willing to tolerate a transmission duration of six minutes, the required transmission bandwidth would be 2.6 kilobits per second (935 kilobits ÷ 360 seconds). For shorter durations the bandwidth would be proportionately broader.

Expression of resolution in elements also provides a universal basis for comparison of various image-reproducing media. Some representative examples are shown in Table 5.A.

Resolution Determinants

1. *Spot Size.* A fundamental factor limiting the resolution capabilities of a fax system is the size of the scan spot. Of course it is a factor only insofar as it is consistent with the scan density, or *pitch* (scan lines per inch), of the system. The optimum relationship (assuming for the moment a round spot) is a spot diameter, in inches, equal to the reciprocal of the number of lines per inch. For example, in a 96 LPI system the optimum spot diameter would be $\frac{1}{96}$ inch.

The greater resolution potential of a smaller spot would be wasted on a system having a scan pitch of significantly fewer lines than could be accom-

QUALITY

TABLE 5.A. Elemental Resolutions of Selected Graphic Reproductions

Image	Resolution in Total Elements (pels or pixels)
$3\frac{3}{4}$ × 5-inch newspaper halftone (about 50 dots per inch)	50,000*
8mm home-movie frame	50,000**
525-line commercial TV	150,000†
$3\frac{3}{4}$ × 5-inch "slick" magazine halftone (about 150 dots per inch)	420,000*
1000-line CCTV	550,000†
96 LPI fax recording of an $8\frac{1}{2}$ × 11-inch document	630,000††
35mm professional movie frame	1,000,000**
Frame of 35mm microfilm (100 cycles/mm resolved in image)	50,000,000

*D. G. Fink, *Television Standards and Practice* (New York: McGraw-Hill, 1943), pp. 62-63.
**Ibid., pp. 5-6.
†Assumes equal horizontal and vertical resolution and various losses (to be discussed later in this chapter).
††Assumes 850 elements per line (100 per inch) and an *actual* vertical resolution of 740 elements.

modated in the same space without overlapping—e.g., a spot of $\frac{1}{200}$-inch diameter in a 100 LPI system. Likewise, an oversized spot, resulting in overlapping strokes, would unfairly limit a system's potential resolution capabilities. In practice no particular rule is applied universally. In some fax systems lines are permitted to overlap so as to effectively render them invisible and at the same time enhance the density of the copy, whereas in TV systems the scan spot is typically somewhat smaller than the optimum for a given pitch.[1]

In existing commercial fax equipment spot size may vary from $\frac{1}{50}$ inch (20 mils) down to $\frac{1}{1000}$ inch (1 mil), and even smaller. Indeed in fax scanners for special applications it can extend down through fractions of a mil, all the way to the *micron* (thousandths of a millimeter) range. Alden's 35-mm microfilm scanner, for example, uses an optically reduced spot measuring about 0.6 mil, which permits the equivalent of 96 LPI scan resolution on a 16X-reduced microimage of the original subject copy. Special microfilm scanners have been developed that use electron or laser beams to produce spots as small as 5 microns (0.2 mil), a size that permits a scan resolution of better than 5000 LPI on the film.

2. *Shape of Scan Aperture.* While a round scan spot, or aperture, will produce images of adequate quality in most cases, the optimum shape for fidelity of image is a rectangle, its long dimension defining the width of the stroke. Why a rectangle and not a square—or a circle—is best explained by referring to Fig. 5.3.

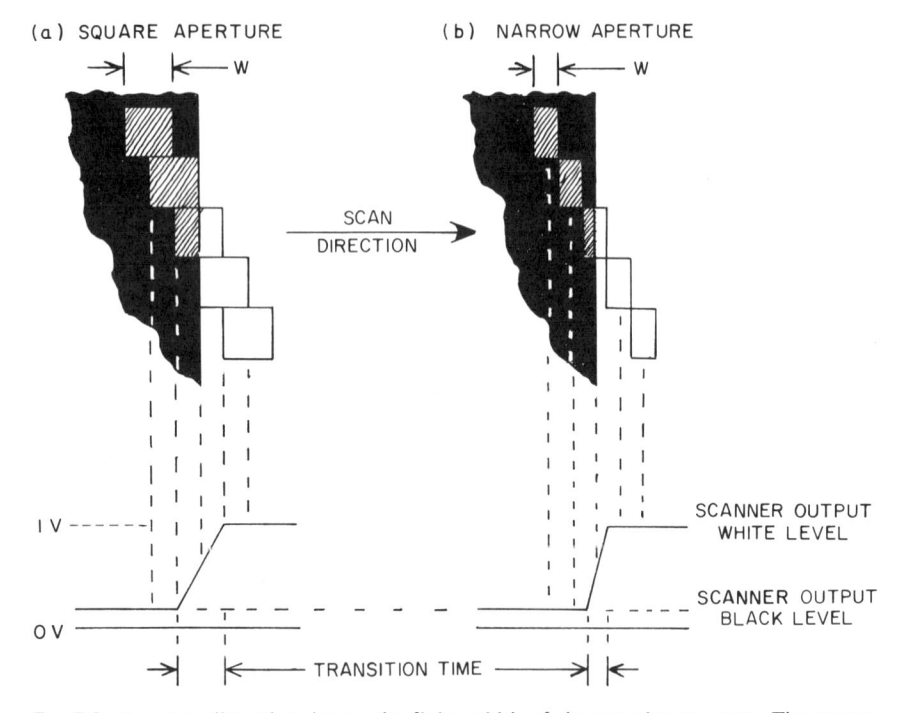

Fig. 5.3. Aperture distortion due to the finite width of the scanning aperture. The greater the width, the less sharp the transition from one contrasting brightness level to another. Note: The descending sequence of aperture positions in each sketch represents successive positions within a stroke.

As illustrated, particularly in 5.3(a), the effect of an aperture of finite width passing across the juncture of dark and light picture elements is somewhat like that of a window shade being raised or lowered (except that here the "window" moves with respect to the "shade"). From the viewpoint of the scanner's photoelectric eye a sharp transition from black to white on the subject copy thus becomes a relatively gradual change because of the finite interval between the aperture's being obscured by the dark detail to its being fully bathed in light.

In other words, the faithfulness with which a sharp tonal transition is reproduced along the scan axis is a function of the relative amount of time it takes for the full width (w) of the aperture to pass the tonal juncture. The wider the aperture, the more sluggish the scanner's response to sharp transitions, and therefore the poorer the fidelity of reproduction. This phenomenon is commonly known as *aperture distortion.*

This suggests that the ideal aperture from a fidelity standpoint would be a slit of infinitesimal width. Obviously such an aperture would be impractical because it would not admit sufficient light. Therefore in practice a compromise is struck, as illustrated in Fig. 5.3(b). The ratio of width to height for a given system will (or should) be governed by the fidelity of the system as a whole.

Where a system operates close to the upper frequency limitations of the transmission channel (as is often the case with systems utilizing the telephone network), it is perhaps just as well that the aperture shape does *not* approximate the ideal. In such cases, it is generally recommended that one limit the transmitted frequencies in the scanning process rather than rely on the response characteristics of the transmission circuitry. The latter alternative may produce distortions even less desirable than the effects of limiting resolution. (See Chapter 4.)

It is also generally true that aperture shape becomes less critical as spot size decreases—consistent, of course, with an increase in scan resolution. Where, for example, an extremely fine scan grain can be achieved through application of a laser or cathode ray device, one should not be greatly concerned if the scan spot is round rather than rectangular or square. (This may depend somewhat on the nature of the scanned original. If it is a microfilm image, for example, scanning distortions are likely to be magnified in the enlarged recording, and due care must therefore be taken.)

3. *Recording Process.* In conventional ink printing and duplicating, quality is governed principally by the process involved and the quality of the paper. There is no doubt that letterpress is superior to stencil duplication, for example, or that glossy paper stock produces better-quality copy than coarse newsprint. The difference is perhaps as much aesthetic as it is physical.

Similarly, in fax recording, where process and print medium are more closely related, the combination of the two is among the factors influencing faithfulness of reproduction. Of the various processes, photographic recording is generally acknowledged as being superior to others from the standpoints of both resolution and tonal latitude. Beyond that, it is difficult to single out any one process as being superior in every respect to another. Each has its distinctive characteristics. (See Chapter 3.)

Figure 5.4 shows a picture segment reproduced by the photographic process and by a direct process. The scan-line resolution is essentially the same—nominally 100 LPI—for both. Bear in mind that these are specimens selected more or less at random, and so do not necessarily represent the best—or worst—that these processes can do.

4. *Axis and Direction of Scanning.* Experimenters have generally concluded that from the resolution standpoint there is no measurable basis for preferring one scan axis, or direction, over another—i.e., horizontal vs. vertical, left-right

168 ELECTRONIC DELIVERY OF DOCUMENTS AND GRAPHICS

Fig. 5.4. Comparison of fax recordings made by the photographic process (lower left), as used by news agencies, for example, and by a popular direct recording process (lower right). Original photograph is shown at top.

vs. right-left, up vs. down. The axis is generally controlled by the relationship of width to height of the subject copy and how that relationship affects orientation of the copy with respect to the scan mechanism.

The prevalence of horizontal, left-to-right scanning—sometimes referred to as "positive," or "normal," scan direction—in TV and flat-bed facsimile systems is best explained as having been influenced originally by the way we read. One may reasonably speculate that had these arts been a product of eastern rather than western cultures, scan orientation and direction might have been influenced accordingly.

5. *Kell Factor.* It was pointed out that different factors control horizontal and vertical resolution in a scan system: within a scan stroke resolution is a function of speed of response, while across the scan grain it is determined *basically* by the relative coarseness of the grain, as measured in scan lines per inch (LPI). The word *basically* is emphasized here because while there should be a 1:1 relationship between scan lines and image elements on the cross-grain axis, effectively the number of elements resolvable per inch is *less* than the LPI.

QUALITY

The loss, generally estimated to be 30 percent of the ideal elemental resolution (equivalent to the LPI), is called *Kell factor*.[2]

In simple terms Kell factor represents a loss inherent in the scanning process. It has to do with the fact that scan lines are of finite width, across which there can be no discrimination between light values. For example, if a stroke of the scan spot just catches the edge of a dark image element on the subject copy, the resulting recording stroke will register a dark mark across its entire width at that point (Fig. 5.5). Thus the scanning process tends to have a spreading effect on elements along the cross-grain axis.

The spreading effect also occurs in the case of lines of image detail that appear parallel to the scan axis but are actually slightly oblique to it. Rarely will an

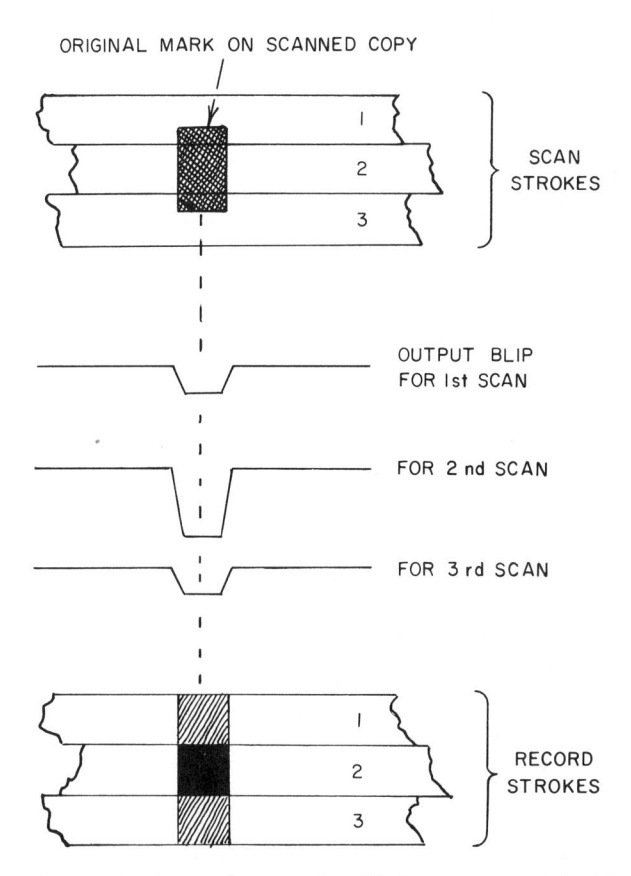

Fig. 5.5. A mark reproduced by a fax recorder will always occupy at least the full width of the recording stroke, regardless of the dimensions of the actual mark on the scanned original.

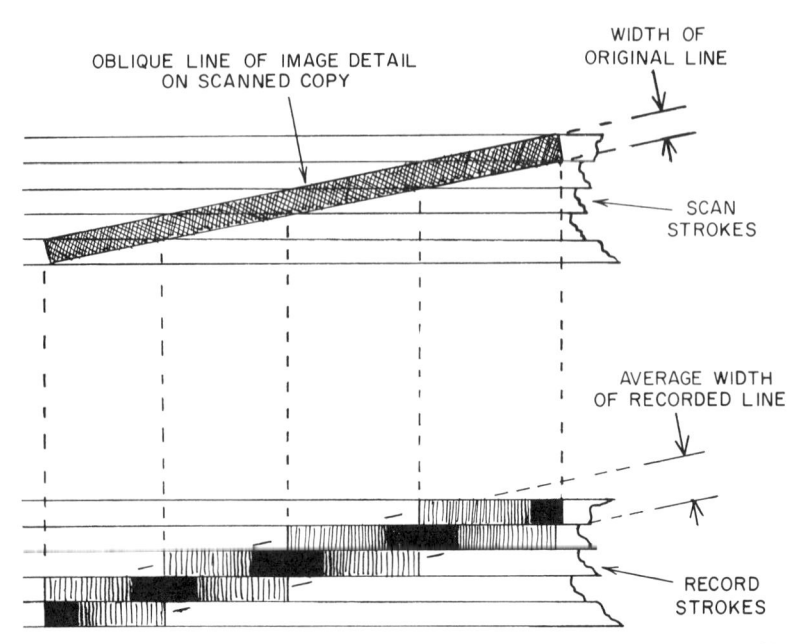

Fig. 5.6. Spreading, or "stepladder," effect in the reproduction of an oblique line of image detail has the effect of increasing the line's width.

image line lie exactly coincident with a scan line. The spreading effect on oblique lines is illustrated in Fig. 5.6.

What has been said of Kell factor so far applies primarily to analog systems. In a digital system, where the scanner output is thresholded and "clocked" so that each scan stroke is reproduced at the recorder as a precisely spaced series of dark and light squares, Kell factor may be said to apply in both the vertical and horizontal directions. Just as the edge of an image element will affect the shading of the full width of a continuous scan stroke at that point, so will it affect the shading of a whole reproduced square within a digital scan stroke.

The degree of loss represented by Kell factor in an analog system has been derived both subjectively and statistically by different researchers, and there is fairly general agreement on a value of 0.70. In a strictly technical sense it may be regarded as a kind of *root mean square* (rms) phenomenon, which simply stated means that it represents an effective value rather than an ideal, and is equal (mathematically) to $\frac{1}{2}$ the square root of 2, or 0.707.

Thus the effective cross-grain output resolution of an analog fax system, in *elements per inch*, is the LPI × 0.70. In a digital system, the same may be assumed to apply to the elemental resolution *along the scan axis* as well.

6. *"Square" Resolution.* For all practical purposes the optimum condition with regard to relative vertical/horizontal resolution in a fax or video system is equality. When the two are equal, we say that the element area resolved by the system is *square*, or that we have *square resolution*.

This should not be taken to mean that where resolution along one axis is fixed and there is margin for improvement along the other, such improvement should not be undertaken. If, for example, the LPI of a fax system is pre-established (possibly for reasons of equipment compatibility) and there is bandwidth to spare, an increase in resolution along the scan axis is bound to result in improved output legibility—up to a point. The increase may be implemented by narrowing the scanning aperture or in a digital system by stepping up the sampling rate (which may also require modification of the aperture). Unfortunately there are no clear criteria for determining the point beyond which increased resolution would be wasted. Subjective tests have placed it anywhere from 25 to 100 percent over that required for square resolution.

There may also be cases where it is advantageous to reduce resolution along the scan axis relative to that along the cross-grain axis. For example, where bandwidth is more than adequate to satisfy the resolution requirements for certain types of documents, it may be expedient to increase the scan rate to obtain a speed advantage without altering the LPI. Depending on how much the balance is shifted, the result could be appreciably less resolution along the scan axis than at right angles to it. How fair an exchange this resolution imbalance would be for the ability to send and receive more rapidly must, of course, be judged on a case-by-case basis.

Another consideration that can affect squareness of resolution is the value we place on *edge gradient*, or the fidelity with which sharp tonal transitions are reproduced. As has already been pointed out, transition fidelity along the scan axis is primarily a function of aperture width, and cannot be divorced from resolution in terms of maximum elements resolvable per scan stroke. Thus the balancing of elemental resolution between the vertical and horizontal axes can impose an imbalance in fidelity.

In practice a compromise is generally struck between square resolution and good fidelity by employing a scan aperture whose width is somewhat less than its height. (See above, 2. *Shape of Scan Aperture.*) Thus, rather than shape the aperture to impose square resolution as a limitation, the system is normally designed to afford resolution along the scan axis *at least* equal to that at right angles to it. In this way allowance can also be made for losses in fidelity that may occur in the circuitry between scanner and recorder.

Technically Kell factor should be considered too in the balancing of vertical and horizontal resolution in an analog system. However, ignoring it will simply result in greater resolution along the scan axis than at right angles to it—which, as noted above, is an acceptable condition. A 96 LPI system, for example, will usually be designed to resolve at least 96 elements (48 cycles) along the scan axis. Applying Kell factor, the nearest the system could come to yielding true square resolution would be a matrix of 67 X 96 elements within a 1-inch square (67 being the product of LPI and Kell factor and therefore the actual cross-grain elemental resolution).

In any event it can be argued that a loss along one axis is no reason for introduction of a compensating artificial loss (as it were) along the other just to preserve true square resolution. The argument is certainly valid, but then a lot depends on where we start—and how we proceed—with the design of a system.

Kell factor is most effectively applied when a system is being designed from the very beginning and the required output resolution has been rigidly specified. Let us assume, for example, that we are designing an analog system and that we start with two specific requirements:

1. *Output Resolution* (minimum)—3.5 line pairs (spatial cycles) per millimeter, horizontally and vertically;[3]
2. *Transmission*—vestigial sideband amplitude modulation via the switched telephone network (effective bandwidth, about 400 to 2400 Hz).

At 25.4 millimeters per inch, 3.5 cycles/mm translates to 89 cycles—or 178 contiguous lines (or elements)—per inch. Taking Kell factor into account, the scan resolution necessary to reproduce 178 elements is $178 \times \frac{1}{0.70}$, or 254 LPI.

For square resolution (ignoring losses other than Kell), the system must be capable of resolving 89 cycles per inch throughout the effective scan length, which we will assume to be 8.5 inches. Thus the system must allow for 8.5 X 89 X 254, or a total of 192,151, cycles to be transmitted (and received) per vertical inch.

Now for bandwidth. For a system of this nature (AM vestigial sideband via the switched network),[4] the carrier frequency will have to be set at about 1800 Hz, so that the maximum frequency transmitted (as determined by how much of a "vestige" of the upper sideband is retained) will be well within the 2400-Hz upper limit. Inasmuch as the carrier has to be somewhat higher than the maximum baseband frequency, the latter should be limited to about 1400 Hz. At 756.5 cycles per scan line (89 cycles per inch X 8.5 inches), 1.85 lines can be transmitted per second, making the scan rate 1.85 X 60, or 111, lines per minute. At 254 LPI, an 11-inch-high document will require a total of 2794 scan strokes and 25 minutes of transmission time.

What we end up with is a system that is painfully slow, may not give a good edge gradient along the scan axis, and is subject to most of the transmission impairments discussed in Chapter 4. But, except for the possibility of being

QUALITY   173

sped up by digital means, from the standpoint of straight elemental resolution it is a system of optimal design. Such are the trade-offs that forever plague the fax system designer.

Effect of "Gaps"

To say that a fax system transmits x number of scan lines within a given interval of time (e.g., 180 LPM, or three lines per second) is not to say that the interval is solidly filled with x lines' worth of picture information. Invariably there are gaps between the end of one line and the beginning of the next. In a conventional system, phasing or synchronizing pulses may occupy the gaps, and in an electronic flying-spot system they represent the *blanking period* in which the scan spot is "flown" back to a new start position at the completion of each line.

To understand the effect that these gaps in signal continuity can have on resolution and bandwidth, it is necessary merely to remember that frequency is a *rate* and not a quantity. For example, if the only activity within a scan line is a burst of ten black-white cycles at some point along the stroke, the quantity of transitions for that line is ten, but the rate is ten per x second. In a 180 LPM system, if the burst is evenly spaced and occurs within a $\frac{1}{4}$-inch segment of an 8.5-inch stroke, the ten cycles represent a frequency of 1020 Hz (ten cycles in $\frac{1}{34}$ of a stroke, at three full strokes per second). (See Fig. 5.7.)

Accordingly, in the preceding example of an 8.5-inch scan line in which 756.5 cycles of spatial frequency are resolved, the 8.5 inches constitutes the *effective*, or *available line, length*, meaning that each line is followed (or preceded) by some finite gap. Assuming the gap to be $\frac{1}{2}$ inch, the 756.5 cycles occur not within a full line, but rather in $\frac{8.5}{9}$ of a line. Thus at 1.85 lines per second the true maximum baseband frequency will be

$$\frac{9}{8.5} \times 756.5 \times 1.85 = 1481.8 \text{ Hz}$$

In order to bring the system back to within the 1400-Hz baseband limitation, the cycles-per-line resolution would have to be reduced from 756.5 to 714.5, or the scan rate from 111 to 105 LPM.

In this example, the difference is slight. But it is plain that as the scan rate increases, the effect of gaps on bandwidth—and therefore on the system's resolution capabilities—can become significant. Consider, for example, a 96 LPI system operating at a scan rate of 360 LPM. We will assume that the total scan-line interval (total length between corresponding points on consecutive lines) is 9 inches, of which $\frac{1}{2}$ inch is a gap, and that the system is designed to resolve 48 cycles per inch along the scan axis. Taking only the effective scan length into consideration, the maximum baseband frequency of the scanner output, at six strokes per second, would be

$$48 \times 8.5 \times 6 = 2448 \text{ Hz}$$

10 SPATIAL CYCLES IN 1/4 INCH OF SCANNED COPY

SCAN STROKE

SCANNER OUTPUT

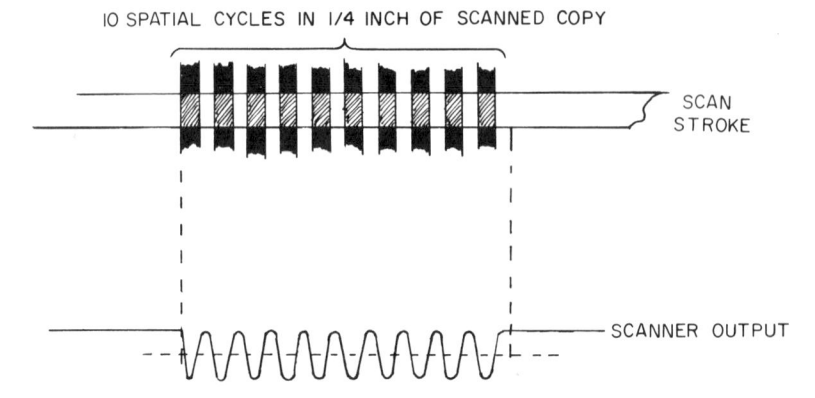

STROKE SPEED: 180 PER MINUTE (3 PER SECOND)
LENGTH OF EACH STROKE: 8.5 INCH
1/4 INCH = 1/34 OF 8.5 INCHES
3 STROKES X 34 X 10 = 1020 ELECTRICAL CYCLES GENERATED PER SECOND (1020 Hz)

Fig. 5.7. Relationship of spatial frequency to scanner output frequency. The former is fixed, whereas the latter will vary with scan rate.

However, taking the gaps into account, the maximum frequency becomes

$$48 \times 9 \times 6 = 2592 \text{ Hz}$$

Unfortunately, although spelled out in various individual standards, there is no *universal* standard governing the length of gaps ("dead sectors") or the percentage of line length they represent. Nor is there one governing whether the gap falls within the page width of the scanned subject copy or beyond it.

The percentage can range from zero to about 16, depending on the particular system. Five to seven percent is common for conventional fax systems, while 16 percent is relatively common for electronic "flyback" systems. As for location with respect to the copy, this too will vary from system to system. In some it is beyond the edge of the subject copy, while in others it falls within either the left or right margin.

Whatever the particular arrangement, the simple rule to follow in computing the resolution, bandwidth, or scan rate of a fax system—whichever of the three is (or are) variable—is that scan length always be taken as the *total* interval between corresponding points on consecutive lines (effective length plus gaps).

CONTRAST, GRAY SCALE, AND POLARITY

There was a time, before photoelectric devices had been developed to their present high state of refinement, when a fax scanner's ability to distinguish be-

QUALITY

tween image details and background of the scanned copy often left much to be desired—particularly if the copy consisted of anything other than black or blue on white. Modern scanners, by comparison, not only are capable of excellent color and gray-scale discrimination, but also are usually designed to compensate automatically for varying background reflectances.

Despite these advances, contrast remains a factor in the faithfulness and clarity with which a fax recorder reproduces the subject copy. While it is true that modern electronics permits the boosting of weak contrast by the mere twist of a knob (or automatically through special compensating circuits), the fact remains that contrast below a certain level can only be boosted at the expense of introducing excessive "noise" into a recording.

A certain amount of noise is always present in the output of a scanner, and a certain additional amount is bound to be picked up in transmission and in amplifier circuitry. The degree to which noise degrades a recording depends largely on the ratio between it and the picture detail component of the signal fed to the recorder, i.e., the *signal-to-noise ratio* (more on this later).

A possible exception to this rule, particularly with regard to transmission noise, is a system in which the scanner output has been thresholded and converted to discrete pulses of uniform height, i.e., a digital system. However, even there, if the copy itself lacks contrast, *thresholding noise* may result from the indecision of the system as it attempts to distinguish image details from background.

The recording process and medium have already been briefly discussed as to their role in reproduction quality. For the most part, the fidelity of tonal rendition and contrast required at the output of a fax system are determined by the particular function the system serves. For the transmission of photographs, gray-scale discrimination is important; for the transmission of messages, it is not.

Measuring Contrast

Contrast may be assessed in various ways. That of a photographic negative, for example, can be measured in terms of *transmission density* on a photoelectric device known as a densitometer. Readings for representative areas of subject and background are compared, and the greater the difference, the better the contrast. Depending on the application, certain minimal standards may be set as a quality control. Scanning-type densitometers make it possible to analyze contrast graphically as a spatial frequency waveform (Fig. 5.8).

Similarly, the contrast of opaque prints is measurable in units of *reflection density*, which correlates directly with transmission density. The instrument used for measurement is a *reflecting densitometer*, or *reflectometer*. Table 5.B shows the relationship among tone, reflectance (or transmission), and densitometer readings.

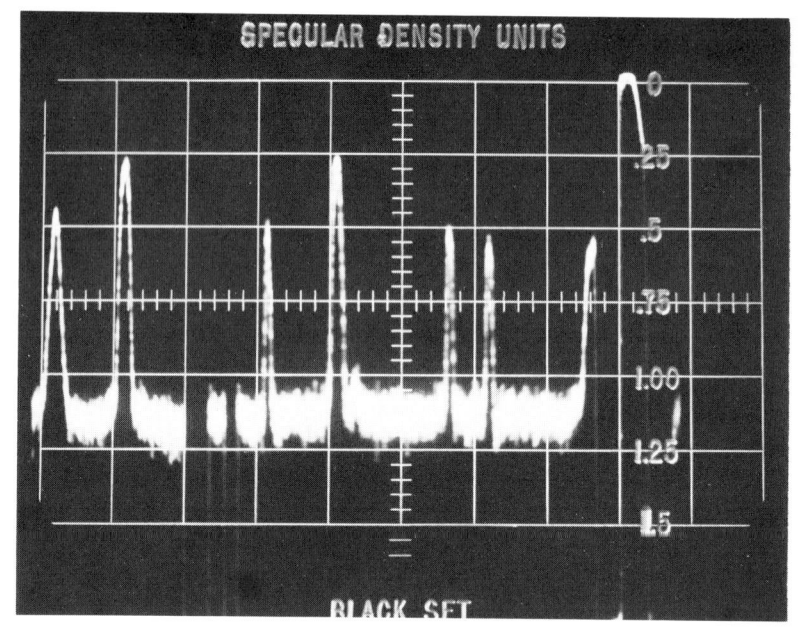

Fig. 5.8. Portion of a negative microimage of a line drawing analyzed for contrast on the screen of a scanning microdensitometer. The analyzed segment consists of "off-white" peaks averaging about 0.5 density unit, against an "off-black" background (broad white areas in the scope image) of about 1.10 density units.

Ordinarily, the contrast of a fax recording is gauged and evaluated by eye. The IEEE Test Chart (Fig. 5.2) contains patterns to aid in the assessment. The principal one appears near the top of the chart between the vertical bar and repeating tri-bar resolution patterns. It is in the form of a *step tablet* containing two rows of 15 discrete tones, each row ranging from nearly pure white to nearly pure black. (Not all 15 are clearly distinguishable in the reproduction.)

TABLE 5.B. Relationship of Tone, Reflectance (Or Transmission), and Densitometer Readings

Tone	Percentage Transmission or Reflectance	Densitometer Reading (Density Units)
black	1	2.00
medium gray	50	0.30
white	100	0.00

The extreme steps have reflection densities of approximately 0.00 and 1.85, respectively.

The continuous "wedge" just above the step patterns represents a linear variation covering essentially the same range. It is useful for testing the quantization of gray levels in a digital fax system.

Gray levels approximating steps 6 and 11 of the step tablet are reproduced at the right-hand side of the chart as an aid in detecting level variations during recording (assuming horizontal scanning of the chart). Paired gray levels, approximating steps 1 and 6 in one case and 11 and 15 in the other, appear in the form of vertical bar resolution patterns at the very top of the chart. The latter are useful in checking the system's response to threshold contrast levels encountered in scanning.

The portrait photograph of the young lady contains approximately the same tonal range covered by the step table. To the discerning eye of one experienced in assessing fax recording quality, its highlights and shadows serve as an effective gauge of a system's tonal capabilities.

Effect of Contrast on Resolution

Although few data exist in the way of specific correlations, there is substantial evidence that within limits, for a given level of legibility, resolution and contrast can compensate each other. It is well established, for example, that the human eye's visual acuity—the ability to discern small objects or the separation between closely spaced objects—increases with an increase in overall subject illumination. The reason is that increased brightness causes a corresponding decrease in the size of the eye's iris opening (equivalent to lowering the f-stop of a camera lens), which results in improved resolving capability. Therefore, to the extent that contrast is improved by an increase in background reflectance of a dark-on-light document reproduction, legibility should also be improved—without the resolution having been altered.

As for empirical data correlating contrast and resolution, the results of a study by Dr. C. E. Nelson are of particular interest. Based on a jury analysis of a large number of paper prints produced from microfilm images containing a variety of contrast-resolution relationships, Nelson's general conclusion was that "satisfactory readability can be achieved with *moderate resolution and high contrast or high resolution and moderate contrast.*"[5]

For the particular prints involved in the analysis, it was possible to establish a contrast-resolution relationship expressible as a simple equation:

$$CR\ index = D_b/D_l \times R$$

where D_b is the background density, D_l the line (or image detail) density, and R the resolution in line pairs (spatial cycles) per millimeter. All values are as read on the microfilm from which the test prints were produced.

According to Nelson an index of 200 represented the threshold of legibility (prints produced from images for which the index was less than 200 proved consistently illegible). He points out, however, that the 200 index applies only to the particular materials and processes employed in this trial and cannot be applied universally.

Tests have also been conducted by H. C. Frey of Bell Telephone Laboratories, the results of which indicate that contrast may influence legibility to an even *greater* degree than does resolution. The tests were of limited scope, however, and the results must therefore be regarded as inconclusive.

Suffice it to say for the time being that there *is* a relationship between contrast and resolution (or contrast and legibility), and that an improvement in the contrast of a fax recording—with resolution remaining fixed—will almost certainly enhance legibility. There is, as yet, simply no sufficiently definitive formula to permit us to predict that if we increase the contrast by a given amount, we can do with x amount less resolution.

Polarity

The tonal polarity of a fax recording—the dark-light relationship of detail to background—may be the same as that of the original (i.e., *positive*) or the reverse of it (*negative*), depending on application.

In systems intended for the production of photographic negatives of the subject copy, the signal to the recorder will have the same electrical polarity as the scanner output prior to amplification. In other words, maximum signal will represent the *white* portions of the original. With a crater tube as the reproducer, maximum signal to the recorder will produce maximum light, which will be interpreted on the processed film as maximum density, or black. Systems designed to reproduce newspaper page proofs as printing masters at a branch plant are frequently arranged to work this way.

In most other applications, regardless of recording technique, the polarity of the output signal to the recorder is *opposite* that of the scanner output. Minimum signal at the scanner, representing black, therefore becomes *maximum* at the recorder, so that black signal components produce maximum marking current.

In a basic analog system, polarity of the transmitted signal may be primarily a function of the number of amplifier stages between the scanner and the transmission facility. In any event, it is determined by the reversals that take place in signal processing between the output of the scanning transducer and that of the transmitter as a unit.

In order to ensure compatibility between terminals within a system, it has been necessary to establish standards on transmission polarity. For message fax and other systems customarily employing direct recording techniques—e.g., weather map distribution systems—*negative modulation*, or *maximum signal on black*, is the standard. However, in systems customarily employing photorecording techniques (with the possible exception of those for the transmission of newspaper page masters), *positive modulation*, or *maximum signal on white*, is usually specified. These tonal-maximum signal relationships obviously assume amplitude modulation. In FM or frequency-shift keying (FSK) systems, black may be represented by a higher or lower frequency than white with no change in amplitude. As with AM, there is no universal standard. In any event these standards are largely a matter of tradition and have nothing directly to do with recording polarity.

The effect of transmission polarity on output quality is difficult to assess. It is primarily a question of whether blemishes due to noise acquired in transmission will impair the background or image detail of a recording. If the blemishes are conspicuous, they are objectionable in either case.

IMPAIRMENTS DUE TO DIGITIZING

The advantages of transmitting fax signals in digital rather than analog form were covered in chapters 2 and 4 in the discussions of fax networks, digital data compression techniques, and digital transmission circuits. In addition, chapter 4 discussed the effects of certain common transmission impairments on digital signals. But the effects of the digital process itself on output copy quality were touched upon only briefly. It is now time to treat this topic specifically.

Where subject copy consists essentially of only two contrasting tonal levels, e.g., black text on a white background, it may be desirable to use a two-level (mark-space; on-off) form of transmission. It is also sometimes practical, and may become more so as digital techniques become more prevalent, to transmit photographs and other halftone material digitally by *pulse-code modulation* (PCM) techniques, in which each of a number of discrete steps in the gray scale is assigned a binary code.

But, whether the digitizing is two-level or multilevel, there are two potential quality impairments inherent in the analog-to-digital conversion process. One is errors in discrimination between one tonal level and another. The other is the displacement of marks that may occur in the recording due to slight differences between the scan rate and the digital sampling rate.

For the purposes of this book it will suffice to confine the discussion to two-level quantization—or, more accurately, *thresholding*. The nature of the impairments is basically the same as for multilevel except that amplitude-quantizing (thresholding) errors are likely to be more conspicuous, and therefore more deleterious, in a two-level system.

Amplitude-Quantizing (Thresholding) Errors

A basic step in two-level analog-to-digital conversion is the setting of a reference level, or *threshold*, for discrimination between image detail and background. For high-contrast originals, there is no problem. For moderate- or low-contrast originals, the probability of discrimination (thresholding or amplitude-quantizing) errors occurring because of the small difference in level between background and detail will depend on how consistent the contrast is. One can appreciate the thresholding problem that might obtain where the subject copy consists of characters of various colors (convertible by the scanner to gray levels) on an off-white background. Figure 5.9 illustrates the nature of the problem.

Sophisticated converters have been developed in which discrimination is governed by a *"floating" threshold*, which changes position with variations in the average between maximum and minimum levels of the analog signal. For most applications, however, it is sufficient for the threshold to be essentially fixed, but adjustable manually as needed.

Figure 5.10 is a segment of a digital fax recording in which the effects of thresholding errors appear.

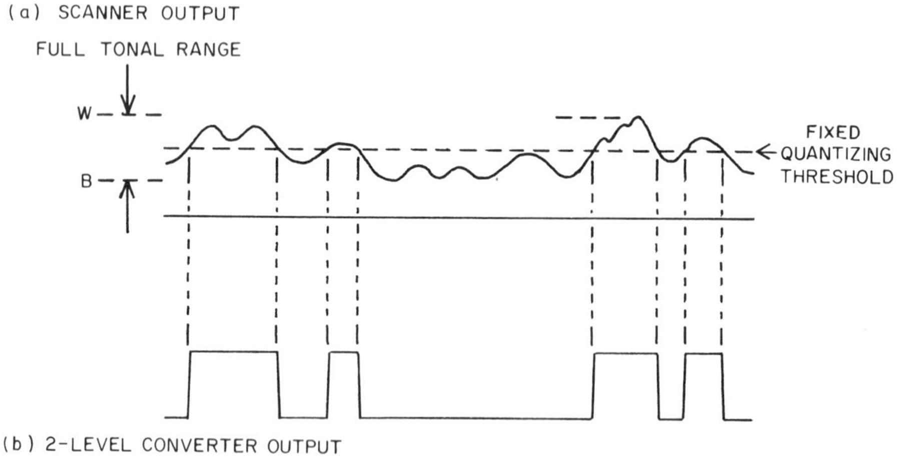

Fig. 5.9. Illustration of the thresholding problem inherent where contrast is lacking between image details and background. Several details are lost because fluctuations in signal strength are too weak to cross the quantizing threshold.

QUALITY

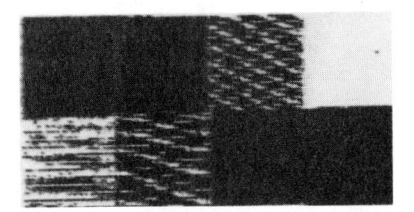

Fig. 5.10. Thresholding errors in a two-level reproduction of a halftone original. Specimen (left) was reproduced by an analog system; specimen (right) by a two-level binary system. Errors are particularly evident in the lower left square of the latter specimen, which on the original, has the least contrast with the white background.

Time-Quantizing Errors ("Jitter")

"Jitter" is an effect occurring at right angles or oblique to the scan axis of a fax recording, the result of slight time and space inconsistencies between scan rate and digital sampling interval or between sampling interval and the angle of a line of picture detail. It can occur in analog systems as well, but is more common and of a slightly different character in digital systems. An actual sample of jitter appears in Fig. 5.11, and the effect is graphically analyzed in Fig. 5.12.

Sampling in a digital system is controlled by what is commonly called a

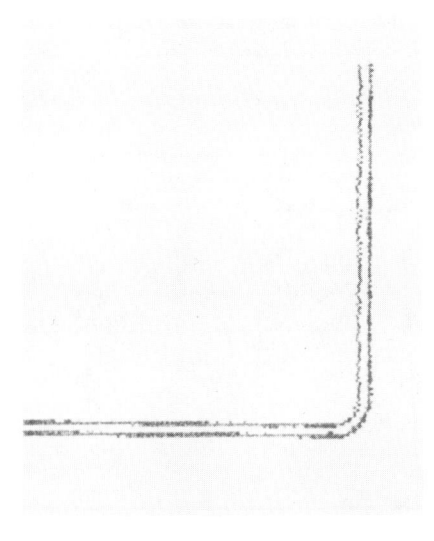

Fig. 5.11. Segment of a two-level fax recording in which jitter is evident. Defects in the horizontal segment are the result of the image line's not having been exactly coincident with the scan axis.

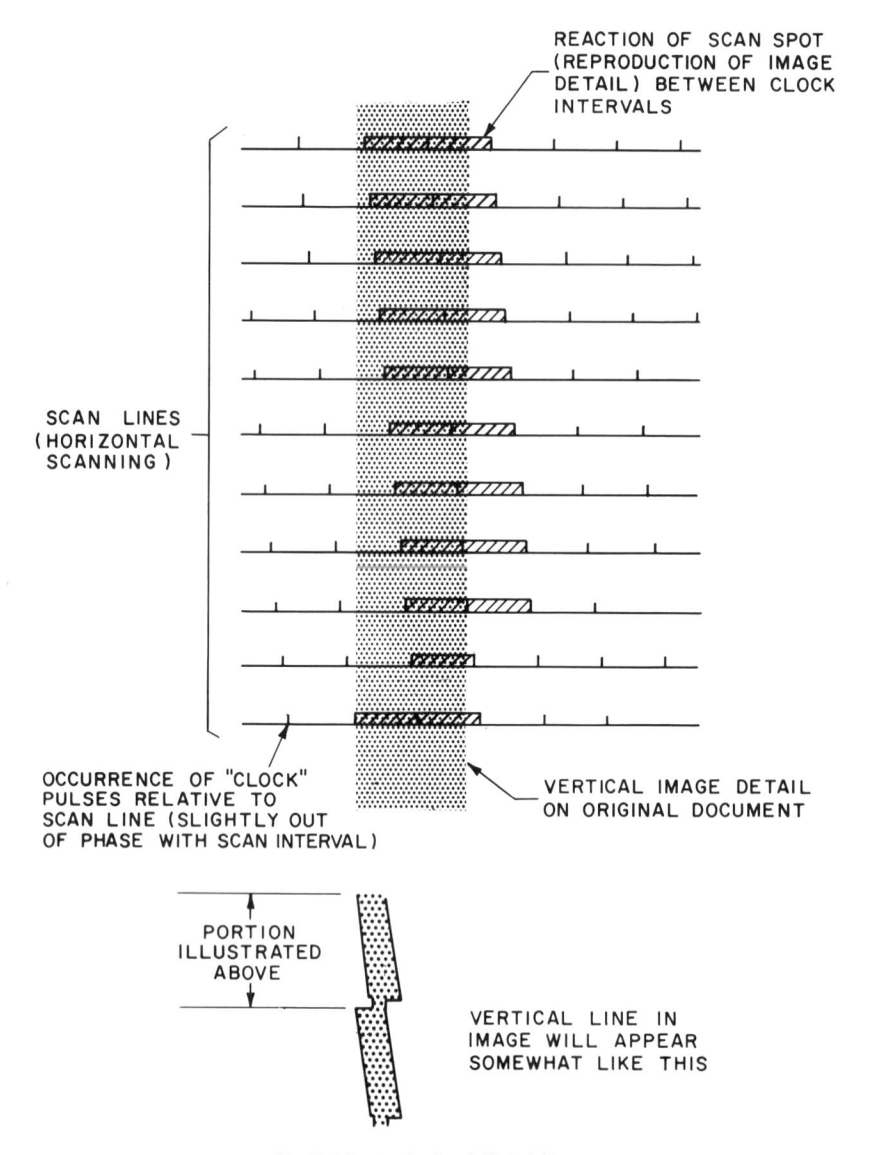

Fig. 5.12. Analysis of digital jitter.

clock, which is essentially a precision oscillator similar to those used for the generation of synchronous power to operate scanner and recorder drive motors. (See Chapter 3.) In effect the clock pulses, occurring at regular intervals, momentarily connect the amplitude-quantized scanner output to a sensing circuit to determine whether an ON (mark) or OFF (space) condition prevails at that moment. If an ON condition is sensed, the carrier is shifted to mark level, where it remains until the sensor detects an OFF condition. Similarly, if an OFF condition is sensed, the carrier is shifted back to space level, where it remains until the next ON sensing. The shifting may be in the form of amplitude, frequency, or phase changes, depending on the modulation technique employed.

The above explanation is perhaps somewhat oversimplified, but it should help to explain Fig. 5.12.

Inasmuch as each *bit* (each ON or OFF pulse) is roughly equivalent to half an analog cycle, the sampling rate in bits per second will usually be at least twice the highest baseband frequency (in hertz) that the system is expected to accommodate. And technically the carrier frequency in hertz should at least equal the sampling rate in bits per second, so that there is a full cycle of carrier for each bit of transmitted data.

Jitter can be minimized by designing the digital system so that the clock frequency, or sampling rate, is somehow synchronized with the drive-motor frequency. One way to achieve this is to use a device called a *shaft-angle encoder*, as described in the section on *Phasing and Synchronization* in Chapter 3. But even then, except for some inprovement in uniformity, the effect will persist for oblique image lines. No degree of precision can entirely eliminate jitter.

But while jitter may be unavoidable, controlled laboratory tests[6] have indicated that there is little if any fundamental difference in character legibility between predominantly black-white recordings produced by a two-level digital system and those produced by straight analog.

LEGIBILITY CRITERIA

The legibility of words or characters is often a key factor in the establishment of output quality requirements for a fax system. But before any specific end requirements can be set, a choice has to be made as to which of the following two basic objectives applies:

Objective No. 1. Individual letters of a word need not be positively identified, so long as the word as a whole is legible; or

Objective No. 2. The identity of the smallest significant individual character must be virtually assured.

Objective No. 1 might apply in the case of a conventional message fax system, the input to which is predominantly text-type material, whereas Objective No. 2 will apply where technical documents—graphs, specifications, etc.—or documents containing significant numerical data are involved. Figure 5.13 illustrates the distinction between word and character identification.

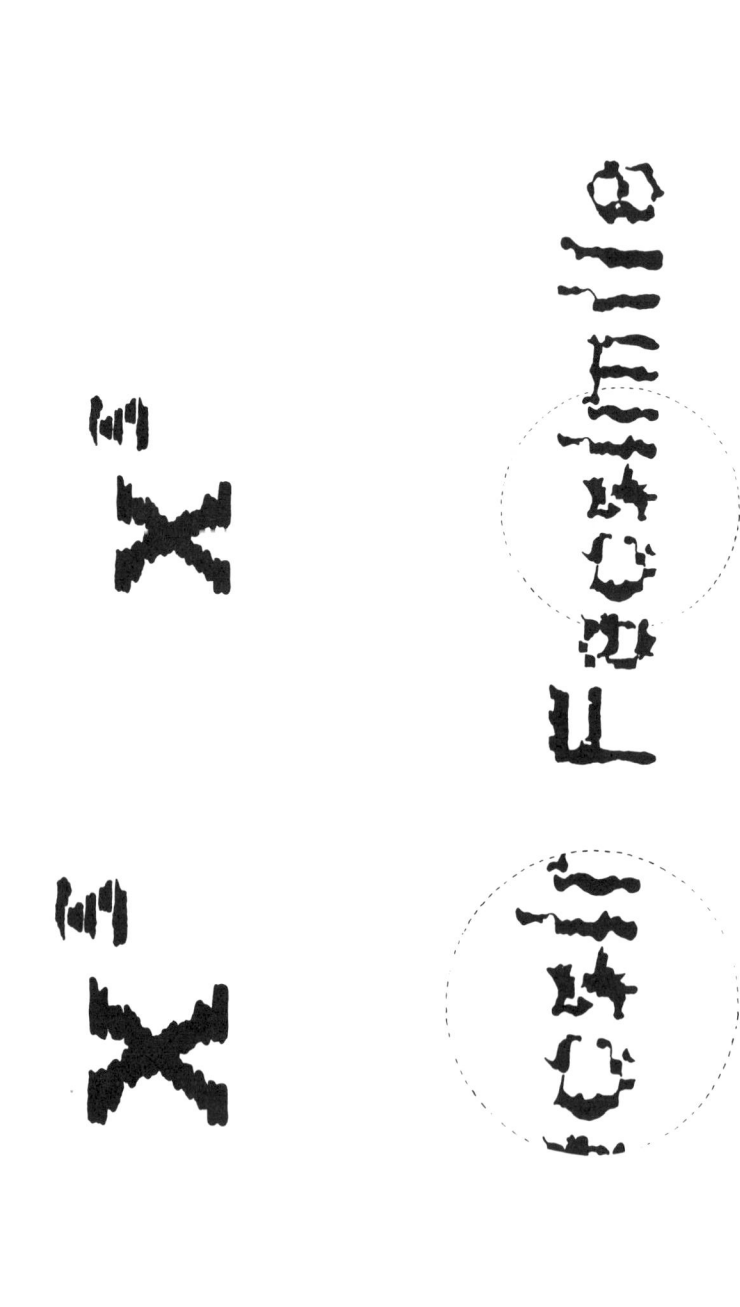

Fig. 5.13. Individual character versus word legibility. On the left, neither the exponent for the x nor the characters in the circle are clearly identifiable. On the right, the characters in the circle become identifiable in context, whereas the identity of the exponent remains in doubt. (These are hand-drafted copies of magnified fax recording specimens.)

QUALITY   185

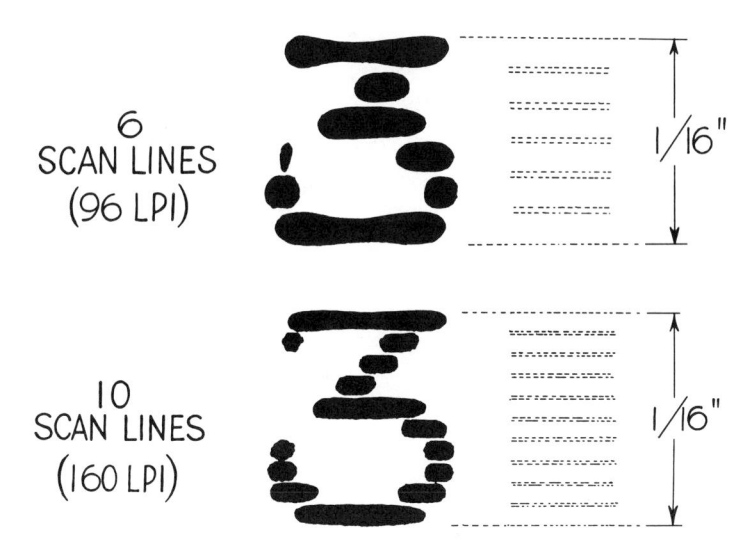

Fig. 5.14. Comparison of the effect of coarse and relatively fine scan resolution on reproduction of the number *3*. (simulated)

Once this basic objective has been decided, there are various experimental criteria by which specific output resolution requirements can be set. Not all of those discussed below apply directly to fax systems, although regardless of the language or the approach, there is general agreement among the various findings as to the fundamental criteria for meeting either of the two basic legibility objectives.

Scan Lines Per Character

The most direct approach to the design (or selection) of a fax system to meet a given legibility objective is to determine the minimum number of scan lines by which an individual character should be dissected. A number of experiments have been performed, by a number of researchers, in which subjects were asked to evaluate the comparative legibility for a varying number of scan lines per character. The collective result has been a more or less unanimous agreement that 10 scan lines is the minimum required for *assured* legibility or an individual character (Objective No. 2).

Supplementing this, experience has shown that approximately half that number, a minimum of five lines, is required to meet Objective No. 1 (word legibility), where individual characters need be identified only by inference from the supporting context. Figure 5.14 illustrates the comparative results of a large and a small number of scan lines per character.

TABLE 5.C. SCAN RESOLUTIONS REQUIRED FOR VARIOUS MINIMUM CHARACTER HEIGHTS, BASED ON FIVE LINES PER CHARACTER TO IDENTIFY WORDS AND 10 LINES TO IDENTIFY INDIVIDUAL CHARACTERS

Min. Character Height	Approx. Equiv. Type Size	Min. Scan Lines Per Inch (LPI)	
		1. − Identify Words	2. − Identify Indiv. Characters
$\frac{3}{64}''$	6−7 pt.	107	214
$\frac{1}{16}''$	8−10 pt.	80	160
$\frac{3}{32}''$	11−12 pt.	53	106
$\frac{1}{8}''$	14−18 pt.	40	80
$\frac{5}{32}''$	20−22 pt.	32	64

Table 5.C interprets the above criteria as scan resolutions required for various minimum character heights. It bears mentioning here that the *point* (1 point = $\frac{1}{72}$ inch) designation for type size is somewhat deceptive. Typically 6-pt. text characters range from about $\frac{1}{32}$ to $\frac{3}{64}$ inch in height; 8-pt. from $\frac{3}{64}$ to $\frac{1}{16}$; 10-pt. from $\frac{3}{64}$ to $\frac{5}{64}$; 11-pt. from $\frac{1}{16}$ to $\frac{3}{32}$; 12-pt. from $\frac{1}{16}$ to $\frac{7}{64}$; 14-pt. from $\frac{5}{64}$ to $\frac{1}{8}$; 18-pt. from $\frac{3}{32}$ to $\frac{9}{64}$; and 22-pt. from $\frac{7}{64}$ to $\frac{11}{64}$. These are based (loosely) on measurements of Caslon Old Style upper and lower case letters. Of the standard typewriter fonts, *elite* falls somewhere between 10- and 11-pt. type, and *pica* between 11- and 12-pt.

Table 5.C shows, among other things, that an LPI of 32 would be sufficient for assured *word* identification if the smallest letter ever to be encountered were $\frac{5}{32}$ inch high, or about 22-pt. upper case. It is difficult to conceive, however, of a text-type application where letter height would be confined consistently to so comparatively large a dimension. A more realistic situation would be one in which characters are consistently $\frac{1}{8}$ inch high and require individual identification. This would be the case for most modern "drafted" documents, such as engineering drawings and specifications.

Drafting standards vary from one company to another, but a recent survey reveals that $\frac{1}{8}$ inch is a fairly widespread minimum for character height, and that for documents larger than 17 × 22 inches in size, the minimum is often upped to $\frac{5}{32}$ inch.

As for text-type material—books, letters, forms, etc.— the minimum character height commonly encountered is likely to be closer to $\frac{1}{16}$ inch, which is a good minimum to allow for in any case, since it will give fair assurance of legibility for the common typewriter fonts as well as most 8-pt. and larger type sizes.

The realistic lower limit for scan resolution, then, for drafted and text-type

QUALITY

material alike is about 80 LPI. However, this should not be interpreted as implying that scan resolutions lower than 80 LPI are to be ruled out as unacceptable. By this particular criterion lower resolutions will merely decrease the *degree of assurance* of output legibility for a given character size.

For individual character identification in text material, the lower-limit figure would be doubled, to 160 LPI. And where items set in 6-pt. or comparable size type—footnotes, exponents, column headings, etc.—are frequently encountered, it may be necessary to go to scan resolutions in excess of 100 LPI (for word identification) or 200 LPI (for individual character identification).

In cases where the material is microfilmed, documents ranging from $8\frac{1}{2} \times 11$ to 17×22 inches in size may be reduced anywhere from 16X to 24X on film, while the larger ones will usually be reduced either 24X or 30X. These are *linear* reductions. In other words, at a 16X reduction, a line of image detail measuring 1 inch on the original will measure $\frac{1}{16}$ inch on the microimage.

Assuming a fax scanner has to be designed to scan a microimage directly at the film plane, the scan resolution (in LPI) required to meet the 5 or 10 lines/character legibility criterion can be derived simply by multiplying the LPIs in Table 5.C by the photographic reduction. These film plane LPIs have been worked out and are shown in Table 5.D.

Microfilm Printback Criteria

In connection with its activities in the area of test standards for the reproduction of microfilmed drawings and documents, the National Bureau of Standards has devised the following equation as a guide in determining the resolution required for a given level of reproduction quality:

$$R = qr/e$$

As applied to microfilm, R is the resolution in spatial cycles per millimeter, as read on the film with the aid of a microscope; r is the linear reduction at which the original document was recorded on film; e is the height in millimeters of a lowercase "e" on the original document, and q is an arbitrary quality index.

According to the NBS, q must be at least 8 when the resolution required for excellent copy is computed, 5 for medium-quality copy, and 3 for marginal quality.

Applying the formula to facsimile, we will want R to represent resolution in scan lines per inch rather than spatial cycles per millimeter. The millimeter-to-inch conversion can be accomplished simply by expressing the character height e in inches; and the conversion from spatial cycles to scan lines simply requires the answer to be increased by a factor of 2. (There are two scan lines to a spatial cycle.) It might be argued that the conversion from resolvable spatial cycles to

TABLE 5.D. SCAN RESOLUTIONS REQUIRED AT THE FILM PLANE TO MEET THE 5- AND 10-LINES-PER-CHARACTER LEGIBILITY CRITERIA IN A MICROFILM-INPUT FAX SYSTEM

Min. Character Height	Min. Scan Lines Per Inch (LPI) At Film Plane							
	1. – Identify Words				2. – Identify Indiv. Characters			
	16×	20×	24×	30×	16×	20×	24×	30×
$\frac{3}{64}''$	1712	2140	2568	3210	3424	4280	5136	6420
$\frac{1}{16}''$	1280	1600	1920	2400	2560	3200	3840	4800
$\frac{3}{32}''$	848	1060	1272	1590	1696	2120	2544	3180
$\frac{1}{8}''$	640	800	960	1200	1280	1600	1920	2400
$\frac{5}{32}''$	512	640	768	960	1024	1280	1536	1920

Note: The headings 16×, 20×, etc., refer to *linear* photoreductions—e.g., a 16×-reduced image means that a detail measuring 1 inch on the original document is reduced to $\frac{1}{16}$ inch on the microimage.

QUALITY

scan lines per inch should also take Kell factor into account by further multiplying the answer by $\frac{1}{0.70}$. The NBS formula, however, already assumes certain magnification and reproduction losses, thereby making further compensation unnecessary.

In addition, since there is no reduction involved, we can simply eliminate r from the equation, which thus becomes

$$R = 2q/e$$

Computing R for a q of 5, which we may assume to represent the minimum quality required for *individual character* recognition, and letting $e = \frac{1}{16}$ inch, we find that

$$R = \frac{10}{0.062} = 161,$$

which agrees with the 10-lines/character criterion previously discussed.

Computing R for a q of 3, which we may assume to represent the minimum quality required for *word* recognition, we find that for an e of $\frac{1}{16}$ inch,

$$R = \frac{6}{0.062} = 97,$$

which is slightly higher than that required to meet the 5-lines/character criterion previously discussed.

Another generally accepted microfilm printback criterion is that to be judged acceptable, a print must be capable of resolving 3.5 to 4 spatial cycles per millimeter. In other words, on a nonreduced print of a photographed NBS Microscopy Test Chart (Fig. 5.15), separate lines and spaces must be discernible in all patterns from the largest down to 3.6 or 4.0.

At 25.4 millimeters per inch, and at two scan lines per spatial cycle, 3.5 cycles/mm works out to 178 scan lines/inch—which is a bit higher than that required for independent recognition of $\frac{1}{16}$-inch characters (by the 10-lines/character criterion) and more than adequate for word recognition where characters are $\frac{3}{64}$ inch, or 6 points. Four cycles/mm works out to 203 scan lines/inch, just short of the minimum needed to ensure independent recognition of $\frac{3}{64}$-inch characters.

So in general the microfilm printback criteria agree with the rules governing the number of scan lines per character for good and for marginal legibility.

Library Tests

In tests conducted in 1966 by the University of Nevada Library (under a grant from the Council on Library Resources) it was determined that a 96/180 message fax system operating over the dial network via phone couplers was able to produce copies of a quality "adequate for most library materials."[7] The ma-

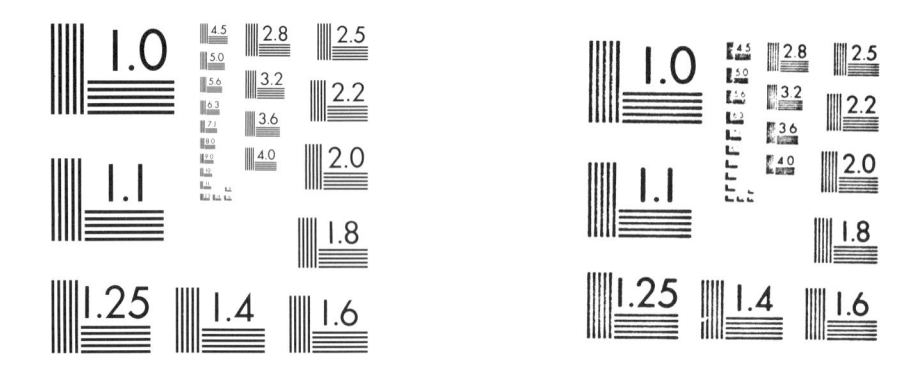

Fig. 5.15. Comparison of an actual NBS Microcopy Resolution Test Chart (left) with a near 1:1 reproduction of an enlarged, first-generation microfilm image of the same chart (right). Allowing for the slight reduction from original to print, there is approximately a 12-target loss from input to output—which is normal for the particular process involved. (These reproductions also, of course, reflect losses inherent in the process involved in the printing of this book.)

terials that were legibly reproduced included printed pages of 8-point and larger type sizes, typed pages (pica and elite), carbons of typed pages, line drawings, graphs, charts, and photos having good contrast.

As for printed matter specifically, the test results further indicated that most 6-pt. typefaces and 8-pt. italics reproduced with borderline legibility, while 6-pt. italics and most type sizes smaller than 6 pt. generally reproduced illegibly. In addition materials of very low contrast resulted in illegible recordings in most cases.

Referring to the previously discussed criteria, we can see that there is a fair degree of correlation with these library tests. By the 10 lines/character criterion, we observe that we can go as low as 80 LPI and still have marginal legibility of 8-pt. characters, and that an LPI of over 100 is necessary for legibility of 6-pt. characters. And, according to the NBS printback legibility equation, marginal legibility of 6-pt. characters requires a minimum of about 97 LPI.

Other Findings

The joint National Micrographics Association/Electronic Industries Association Microfilm-Facsimile Standards Committee, TR 29.1, conducted subjective tests in 1969 to determine minimum scan resolution requirements for engineering drawings and other technical documents requiring individual character identification. The 14 committee members (including the author) themselves constituted the jury that analyzed the two specimens used in the trial—a segment of an engineering drawing that was judged "typical" and a slightly modified IEEE facsimile test chart—each of which was scanned at nominal resolutions of 100, 135, 150, 165, 180, and 200 LPI. The object was to evaluate subjectively the threshold of acceptability, both intuitively and with the aid of resolution test charts.

The results of the analysis were (*a*) a general conclusion that type sizes smaller than 6 pt. require a scan resolution in excess of 200 LPI for independent character identification, and (*b*) a specific recommendation of 165 LPI as the standard scan resolution for engineering drawings.

In research conducted during the early 1970s Singer Simulation Products of Sunnyvale, California, concluded that for "good" legibility reproduction in a digital data compression storage and transmission system, scan resolutions of 130 LPI and 195 LPI are required for "ordinary" documents and engineering drawings, respectively.

One further consideration regarding required resolution for character legibility is that there are instances in which one may wish to *re*transmit a fax-received document. In such cases, any impairment of legibility by the transmission process will be compounded somewhat, and it may therefore be advisable to choose a scan resolution sufficient to provide an adequate margin for that eventuality. To the best of this writer's knowledge, there are no published study results prescribing what that margin should be (as a general rule), or how it is best calculated in a given situation.

PICTORIAL CRITERIA

So far the discussion has been confined largely to alphanumerics, or symbols and characters of a more or less uniform construction, normally isolated from one another against a contrasting and uniform background. But what of graphics that are not so conveniently modular in character, such as handwriting and photographs? What sort of acceptability criteria applies to them?

Obviously, this category of graphics does not as readily lend itself to the establishment of acceptability thresholds. Nevertheless, there has been a certain amount of quasi-standardization that deserves mention.

Handwriting

The nearest thing to a resolution standard applying to handwriting is the establishment of 80 LPI as adequate scan resolution for the identification of signatures.

Fig. 5.16. Handwriting specimen reproduced by a nominal 100 LPI fax system.

This was the consensus of engineers and others who participated in the development of an experimental facsimile signature verification system by Bell Telephone Laboratories in the late 1950s.[8] The system was developed primarily for use by banks, where customer signature records are frequently filed at a central location and are therefore not readily accessible at branches. It was designed to transmit an area of $1\frac{1}{4} \times 5$ inches in 20 seconds via an ordinary telephone circuit, and in five seconds via a special 5 kHz circuit.

One commercially available fax system for this purpose, produced by the Alden Company, uses a scan resolution of 100 LPI, which, of course, is preferable to 80. (Higher resolution is always preferable from a quality standpoint in any graphic system.) A signature verification system is unique in that the area to be scanned is very limited, thus permitting a favorable trade-off of time and bandwidth for quality.

The Alden 400 System is able to transmit the contents of a $3\frac{1}{2} \times 6$-inch record card in 1.7 minutes—or the signature alone in only 11 seconds—via ordinary dial-up phone circuits. An earlier version of the present system was able to send the entire card in only 10 seconds at the same resolution, but required a special transmission circuit with a 10 kHz bandwidth.

A nominal 100 LPI fax recording of a handwriting specimen is reproduced in Fig. 5.16.

Fingerprints

In law enforcement, the expediting of fingerprint records between outlying precincts and police record centers has proved a definite aid in the speedy identification of a criminal suspect and in the limiting of detention time. The fact that no two fingerprints are exactly alike is an indication of their complexity. Besides the three general pattern types—arch, loop, and whorl—there are numerous subclassifications, within each of which are subtle variations.

Experience has shown that a scan resolution of about 200 LPI is required to preserve all of the essential details of a fax-transmitted fingerprint. In addition, a reasonably good gray-scale capability is desirable to ensure reproduction

QUALITY

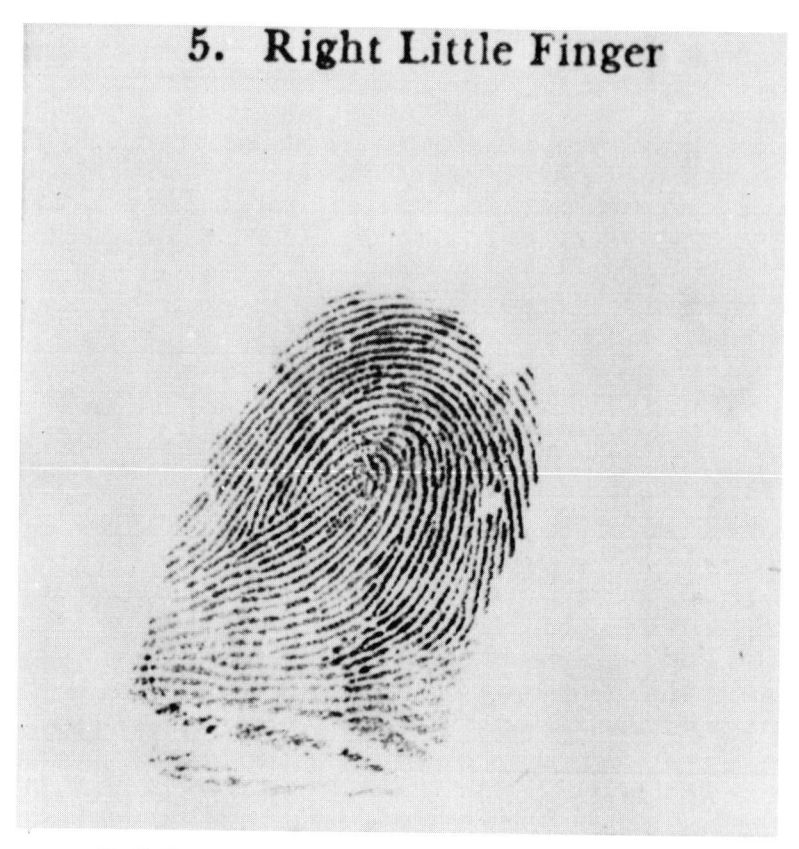

5. Right Little Finger

Fig. 5.17. Fax-transmitted fingerprint. (Courtesy Litton/Datalog)

of very light or smudged prints. A 200 LPI fax recording of a fingerprint is reproduced in Fig. 5.17.

Maps

It is characteristic of maps to vary widely in complexity, making it somewhat impractical to generalize on fax resolution criteria. However, one type of map, the traditional weather map (Fig. 5.18), has been married to facsimile long enough now for its users to have agreed on what constitutes acceptable quality. The standard that has evolved is 96 LPI, the same as that for conventional message fax.

Limited use has been made of lower resolutions—as low as 48 LPI—where the maps are relatively simple and contain no fine detail. The reduction permits

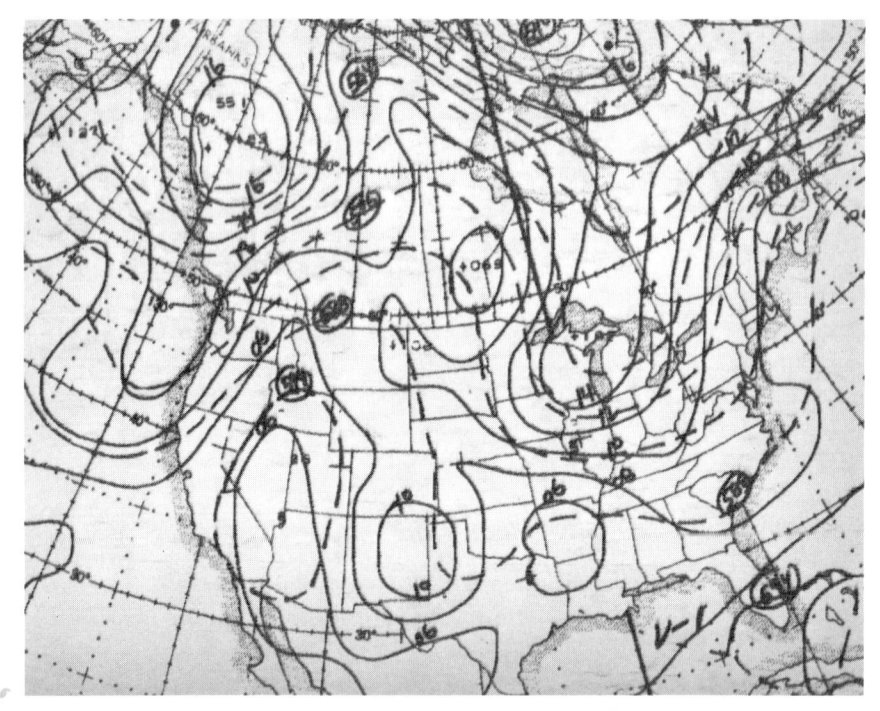

Fig. 5.18. Weather map reproduced by facsimile at 96 LPI. (Courtesy National Weather Service)

speedier transmission or use of lower-grade (and therefore less expensive) transmission facilities. Conversely, higher resolutions are sometimes employed for particularly "busy" maps.

As for other types of maps, required resolution will depend on relative complexity and fineness of detail. A typical road map, for example, with its extensive use of comparatively small characters and symbols, would probably require a scan resolution in the neighborhood of 200 LPI. In addition, the liberal use of color in such maps would require that the system have a good gray-scale capability in order to maintain the distinction between background and detail.

Photographs

Photographs, like maps, can differ widely in complexity. While scan resolution requirements will usually vary according to the level of output quality required in a given application, anything from about 110 to 135 LPI has proved adequate for the fax distribution of news photos. The AP's current standard is 111 LPI,

QUALITY

whereas the UPI transmits at 135 LPI domestically and 109 LPI internationally. The generally increasing quality of newspaper photos (resulting from the use of offset printing techniques) may drive these figures slightly upward in the near future.

Press photos are usually transmitted as 8 X 10-inch enlarged prints of the photographers' camera negatives. When these prints are scanned at 135 LPI, the recorded result is a picture having more than twice the clarity of the same picture filling a commercial TV screen, viewed at a proper distance. The received fax-recorded picture is usually reduced in size for publication, which has the effect of increasing its resolution somewhat (just as a smaller TV image appears sharper than a larger one). Much resolution is subsequently lost, however, in the halftone printing process.

In police facsimile systems, "mug shots" of criminal suspects are usually transmitted at a resolution of 200 LPI; but that is only because these systems are geared to the transmission of fingerprints, which, as already pointed out, requires a high resolution to preserve essential details.

Fax systems designed for the transmission of full-size newspaper page proofs to satellite printing plants, and which must be capable of faithfully reproducing halftone photos as well as text, employ scan resolutions as high as 1000 LPI. The choice is influenced primarily by the need to minimize the occurrence of *moiré*, or interference, patterns caused by the interaction of scan lines with the dot arrays constituting the halftone pictures.

Experience has shown that to avoid moiré patterns completely in the fax reproduction of halftones, the scan resolution must be at least 10 times the halftone screen density in lines (or dots) per inch. Thus, if 65-line halftones are the finest used, a scan resolution of 650 LPI will suffice for moiré-free reproduction. Likewise, use of 100-line halftones will require 1000 LPI.

However, experience has also shown that despite some occurrence of moiré patterns, a scan resolution of only seven times the halftone screen density will afford acceptable reproduction. By this criterion, the scan resolutions for 65- and 100-line halftones can be reduced to about 450 and 700 LPI, respectively. These page-proof fax systems are offered commercially with resolution as low as 300 LPI, which, while offering no guarantee against the appearance of interference patterns in halftones, at least makes it possible for the system to operate at a reasonable speed over a comparatively inexpensive transmission facility: approximately six minutes per 16 X 24-inch page via a 48-kHz group circuit. The higher-resolution systems to be competitive in terms of speed must either transmit via channels in the megahertz range or resort to digital data compression techniques requiring more expensive terminals.

With the growing stress on the use of digital techniques in fax systems, it is well to be acquainted with a couple of the methods by which at least the semblance of continuous tone can be preserved in the transmission of photos and other halftone material by such systems. The best-known method is, of course, pulse-code modulation (PCM), in which a digital code "word" is assigned to each of a selection of predetermined gray-scale threshold levels. Experience has shown that a minimum of from 14 to 20 gray levels (including black and white) is required for reasonably "smooth" reproduction of tonal gradations. A large number is, of course, preferable. However, since each level may require as much as a 5-bit code for transmission, bandwidth requirements can be quite high. Figure 5.19 illustrates the effect when only eight levels are quantized.

An interesting and relatively new development that offers a way around the characteristic quality-bandwidth trade-off of PCM is a process called *dither*. The basic principle is the assignment of various *response thresholds* to the individual "addressable points" constituting the ordered matrix of picture

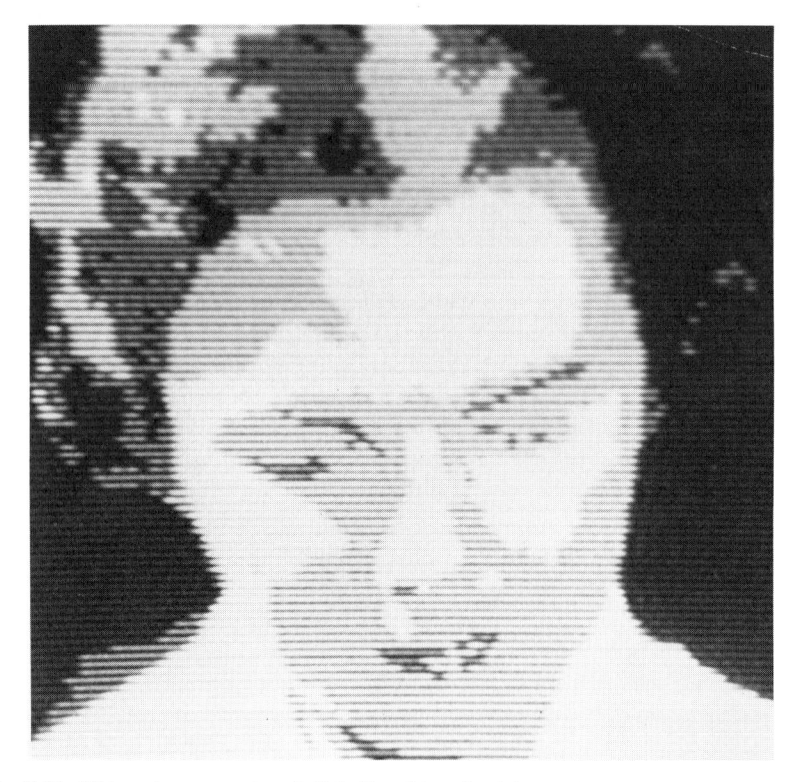

Fig. 5.19. Video picture produced digitally with only eight brightness levels, resulting in a "paint-by-numbers" effect. (Copyright 1970, Bell Telephone Laboratories, Inc.; reprinted by permission of the Editor, Bell Laboratories *Record*)

QUALITY 197

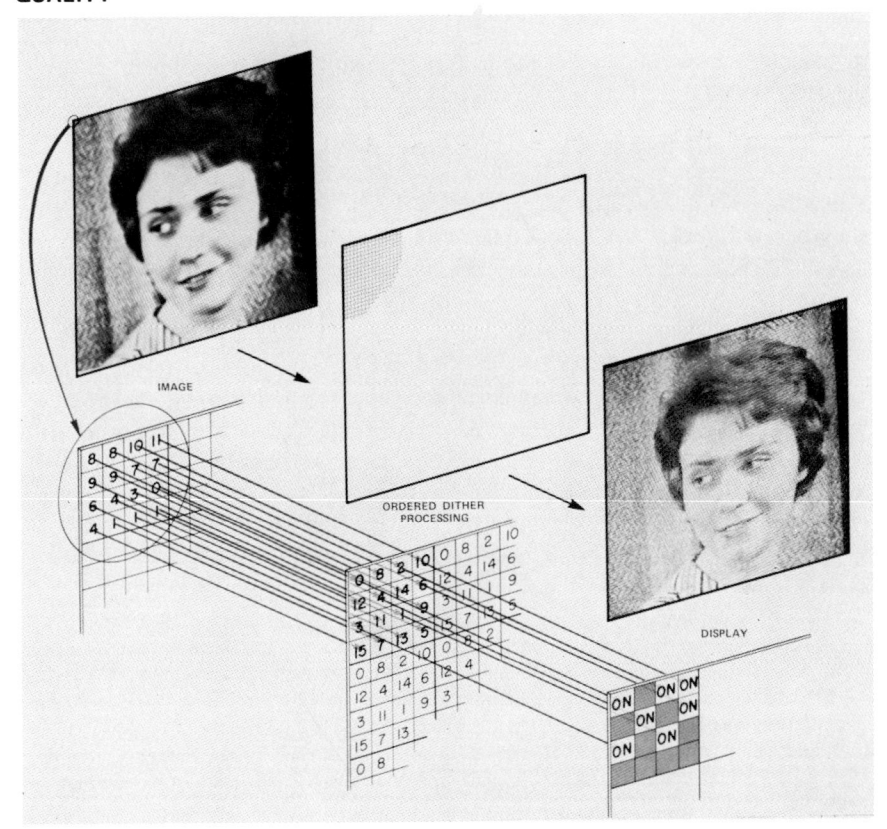

Fig. 5.20. The "dither" process creates the illusion of continuous gray scale in an electronically reproduced photo by segmenting the picture into a grid matrix and subdividing each grid square into a cluster of black and white elements. The ON or OFF (white or black) state of each element of the submatrix in the reproduced picture (right) is determined by whether the corresponding portion of the original photo (left) exceeds a predetermined brightness threshold (center). Where the same threshold array is repeated from one submatrix to the next, as in this illustration, the process is called *ordered* dither. The array can also be random over the entire picture area. The eye tends to blend the black-white clusters into shades of gray. (Picture copyright 1976, Bell Telephone Laboratories, Inc.; reprinted by permission of the Editor, Bell Laboratories *Record*)

elements into which the sampling scanner divides the subject copy. This threshold assignment may be random, or it may be "ordered" in the sense that a given submatrix of, say, 16 predetermined thresholds is repeated throughout the full picture matrix. This ordered dither approach is illustrated in Fig. 5.20.

Referring to the figure, we can see how this technique might be applied to digital fax scanning. A relatively simple algorithm would instruct the scanner to generate a pulse only when the picture element at the sampled point exceeds the pre-assigned threshold as follows:

1st scan line–0, 8, 2, 10, 0, 8, 2, 10, 0, 8, . . . (etc.)

2nd scan line–12, 4, 14, 6, 12, 4, 14, 6, 12, 4, . . . (etc.)

and so on, through the fourth scan line, after which the entire sequence is repeated. The result, as may be discerned in the figure (right-hand picture), is a reasonably faithful reproduction of the continuous tones in the subject copy.

One disadvantage of dither is that it reduces the resolution of relatively low-contrast image elements. An important compensating advantage, however, is that compared with a PCM system, "contouring" (the "paint-by-numbers" effect depicted in Fig. 5.19) is virtually eliminated without having had to resort to an impractical number of quantizing levels.

FOOTNOTES

1. However, according to Mertz and Gray in their classic "A Theory of Scanning . . . ," (*The Bell System Technical Journal*, July 1934), there is something to be said for over-lapping. While it reduces resolution, it may also minimize the occurrence of certain "extraneous patterns" such as the moiré patterns and the serrations that appear on oblique details of an image.
2. Named for R. D. Kell, an RCA engineer who introduced it as the "K factor" in a co-authored 1934 technical paper, "An Experimental Television System" (*Proceedings of the IRE*, vol. 22, p. 1247).
3. A recommended minimum for assured legibility of $\frac{1}{16}$-inch-high characters is 3.5 cycles/mm of photographic resolution.
4. See Chapter 4.
5. See C. E. Nelson, *Microfilm Technology* (New York: McGraw-Hill, 1965), pp. 303–304.
6. Specifically those conducted by R. L. Erdmann and A. S. Neal and reported in their article "Character Legibility of Digital Facsimile Resolution," *Human Factors*, Oct. 1968, pp. 465–473.
7. H. G. Morehouse, "Telefacsimile Services Between Libraries With the Xerox Magnavox Telecopier," Dec. 20, 1966 (study prepared for the Council on Library Resources, Inc., published by the University of Nevada Library, Reno).
8. F. K. Becker, et al., "An Experimental Visual Communication Systems," *The Bell System Technical Journal*, vol. 38 (Jan. 1959), pp. 141–176.

6
ECONOMICS

Considering the combined effects of the accelerating pace of technological advances, the various new forces at work in the communications services market,[1] and the changing economic picture in general, it is virtually impossible to set down specific guidelines on facsimile economics that are likely to remain valid for any length of time. About the best this chapter can do is to outline the basic elements and general philosophies governing fax system costs *at the time of writing*. Specific current cost data will be included only for the purpose of illustrating comparative weights of the various cost elements and the comparative costs of different basic system configurations.

In addition, certain current trends will be analyzed and projected in an effort to determine what their impact may be on facsimile's economic future.

Throughout the chapter, unless otherwise noted, references to "current" or "present" conditions, or to typical or specific costs, pertain to roughly the first half of 1977.

COST FACTORS

The various factors influencing fax economics can be grouped into three basic categories: (1) *terminal costs* (the actual fax terminal machines and accessories); (2) *material costs* (recording paper, chemicals, etc.); and (3) *transmission costs* (common-carrier charges).

There are also certain labor costs associated with the operation of fax systems, such as those of routine maintenance (e.g., reloading of recording paper) and the placing of calls. But these probably amount to no more than the cost of filling and addressing envelopes if the same documents were sent by mail, and can therefore be generally disregarded.

Terminal Costs

In the United States, the vast majority of fax terminals—transmitters, receivers, and transceivers—are rented rather than purchased, although both alternatives are open to the customer. (It is the author's understanding that the opposite is true in a number of other countries.) At present, a desk-top message (or "convenience") fax transceiver can be purchased for anywhere from $1000 to about $8000, and rented for as little as $29 a month and for as much as $150 a month. Typically, a send-only or receive-only terminal costs about as much to purchase or rent as a transceiver with comparable features. Thus, a full-duplex transceive capability, accomplished with two separate terminal machines, runs approximately double the cost of a half-duplex transceiver.

Systems capable of higher speeds and/or better resolution, or designed for special applications, may range from a few dollars to several thousand dollars higher in price than conventional message fax equipment. Telautograph's 900-line-per-minute *Copyphone** system, for example, sells for about $6000 per send/receive pair (one transmitter, one receiver), whereas a typical digital data compression transceiver runs about $10,000 to purchase (or about $300 a month to rent). At the top of the scale is the type of system used by periodical publishers to transmit high-resolution page proofs between printing facilities. A single one-way terminal for this system can run as high as a few hundred thousand dollars.

The question of whether to lease or purchase is largely one of the individual customer's preference. There are pros and cons for each choice. It is true, for example, that for perhaps only two and a half years of rental payments a system could be purchased outright, but there is also the matter of maintenance costs to consider. Where complex and highly specialized electronic or electromechanical apparatus is concerned, servicing coverage can be a primary inducement for leasing.[2]

Another factor to be considered is the fast pace of technological change. Electronic equipment in particular is vulnerable to rapid obsolescence. There is always the chance, therefore, that a purchased fax system will become antiquated before its cost has even been amortized.

Beyond these practical considerations, the decision to lease or buy may be dictated simply by accounting practices, which vary from one company to another.

The choice between a half-duplex transceiver and separate transmit and receive units is often solely an economic one, but it may also be governed by system objectives. There are systems, for example, in which communication is strictly one way and which therefore require only a transmitter at one end and a receiver at the other (or multiple units, as the system requires). In such systems, it would seem only logical to use separate transmitters and receivers exclusively. However, from a purely economic standpoint, it might be expedient to use transceivers, even though half their capability may be wasted.

Copyphone is a registered trademark of Telautograph, Inc.

Other advantages traditionally cited for separate send and receive units are the abilities to send and receive simultaneously at a given station, to produce a local copy of the transmitter output, and to help isolate trouble by back-to-back hookup of the separate units. In recent years, however, duplex transceivers offering these same capabilities in a single unit have become generally available.

Terminal costs may also include certain accessories. For example, where opposite ends of a system cross a commercial power grid boundary, a precision power supply may be required at each terminal (see Chapter 3). These can cost as much as a few thousand dollars per unit and rent for $10 to perhaps $50 or more per month per terminal. Other typical accessories are automatic-answer devices for use with acoustic-coupled systems (typically $5 to $15 a month), and automatic document feeders ($300-$1000 to purchase, $10-$30 a month to rent).

Separate acoustic/inductive (phone handset) couplers, sometimes included in the rental of analog transceivers for the convenience of "wireless" hook-up to a phone line, are also separately available for $80 to $200 each, or about $5 a month. Separate modems for digital fax systems can range from $800 to over $20,000 a terminal (or roughly $20 to over $500 a month), depending on the bit-rate, error-correcting capability and other factors relating to system requirements and the degree of sophistication desired. The typical cost for a voiceband system is $5000, or $100 a month per terminal. (Most digital fax systems currently available have built-in modems.)

Material Costs

In many if not most fax systems in operation today, the only significant continuing material cost is that of recording paper. A possible exception is photographic recording, widely used in the distribution of news photographs and the recording of cloud cover pictures from earth-orbiting weather satellites. To the extent that the recording requires processing chemicals, sensitized paper, and special recording lamps, photographic material costs can be substantially higher than for those systems that consume paper alone. However, in recent years, the trend has been toward much simpler and less expensive processing of fax-received photographs.

Of the more popular direct recording processes, electrolytic is generally the least expensive, costing from three to four cents a copy. Electrostatic ranks next with a range of from four to six cents a copy, and electroresistive ("burn-off") recordings cost from six to nine cents apiece. Thermal, which at this writing is still not widely used in fax systems, also ranks among the less expensive direct processes.

An additional continuing material cost peculiar to the electrolytic process is that of printing electrodes. These tend to erode as a result of electrochemical action and must be periodically replaced. Printing blades can cost from a few cents to several dollars apiece, and the useful life of each can range from two hundred to a few thousand copies, depending on the specific process involved and the nature of the copy (the amount of "blackness"). But at worst the consuming of printer electrodes adds only a fraction of a cent to the cost per copy (as does that of styli in other processes) and can therefore be generally disregarded.

Of the indirect processes, the least expensive in terms of material cost is transfer electrostatic, which uses ordinary paper. The cost is about a penny a page. The most expensive is, as indicated above, the "wet" photographic process, the cost for which can easily reach 25 cents a page and higher. By contrast, the newer heat-processed, dry-photo recordings cost between seven and ten cents a page.

Transmission Costs

The first consideration in determining transmission cost is the required bandwidth. Because of modulation, the maximum frequency to be transmitted is invariably higher than the maximum baseband frequency. The amount of margin needed depends primarily on the transmission scheme employed—FM, AM single sideband (SSB), AM vestigial sideband (VSB), or AM double sideband (DSB). (The various aspects of bandwidth determination have been covered in preceding chapters.)

With bandwidth requirements determined, the next question is what facilities or services are available to meet those requirements. These could range from the readily accessible telephone dial network to a specially installed private microwave facility. Without resorting to data compression techniques, the dial network (or public switched network) is limited to an effective maximum frequency of about 2400 Hz. For broader bandwidths or for better assurance of consistent transmission quality, a leased private line is usually required.

A limited selection of higher-than-voiceband "switched" services is available from various common carriers: Western Union's *Broadband Exchange* service, for example, which offers switched channels slightly above voiceband. In the offing are switched digital services covering the range from 4800 bps, through 56 kilobits, to as high as 1.5 megabits.

The choice between switched service and a dedicated leased line will usually be made on the basis of the volume of "transactions" (transmitted or received pages, or both) within a given time frame. Inasmuch as the tolls paid for switched service apply *only for the duration of a connection*, the cost naturally increases

ECONOMICS 203

with volume. By contrast, a leased circuit is generally paid for on a *predetermined per-month-per-mile* basis, and therefore costs less *per transaction* as the volume increases. This basic distinction is important to keep in mind. It is illustrated graphically in Figs. 6.1 and 6.2.

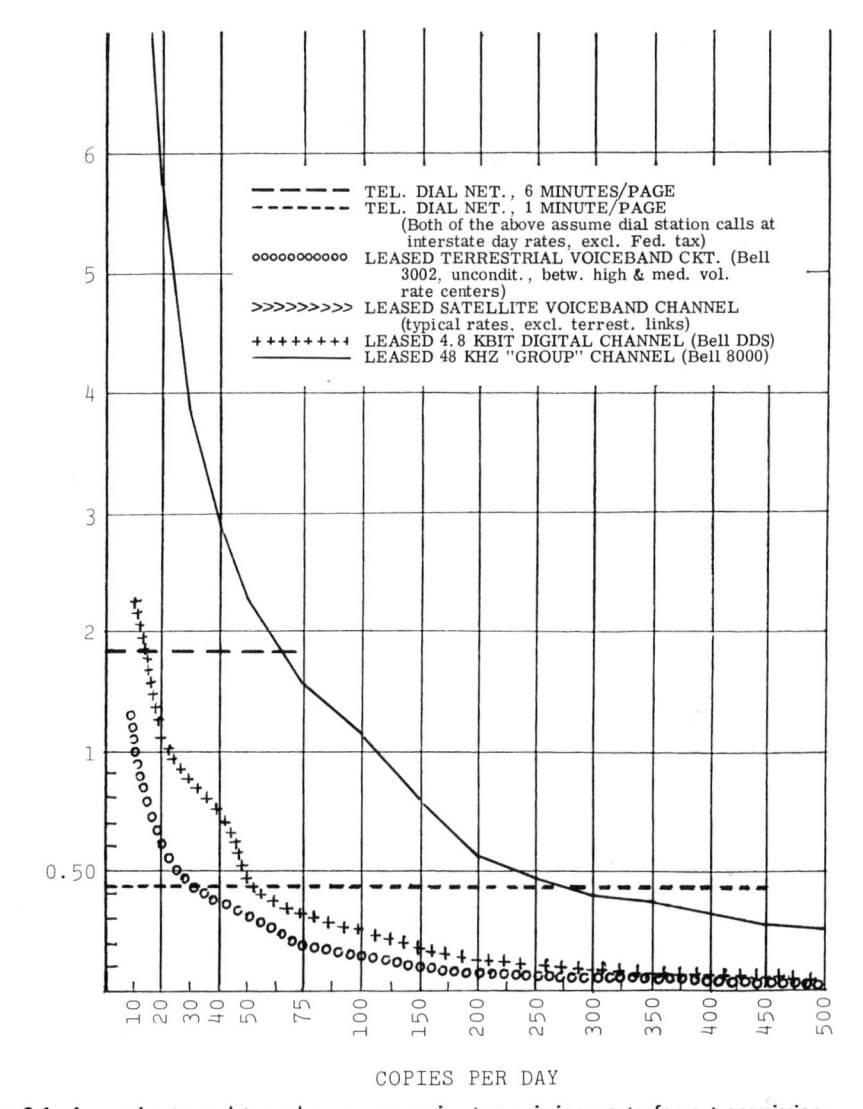

Fig. 6.1. Approximate end-to-end common-carrier transmission costs for a transmission distance of 100 airline miles. Copies-per-day volumes assume 22 working days per month; single transmission channel. (Note that satellite channels are not generally tariffed for distances of less than 500 miles.)

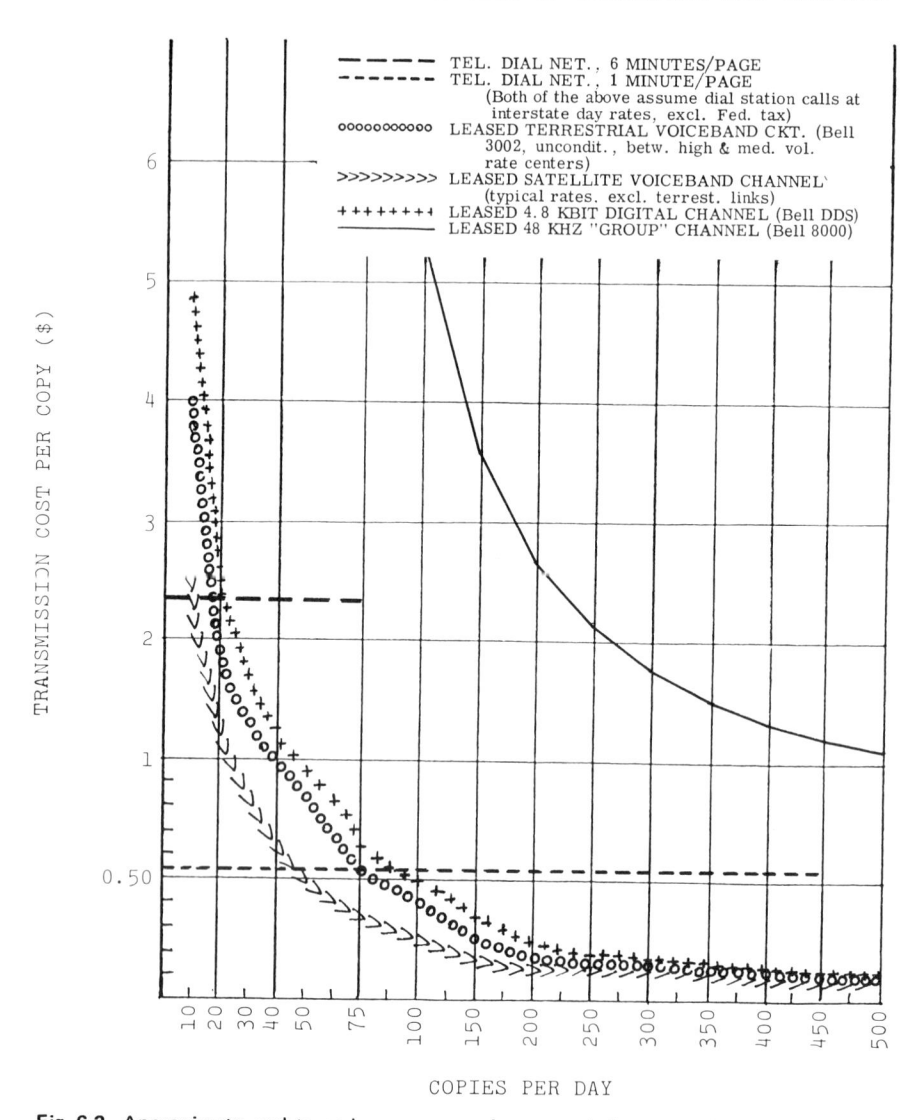

```
                    ─ ─ ─ ─   TEL. DIAL NET., 6 MINUTES/PAGE
                    ───────   TEL. DIAL NET., 1 MINUTE/PAGE
                              (Both of the above assume dial station calls at
                              interstate day rates, excl. Fed. tax)
                    ooooooooooo LEASED TERRESTRIAL VOICEBAND CKT. (Bell
                              3002, uncondit., betw. high & med. vol.
                              rate centers)
                    >>>>>>>>>> LEASED SATELLITE VOICEBAND CHANNEL
                              (typical rates. excl. terrest. links)
                    ++++++++  LEASED 4.8 KBIT DIGITAL CHANNEL (Bell DDS)
                    ─────────  LEASED 48 KHZ "GROUP" CHANNEL (Bell 8000)
```

Fig. 6.2. Approximate end-to-end common-carrier transmission costs for a transmission distance of 1000 airline miles. Copies-per-day volumes assume 22 working days per month; single transmission channel.

It is true, of course, that regardless of volume, switched telephone service (the public dial—or DDD—network) is often selected as the transmission medium for voiceband fax systems simply because of the convenience of its availability.

Another consideration in choosing between a switched network (voice or broadband) and a private line is whether transmission is to be in two directions

ECONOMICS

simultaneously—full duplex (or just plain *duplex*)—or one direction at a time—*half-duplex*. Switched networks are customarily confined to the latter mode, whereas private lines can usually be leased in either mode.

If the dial network is to serve as the transmission facility, an interface may be required between the fax terminal and the line at each end of the system. The intent is to ensure that certain characteristics of the signal—particularly its level—are controlled so as not to cause interference with other circuits or damage to the network's switching gear and amplifiers. Although recent legislation has made it possible to hook up directly with the line, provided the terminal has been properly certified and registered (see Chapter 4), there may continue to be cases in which a telephone company Data Access Arrangement may still be required. Alternatively, connection may be made indirectly via a customer- (or vendor-) provided acoustic/inductive coupler. Both of these interface options were described in Chapter 4.

But whatever the hook-up arrangement, a separate interface may still be required in some cases to tailor the fax terminal output to the signalling requirements of the line. The system requiring it will most likely be of the digital data compression variety, and the interface will be in the form of a baseband/voiceband modem. Typical costs of such devices were covered above under (1) Terminal Costs.

Where leased lines are concerned, the customer can connect fax terminals and noncommon-carrier interfaces directly to the line, provided the output of these devices is not such that it could pose a safety hazard to maintenance personnel or cause interference in adjacent lines. (Criteria are stated in FCC Tariff 260.)

A big factor in transmission cost, regardless of whether the medium is a switched network or a leased line, is mileage. Some sample costs illustrating the relationship of dollars to distance are shown in Table 6.A.

Note that the costs shown reflect the *base rates* only. To complete the picture on transmission costs, various miscellaneous charges must also be taken into account. For example:

- Federal excise tax on all interstate toll calls: 5% as of January 1, 1977, and declining 1% per year until total repeal in 1982;
- Monthly termination charge of $25 (each terminal) for a Bell System unconditioned 3002 voiceband data channel;
- Termination charge of about $70 a month at each end of a half-duplex Telephoto link;
- Minimum termination charge of approximately $124 a month (each end) for Bell System Digital Data Service (DDS) at the 4800 bps rate;
- Approximately $460 a month per termination for full 48 kHz "Group" service from the Bell System.

TABLE 6. A Sample Costs, Illustrating Relationship of Dollars to Distance, for a Selection of Transmission Facilities and Services. The Costs Shown Reflect Base Rates Only

Transmission Service	Mileage (Air Miles Between Terminals)								
	10	25	50	100	250	500	1000	2000
1-minute toll call on dial net.*	$0.16	$0.29	$0.37	$0.41	$0.46	$0.49	$0.52	$0.56
Unconditioned 3002 leased line (per mo.)†	81.70	129.50	172.70	240.20	339.20	504.20	834.20	1234.20
Half duplex "Telephoto" leased line (per mo.)	43.30	108.25	184.00	335.50	659.50	1037.00	1577.00	2657.00
4800 bps DDS ckt. (per mo.)	69.00	93.00	121.00	177.00	194.00	441.00	771.00	1171.00
Full 48 kHz "Group" service (per mo.)	162.00	405.00	810.00	1620.00	4050.00	6900.00	10,975.00	19,125.00
Domestic Satellite channel-voiceband‡ (per mo.)	$500 to $1100, depending on terrestrial mileage						
Domestic Satellite channel-48 kHz‡ (per mo.)	$5000 to $9000, depending on terrestrial mileage						

*Based on interstate day rates.
†Rate shown assumes Bell System Multischedule Private-Line Service (MPLS), with one terminal in a low-density rate center and the other in a high-density center. (Rates would be lower between two high-density centers higher between two low-density centers.)
‡Figures shown assume single channel. (There are quantity discounts for multiple channels.)

ECONOMICS

In addition, circuit conditioning may be required to ensure undistorted transmission of fax signals over long distances via a leased voiceband circuit. That requirement will impose additional charges of approximately $5 to $40 (or more) per channel end per month, depending on the types and degree of conditioning needed.

In spite of the many separate cost elements that can enter the picture, there are conditions under which there is—ostensibly, at least—*no* transmission cost whatever. They are basically four:

1. when the transmitter is indirectly coupled to an existing phone, and transmission is confined *within* an exchange area;
2. when the transmitter is indirectly coupled to an existing phone, and transmission is via an existing tie line;
3. when the transmitter is indirectly coupled to an existing phone, and transmission is via the Bell System's Wide Area Telephone Service (WATS);[3]
4. when the transmitter and receiver(s) are interconnected via a customer-owned, solid-wire circuit.

Obviously, there are hidden costs in all of these. For example, if an existing phone installation is used for fax transmission to the extent that an additional circuit becomes necessary to maintain the normal volume of voice communication, the cost of the first circuit—or at least part of it—must be regarded as a fax transmission cost. Thus, the first three of the above conditions assume a modest fax transmission volume.

One final point regarding transmission cost: There are situations (not very common) where cost can be incurred in the form of a service charge for an occasional point-to-point transmission. A mundane example, just to illustrate the point, is the sending of a telegram. A more exotic one—and one more closely related to fax—is the international transmission of photographs by one of the "record" carriers on a one-time service basis. Charges are based on document size (in square centimeters), and they vary somewhat depending on the particular cities between which transmission takes place. RCA GlōbCom, for example, has a minimum 150 sq. cm charge, plus an extra charge for each additional square centimeter. At the rates in force as of October, 1974, transmission of an 8 × 10-inch photo from New York to Paris cost $60, and from San Francisco to Hong Kong approximately double that amount. This obviously goes beyond simple transmission cost in that the charge is computed to recover various costs incurred by the common carrier, including terminal operation and consumption of expendable materials.

TRADE-OFFS

There is not much that can be said about the trade-offs among economy, speed, and quality that has not already been said elsewhere in this book. But perhaps this is a good point at which to review the basic considerations, with the emphasis on economics.

The general rules governing trade-offs can be summarized as follows:

To Increase...	Must Sacrifice...
quality[4]	speed and/or economy
speed	quality and/or economy
economy	quality and/or speed

Expanding on these slightly, we note that increasing either speed or output resolution, without sacrificing one for the other, will require either an increase in channel bandwidth or the adoption of redundancy reduction (data compression) techniques to obtain more efficient utilization of the available bandwidth. Either one will result in increased cost.

Supplementing these technical fundamentals are certain economic facts of life that one should be aware of in planning or modifying a system:

1. In general, the two elements impacting most heavily on the cost of a fax system are *transmission distance* and *required signalling speed*;
2. In terms of common-carrier standard service offerings, the steps in which channel bandwidth or bit rate may be increased are likely to be large and represent substantial increases in cost.
3. The volume of copies transmitted in a system already operating at capacity may be increased in one of three ways:
    (a) by adding additional circuits and terminals,
    (b) by adopting data compression techniques to permit increased transmission speed (new, more sophisticated terminals), or
    (c) by converting to a broader bandwidth facility to permit increased transmission speed (which will probably also mean the acquisition of new, probably more expensive terminals).

   The most economical choice of the three will vary with volume, distance, and comparative costs of terminal gear.

About all that can be said further about trade-offs from the economic standpoint is that once speed and quality requirements—and volume requirements to the extent that they can be predicted—have been tied down, the most economical system configuration can be determined only by an analysis of various combinations of the cost elements previously discussed. The charts in Figs. 6.1 to 6.4 will be helpful in assessing comparative transmission costs.

COST PER COPY

A good basis for comparing the economics of different fax systems is to compute the *cost per copy* for each—assuming, of course, that the same size original material is to be used in all systems. In the case of a leased private line, the cost per copy will be the prorated share of the total cost of the system within a given period. And where the system utilizes an existing switched network, the cost per copy will be the prorated share of the fixed cost, *plus* the toll charge per "call."

Where terminal equipment is purchased outright, the period of time over which the total purchase is to be amortized must be determined. (The purchase price is thus convertible to a continuing annual or monthly cost.) In addition, a reasonable allowance must be made for the servicing costs that would normally be covered automatically were the same equipment rented rather than purchased. Eight to nine percent of the purchase price is a fair rule of thumb for annual maintenance.

To avoid inconsistent results in the distribution of total costs, certain guidelines as to accounting procedure must be established before any computations are made. A question that must be resolved at the outset is whether all terminals of a system are to be financed from the same budget, or each from a separate budget.

Where the first condition—a single budget—applies, there is no problem. However, where each terminal is financed from a separate budget, which might well be the case where transmission is via a switched network, the total cost of a terminal must be confined to the expenses incurred *by that terminal only*. Computation is complicated in the latter case by the question of how the send and receive costs per copy (assuming a customer both sends and receives) are to be distributed. The easy way would be to simply ignore receive costs, since at low to moderate volumes they are negligible anyway. But a better way is to base cost per copy on *copies sent only*, and to include receive (material) costs as an *average* under continuing monthly costs.

Computation of cost per copy may be further complicated by the existence of a "meter plan" in which the *terminal cost* will vary with the volume of pages sent or received. One fax vendor is currently charging five cents per "transaction" (transmission *or* reception) on top of terminal rental and material costs. Another offers an optional meter plan which affects transmissions only and is applied on a sliding scale.

Let us examine two hypothetical systems: System #1 a single transceiver interfaced with the dial network by a phone coupler, and System #2 has two transceivers interconnected by a leased, voice-grade private line. Actually, System #1 is merely a terminal, which qualifies as a system only when it has linked up with a compatible terminal elsewhere on the dial network. For the sake of this example, we shall assume that it customarily communicates with a terminal (or terminals) located 50 air miles away.

(It should be understood that although the costs cited in these examples are based on those for actual products and services at a given point in time, they are meant to be merely illustrative.)

System #1, Terminal A

1. Terminal costs: leased transceiver, $130/mo. (incl. phone coupler)
2. Material costs: recording paper, $0.04/copy
3. Average daily copies sent: 10
4. Average daily copies received: 20
5. Transmission—
    Facility: public telephone (dial) network
    Distance: 50 miles
    Duration, per copy: 4.5 minutes
    Cost per call: $1.33 (interstate day rate as of mid-1976, incl. 6% federal excise tax)
    Cost per copy (assuming 22 working days per month):

$$\frac{\$130 + (22 \times 20 \times \$0.04)}{22 \times 10} + \$1.33 = \$2.00$$

System #2

	Terminal A	Terminal B
1. Terminal costs (leased transceivers)	$130/mo.	$130/mo.
2. Material costs (recording paper)	$0.04/copy	$0.04 copy
3. Average daily copies sent	10	20
4. Average daily copies received	20	10

5. Transmission—
    Facility: voice-grade leased line (Bell System MPLS)
    Distance: 50 miles
    Duration, per copy: (does not apply—unless the volume of copies is sufficient to require additional circuits)
    Base rate (for 50 miles): $172.70/mo.
    (*Note:* This assumes "Schedule II" service, namely one terminal in a low-density rate center and the other in a high-density center.)
    Termination costs: $50.

Fixed monthly system cost:

($130 × 2) + $50 = $310

$310 + $172.70 + (22 × 30 × $0.04) = $509.10

ECONOMICS

(Note: 22 is the working days per month; 30 is the total copies per month, both terminals.)

Terminal A share of fixed monthly cost:

$$10/30 \times \$509.10 = \$169.70$$

Terminal B share of fixed monthly cost:

$$20/30 \times \$509.10 = \$339.40$$

$$\text{Cost per copy, Terminal A} = \frac{\$169.70}{22 \times 10} = \$0.77$$

$$\text{Cost per copy, Terminal B} = \frac{\$339.40}{22 \times 20} = \$0.77$$

Obviously, the cost per copy for System #2 can more readily be computed for the system as a whole, as follows:

$$\frac{\$310 + \$172.70}{22 \times 30} + \$0.04 = \$0.77$$

The trap to avoid is that of splitting the total operating cost evenly between Terminals A and B, and then computing the cost per copy for each terminal on that basis, thus:

$$\$509.10 \div 2 = \$254.55$$

$$\text{Cost per copy, Terminal A} = \frac{\$254.55}{22 \times 10} = \$1.16$$

$$\text{Cost per copy, Terminal B} = \frac{\$254.55}{22 \times 20} = \$0.58$$

Average cost per copy (unweighted) = $0.87

Unfortunately, the trap may be difficult to avoid where separate terminals connected to the dial network are lumped together and treated as one system—especially when each sends or receives a different volume of copies within a given period. Assume, for example, that System #1 is completed by a Terminal B, which sends the 20 copies that Terminal A receives and receives the 10 that Terminal A sends. It would seem a simple enough matter to split the total cost of the system equally between terminals A and B, in which case cost per copy might compute as follows:

Total system cost per month:

$$(2 \times \$130) + 22 \times 30 (\$1.33 + \$0.04) = \$1164.20$$

Each terminal's share:

$$\$1164.20 \div 2 = \$582.10$$

$$\text{Cost per copy, Terminal A} = \frac{\$582.10}{22 \times 10} = \$2.65$$

$$\text{Cost per copy, Terminal B} = \frac{\$582.10}{22 \times 20} = \$1.32$$

Average cost per copy (unweighted) = $1.99

Alternatively, the cost per copy can be computed on the basis of each terminal's own cost:

$$\text{Cost per copy, Terminal A} = \frac{\$130 + (22 \times 20 \times \$0.04)}{22 \times 10} + \$1.33 = \$2.00$$

$$\text{Cost per copy, Terminal B} = \frac{\$130 + (22 \times 10 \times \$0.04)}{22 \times 20} + \$1.33 = \$1.65$$

Average cost per copy (unweighted) = $1.83

But, as in the case of the leased-line system, the *true* cost per copy is obtained by allocating a prorated share of the total system cost to each terminal—or simply by computing for the system as a whole, as follows:

$$\frac{2 \times \$130}{22 \times 30} + \$1.33 + \$0.04 = \$1.76$$

For comparison purposes, it would be well to select one method of computing cost per copy and apply it consistently, even if it does not produce absolute true costs. In comparing a leased-line system with a "detached" dial network terminal, for example, it is imperative that we treat each terminal as a separate entity, except that for the leased-line system, we will have to allocate a share of the fixed transmission costs to each terminal under study.

So much for the mechanics of cost computation. With a little additional manipulation of the foregoing data, we can also illustrate a couple of basic economic distinctions between switched-network and leased-line systems in general (Table 6.B).

First, by reducing the average daily copy volume from 15 to 3 per terminal for our two systems, we can demonstrate the relationship of copy volume to economic advantage. Both systems will naturally be less economical at the

ECONOMICS

TABLE 6.B Comparative Costs per Copy for a Basic Dial Network Fax System (System #1) and a Basic Leased-Line System (System #2) at Two Mileages and Various Copy Volumes

	Avg. 3/day per Term.	Avg. 4/day per Term.	Avg. 5/day per Term.	Avg. 10/day per Term.	Avg. 15/day per Term.
1. 50 Miles					
System #1	$3.34	$2.84	$2.55	$1.96	$1.76
System #2	3.70	2.78	2.23	1.13	0.77
Percent Difference	+11	−2	−12	−42	−56
2. 100 Miles					
System #1	$3.55	$3.05	$2.76	$2.17	$1.97
System #2	4.21	3.17	2.54	1.29	0.87
Percent Difference	+19	+4	−8	−40	−55

Note: Percent difference figures are the approximate percentage by which the lower (System #2) figure either is greater than (+) or is less than (−) the upper (System #1) figure.

lower volume because of the larger share of the fixed terminal cost apportioned to each copy. But whereas System #2 showed a decided economic advantage at the higher volume, System #1 will now be the more economical by a slight margin: $3.34 per copy versus $3.70 for System #2.

The general rule that reveals itself here is that in a switched system, transmission *duration* is generally the controlling economic factor, whereas in a leased-line system, the greater the number of "calls," the more economical the system becomes. This assumes, of course, that the "calls" made in the switched system are toll calls, so that the longer their duration or the greater their number within a given period, the higher the total cost. In a nontoll situation, such as the use of existing tie lines, the rule would obviously not apply.

Now the influence of distance can be demonstrated by increasing the terminal separation from 50 to 100 miles for both systems. The effect on rates is an increase from $1.33 to $1.54 per five-minute call in the switched system, and from $172.70 to $240.20 per month for the leased-line system. At these higher rates, we find that the copies a day per terminal has to be increased to at least *five* in order for System #2 again to become more economical than System #1. What is indicated here is that increasing distance tends to have a somewhat greater economic impact on leased-line systems than on switched systems.

The charts in Figs. 6.1 to 6.4 will provide a more graphic picture of the ways in which copy volume, distance, bandwidth, and "call" duration interact to affect cost per copy in fax systems. Note that only transmission costs are

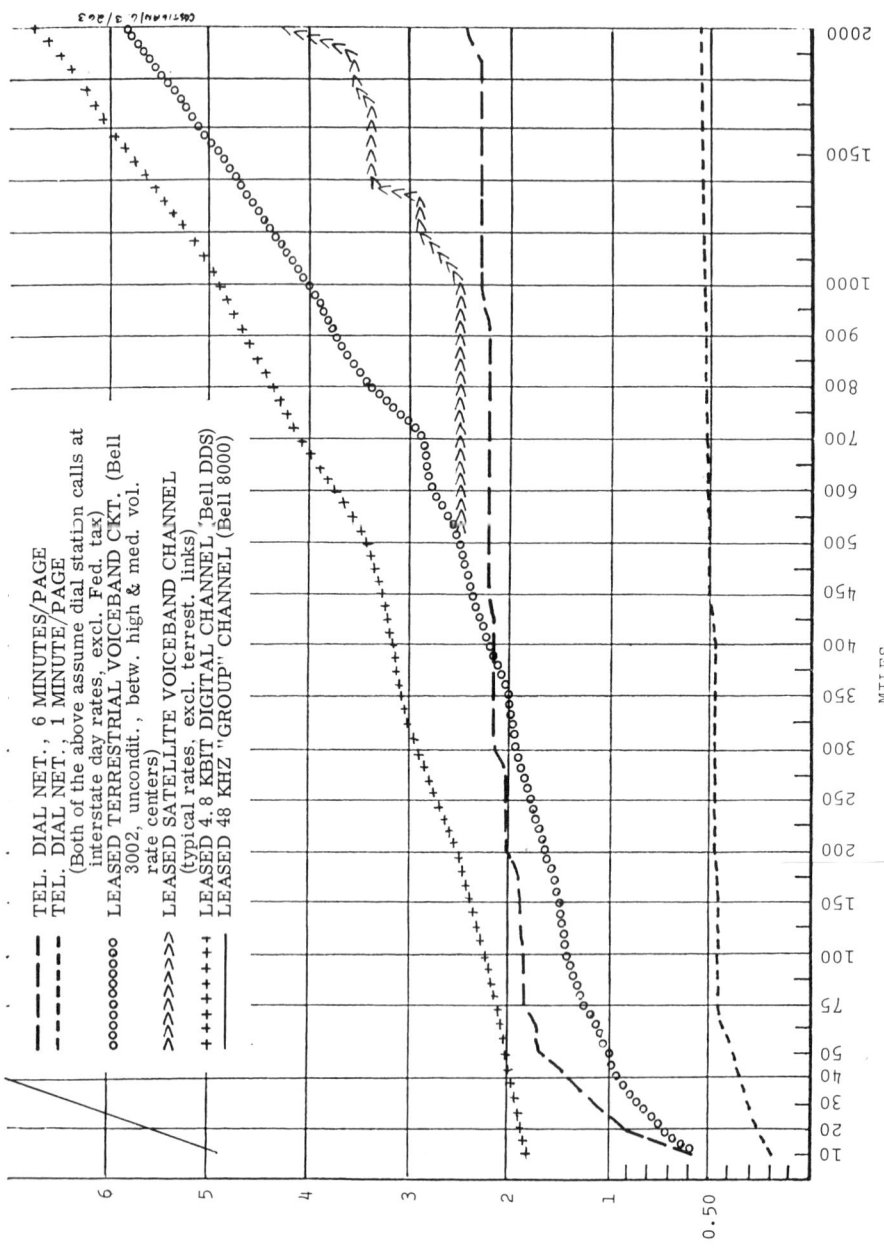

Fig. 6.3. Approximate end-to-end common-carrier transmission costs for a transmission volume of 10 pages per day via a single trans-

ECONOMICS

Fig. 6.4. Approximate end-to-end common-carrier transmission costs for a transmission volume of 100 pages per day via a single transmission channel, assuming 22 working days per month.

Legend:
- ─── TEL. DIAL NET., 6 MINUTES/PAGE
- ─ ─ ─ TEL. DIAL NET., 1 MINUTE/PAGE
  (Both of the above assume dial station calls at interstate day rates, excl. Fed. tax)
- ooooooooooo LEASED TERRESTRIAL VOICEBAND CKT. (Bell 3002, uncondit., betw. high & med. vol. rate centers)
- >>>>>>>>> LEASED SATELLITE VOICEBAND CHANNEL (typical rates, excl. terrest. links)
- +++++++ LEASED 4.8 KBIT DIGITAL CHANNEL (Bell DDS)
- ─── LEASED 48 KHZ "GROUP" CHANNEL (Bell 8000)

X-axis: MILES
Y-axis: TRANSMISSION COST PER COPY ($)

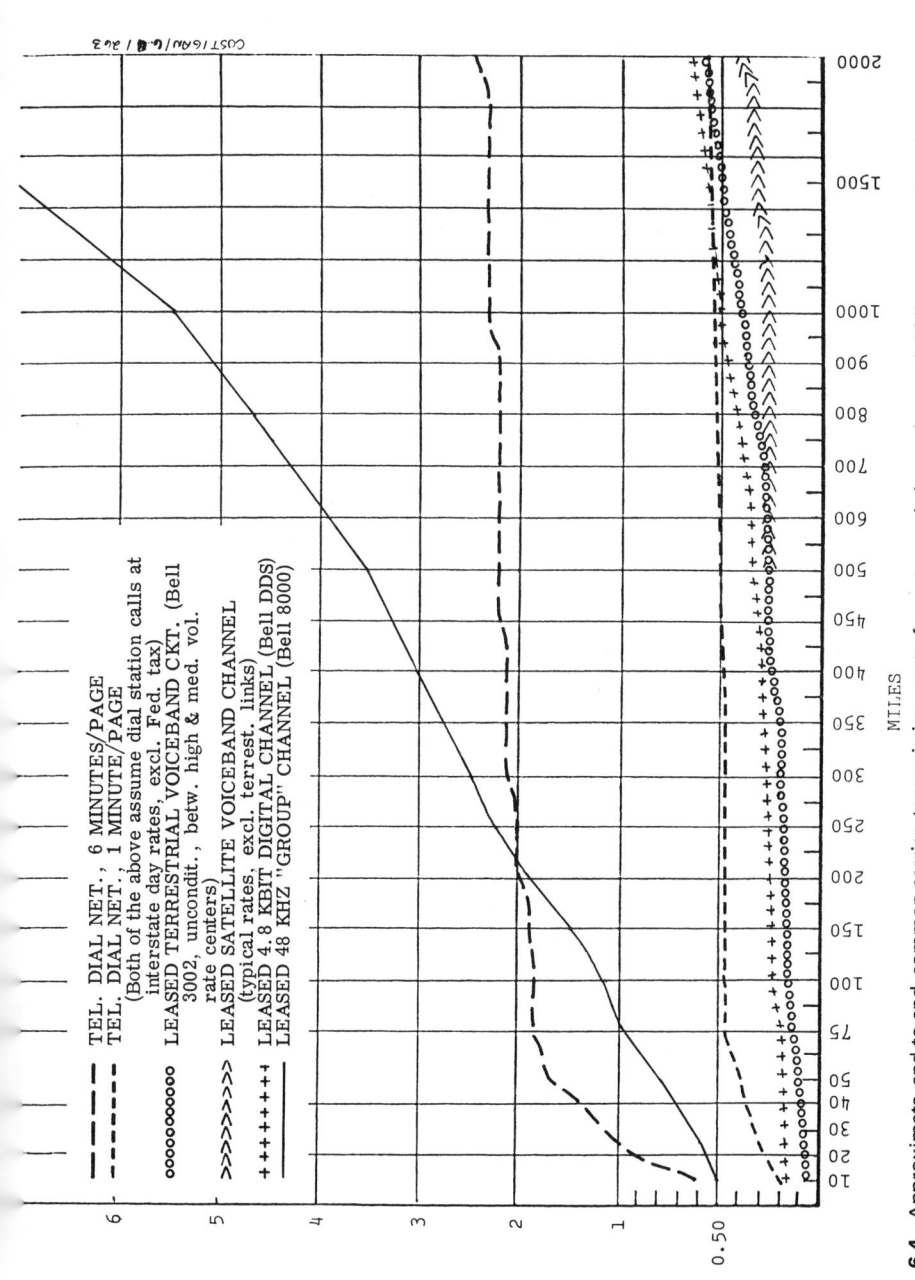

reflected—including termination and conditioning costs where applicable—but *not* the cost for customer-furnished interfaces, such as noncommon-carrier modems. Addition of terminal and material costs will affect the shape and crossover points of the various curves to a certain extent, but the basic interrelationships will remain the same.

Regarding the 48-kHz Group service, the customer has the option—at added cost—of splitting the full channel into various subchannel configurations, thereby possibly sharing the service among two or more different systems operating between the same two points. This would, of course, offset the cost of each individual system. On the charts, however, it is assumed that each of the leased services is devoted exclusively to a single fax system, even though the full bandwidth may not be utilized in all cases.

One-Time Charges

One cost not included in any of the above analyses is the one-time charges for installation of terminal units and for the terminating of the transmission link. Although normally incurred only once in the life of a given system configuration and usually not of sufficient weight to influence the choice of one system over another, these are costs that should be accounted for in any thorough economic analysis. Table 6.C provides a representative sampling of such costs, as of mid 1976.

TABLE 6.C Representative Sampling of Installation Charges for Transmission Services and Fax Terminals

Service or Terminal	Approx. Installation Cost (Per Main Sta. for Transmission Service; per Unit for Terminals)
Bell System Type 3002 leased voice-grade	$55
Bell System Digital Data Service	$130 at 2.4–9.6 kb/s
	$180 at 56 kb/s
Western Union Broadband Exchange Service	$25
Satellite channels	$50–$100
Qwip 1000 transceiver	$10
GSI dex 180 transceiver	$28 (delivery charge)
S-W Datafax 240 transceiver	$60
Xerox Telecopier 200 Transceiver	$75 placement
	$25 removal
Rapifax 100 transceiver	$100 placement
	$35 removal

ECONOMIC TRENDS

In the light of current trends, what is facsimile's economic future? And what are some of the trends that particularly bear watching?

From the author's point of view, there is every indication that the economics of facsimile will continue to improve. Looking at the terminal market alone, we see new names constantly cropping up, enlarging the circle of competitors. Each new entrant contributes some attractive improvement at a bargain price. The stimulation of competition among suppliers is naturally beneficial to fax customers by itself, but it is additionally encouraging that the improvements offered by each new competitor have not been superficial. The emphasis appears generally to be in the right place—on improved operating efficiency.

The first commercial device to compress fax signals for transmission economy became available around 1969. Today there is a proliferation of analog terminals with 2:1 speed compression capabilities, and worldwide there are at least six vendors actively (and successfully) marketing fully self-contained digital data compression fax terminals for use on voice-grade circuits, such as the dial telephone network. The digital machines have average speed compression ratios of about 6:1.

In addition, there are some highly specialized fax terminals using advanced compression techniques, and several vendors are currently developing compression-type terminals for commercial introduction in the near future. Although such terminals cost more than their slower analog counterparts, the improved transmission efficiency they offer will yield a net economic advantage at relatively high transmission volumes, particularly where the terminals are separated by a considerable distance. Moreover, the shorter durations per page become particularly meaningful in situations where a given transmission circuit is being shared among various types of terminals (e.g., fax, data, and voice).

It seems only natural that in order to compete, each new supplier to introduce a compression (or redundancy reduction) system will have to prove his product's superiority primarily in terms of *bit rate per dollar*. In other words, whether the customer wants increased speed for its own sake or more economical utilization of transmission facilities, the supplier who gets the order in most cases will be the one who can meet either requirement most economically.

Reformatting the fax signal at the terminals is, however, just one way of effecting operational economies, and it has its limitations. A more important potential source of transmission economy is the transmission area itself.

One of the significant changes the communications industry has undergone in the past few years has already resulted in certain economies. That change was the easing of restrictions on the connection—or interconnection—of non-common-carrier devices to the common carriers' lines. The immediate result was the substitution of relatively inexpensive Data Access Arrangements for the Data Sets formerly required, and now it appears that even the DAA may be eliminated as a required interface for hardware interconnection. Furthermore, competition has been stimulated among commercial suppliers of modems aimed at increasing transmission efficiency.

Another change, perhaps better characterized as a "latent force," has been the increasing potential for competition in this lucrative market. The development of new radio communication technologies—notably microwave and space borne satellites—during the past two decades has opened the door to the formation of private communication networks with the potential for competing with the established common carriers. Here, as in the terminal market, increasing competition could have a favorable long-range effect on transmission costs. However, the presumption that the user automatically benefits from competition among communications carriers warrants very careful analysis. This is a complex issue.

Meanwhile, there are certain changes self-imposed by the common carriers that should result in significant transmission economics for customers. One is adoption of the 1-minute charge basis for toll calls, which is presently in force for *all* direct-dialed *inter*state calls and for about half of all *intra*state day calls as well. In addition, *most intra*state calls made after 11 p.m are on a 1-minute basis. This relatively new rate structure further enhances the transmission economies possible with digital data compression fax terminals, the average speed of which is such that a "typical" letter-size document page can be sent in a minute or less.

Another common-carrier action that bodes well for the customer is the strong present prospect that international fax transmissions will eventually be permitted on overseas telephone circuits, at substantial reductions over present methods. At this writing (mid-1977), the phone network is still officially off limits to any international communications other than voice. Fax has to go via the international "record" carriers: RCA Glōbcom, ITT WorldCom, Cable & Wireless, WUI, and TRT. However, the FCC has recently begun to reassess that position, and, besides having given a tentative green light to the Bell System, has authorized two digital packet-switching carriers (Graphnet and Telenet) to extend their respective networks beyond domestic limits.

Still another important change affecting the efficiency with which the common carrier gets the customer's message from point A to point B is the strong trend toward geostationary satellite systems (which were discussed in Chapter 4.) The growing use of this relatively new mode of radio relaying promises substantial communication economies.

But perhaps the most important change on the common-carrier front is the growing emphasis on digital facilities. The digital mode of transmission affords

ECONOMICS

Fig. 6.5. This compact, phone-coupled fax transceiver is representative of the economies possible with modern electronic technology. The unit rents for less than $30 a month: about half what a comparable unit would have cost ten years ago. (Courtesy QWIP Systems)

the common carrier certain internal economies, which can, in turn, be passed on to the customer. (See chapters 2 and 4.) It also enables elimination of the digital/analog/digital modems now required for operation of digital systems on analog circuits, and thus makes possible a substantial reduction in the terminal cost. It has already been determined that, ultimately, common-carrier facilities will be almost exclusively digital.

Another current trend, one that has been making fax increasingly attractive to the low-volume user, is the availability of analog transceivers for under $30 a month. Exxon's recently formed QWIP Division has set the pace with its compact Model 1000 machine (Fig. 6.5), offering—for $29 a month—2-speed send/receive operation and at least partial compatibility with some other makes of terminals. In terms of cost and simplicity, it is somewhat reminiscent of the Western Union Desk-Fax system. (See chapter 1.) The *QWIP** machine is perhaps even simpler in design than its once-ubiquitous predecessor—the QWIP electronics are confined to a single, easily replaced solid-state circuit board—yet offers considerably greater utility.

Similarly, some earlier-generation analog machines, such as the Magnavox-

QWIP is a registered trademark of Exxon Corp.

developed Magnafax line (now handled by the 3M Company), are available at attractively low rentals that may or may not be tied to a lease-purchase deal. Such bargains—the machines offered are usually highly effective and quite reliable—may become increasingly common as the newer digital systems continue to catch on and become more widely available.

Despite inflation, the prospects of a fax terminal in the same purchase-price class as, say, a moderately priced electric typewriter ($500 or less) are growing. The necessary technology is already available. The main obstacles at the moment are the economics of fabrication and the need to recover high development costs.

These, then, are the obvious current trends that are likely to influence facsimile's economic future. Others will be emerging that could either enhance or blemish this optimistic picture. Things can change fast, particularly in a field as competitive as fax has become.

FOOTNOTES

1. Easing of interconnection restrictions, emergence of new sources of competition, changes in public attitude toward utilities, etc.
2. The words *lease* and *leasing* are used here in the sense of a customer leasing or renting from a supplier. The words *lease* and *rent* are viewed as interchangeable.
3. A long-distance service that permits the customer unlimited calls (or 10 hours of calls, depending on the arrangement chosen) within a given zone at a fixed monthly rate.
4. "Quality" pertains primarily to output resolution. (See Chapter 5.)

7
STANDARDS

Whenever a large number of like components are linked to form a system, some degree of standardization to ensure compatibility is necessary. Home TV receivers, for example, may differ widely in size, layout of controls, circuit design, and overall appearance, but all are designed to accept the same input signal and to produce from it a 525-line, sound-accompanied picture, renewed at the rate of 30 frames a second. Needless to say, commercial TV could not exist as a system without this common bond.

Closed-circuit television, on the other hand, need not be similarly constrained. In fact, there can be any number of different CCTV standards, just so long as one set of standards is applied throughout a given system.

The same goes for facsimile. Each apparatus manufacturer is free to design to his own standards, but if he wants to sell to the U.S. National Weather Service or to a big news agency, or to any individual customer who has to rely on common-carrier facilities for transmission, he has to observe the standards that have been set for the particular realm within which the apparatus is to function.

Within weather forecasting and telephotography networks, for example, even though various types and makes of equipment may be used, there are specific standards—strictly enforced—governing input, output, and transmission parameters. The systems could not otherwise function.

But within other realms, such as the exchanging of documents and sketches on a private basis ("message fax"), there is somewhat less incentive for strict adherence to a given set of standards. Recommended standards that do exist may be applied or ignored, as the manufacturer or customer sees fit.

There are, in fact, established standards affecting practically every facet of facsimile communication. There are separate domestic and international standards, and, domestically, there are both industry and government standards. There are standards outlining engineering specifics such as the duration of a phasing pulse, and there are standards establishing rates and classes of service. Common carriers offer certain standard transmission services that many facsimile systems have come to rely upon and that exert a strong influence on the design of most manufactured fax equipment.

This chapter will concern itself chiefly with technical standards relating to engineering and transmission, and only with those that are established and have discernible influence. Standard common carrier transmission services were covered in Chapter 4, and quality standards (such as they are), in Chapter 5. Standards regarding rates for "message" services or for the leasing of transmission facilities will not be discussed, since they are periodically subject to change.

Note: Throughout the remainder of this chapter, wherever the contents of specific standards or recommendations are reproduced, the wording and the arrangement may not be precisely that found in the actual documents. In the interests of consistency and of minimizing detail—and, in some cases, of clarification—the author has taken the liberty of abridging and rearranging as required. Words and data in brackets [ ] are, in all cases, the author's.

DOMESTIC STANDARDS

EIA Standards

The Electronic Industries Association (EIA) is a trade association composed mainly of component and equipment manufacturers. Oldtimers may remember it as the RMA (Radio Manufacturers' Association), its name from 1924 to 1950, or as RETMA (Radio-Electronics-Television Manufacturers' Association), which it had become briefly before assuming its present identity in 1957.[1] Its Engineering Department publishes a number of recommended standards, among which are three that deal specifically with facsimile.

RS-328, titled *Message Facsimile Equipment for Operation on Switched Voice Facilities Using Data Communication Terminal Equipment*, is the standard that gives official recognition to the popular 96/180 combination for scan resolution and rate: the so-called 6-minute speed of the majority of analog fax terminals in use today (mid 1976). It also standardizes eight additional design parameters, as follows:

Index of Cooperation: 805 to 889 (829 recommended)
Total Line Length: 8.5 to 9.2 inches
Available Line Length: 8.0 inches minimum (8.5 inches recommended)
Phasing Time: 15 seconds
Signal Sense: Maximum on black
Synchronization:
   (a) 60 (±0.5) Hz transmitted signal
   (b) frequency standard accurate to one part in 100,000

STANDARDS

Paper Speed: $1\frac{7}{8}$ inches/minute (nominal)

Phasing Pulse: 25 (±3) mseconds; black signal interrupted by white pulse once per scan line, black-to-white transition occurring at right-hand edge of $8\frac{1}{2}$-inch-wide sheet

The standard defines message facsimile as relating to "equipments used for transmission of documents such as letters, charts, forms, drawings, graphs, etc., which do not require accurate reproduction of the grey tones between black and white." In its present form, the standard dates back to 1966. Plans for its reissue are currently being considered by the EIA's Facsimile Standards Committee, TR-29.

RS-357, *Interface Between Facsimile Terminal Equipment and Voice Frequency Data Communication Terminal Equipment*, is the second EIA fax standard. As the title implies, it concerns itself with the various electrical and signalling considerations associated with the interconnection of an analog fax system with voice band transmission facilities by way of a so-called data set, or modem. It specifies signal voltage levels, impedances, and general wiring arrangements. It is applicable to both switched-network and private-line systems.

Finally, there is RS-373, *Unattended Operation of Facsimile Equipment*, which specifies the minimum requirements for unattended operation of fax terminal equipment conforming to the preceding two standards. The requirements concern the sequence and duration of control signals necessary to permit one terminal to "call" another, send or receive a "message" (a document or whatever), and terminate the call—in that sequence. Because of recent technological advances and changing customer requirements, this standard has not been widely adhered to by terminal manufacturers.

In recent years, the EIA's fax standardization efforts have been largely influenced by related international standardization objectives, namely those of the International Telecommunications Union, and specifically the CCITT (the activities of which will be discussed later in this chapter). The two organizations have cooperated closely in the past, and, at this writing, their top priority is to arrive at some practical and universally acceptable basic standard controlling the compatibility of digital data compression fax systems (so-called "group 3" systems).

Among the EIA's better known and more widely applied standards is RS-232-C, *Interface Between Data Terminal Equipment and Data Communication Equipment Employing Serial Binary Data Interchange*. It relates to fax only to the extent that it affects the use of digital fax systems on analog circuits—and then it normally applies only when the modem is separate from the fax terminal. As indicated by the title, the standard applies to the interfacing of a terminal with its associated modem. It is one standard with which modem manufacturers have widely complied (though in varying degrees.)

In this same general category—standards concerning digital signalling—are RS-247, *Analog-to-Digital Conversion Equipment*; RS-269-A, *Synchronous Signaling Rates for Data Transmission*; and RS-334, *Signal Quality at Interface Between Data Processing Terminal Equipment and Synchronous Data Communication Equipment for Serial Data Transmission.*

Other published EIA standards in the RS- series that relate (either directly or indirectly) to document and graphic transmission systems in general are RS-343-A and RS-412-A, both of which deal with performance standards for high-resolution closed-circuit TV systems (see the next chapter), and RS-210 and RS-252-A, which concern microwave communication systems. All of these standards are available for a small fee from the Engineering Department, Electronic Industries Association, 2001 Eye Street, N.W., Washington, DC 20006. Some may also be available from the American National Standards Institute, 1430 Broadway, New York, NY 10018.

IEEE Standards

The Institute of Electrical and Electronics Engineers (IEEE) has worked closely with the EIA in the establishment of facsimile standards. Like the EIA, it publishes two standards dealing specifically with fax: No. 167, *Test Procedures for Facsimile* (formerly IRE Standard 43-9 S1) and No. 168, *Definitions of Terms for Facsimile* (formerly IRE Standard 56-9 S1).

In addition, there is the IEEE Facsimile Test Chart, the 1970 update of which was largely the result of recommendations by a joint EIA/National Micrographics Association Committee (EIA Subcommittee TR-29.1) formed in the 1960s to set standards for microfilm-input facsimile systems. The latest (1975) update of the chart, illustrated in Chapter 5, is identical in content to the 1970 issue. It is differently designated mainly to identify it as having come from a separate production run, and therefore having slightly different tonal properties than the earlier issue.

IEEE No. 168 is the basis for terms used in EIA Standard RS-328, and the test procedures outlined in No. 167 are, in turn, arranged to reflect the various parameters standardized in RS-328.

No. 167 is divided into three sections, each of which is quite comprehensive. The three are "Equipment Specification," "Distortions from Facsimile Equipment," and "Distortions from Transmission Medium." Among the specific items for which tests are outlined within the sections are scanning line length, stroke speed, spot dimensions, halftone characteristics, effective band, index-of-cooperation distortion, skew, jitter, delay distortion, echo, noise (random, impulse, and single-frequency), and Kendall effect. While this standard is still largely applicable to the majority of present fax systems, it is somewhat wanting with respect to the newer digital data compression systems. In time a whole new testing standard, more widely applicable to the present technology, will no doubt be issued.

No. 168, in its present edition (January, 1971), defines 117 facsimile terms. However, having been compiled in the mid 1950s, it lacks several terms that one would expect an up-to-date facsimile glossary to include—terms like data set, digital transmission, Kell factor, modem, quantizing (amplitude and time), and run-length coding. Some if not all of these are expected to be included in future updates of this standard, the latest of which is currently being finalized.

Two other IEEE standards that may be of interest inasmuch as they relate indirectly to fax transmission are No. 170, *Definitions of Terms for Modulation Systems* (Formerly IRE Standard 53-11 S1) and No. 269, *Method for Measuring Transmission Performance of Telephone Sets.* Copies of all of these standards, including the Facsmilie Test Chart, can be purchased from the Institute of Electrical and Electronic Engineers, either through its headquarters at 345 East 47th St., New York, NY 10017, or, more directly, from the IEEE Service Center (Order Dept.), 445 Hoes Lane, Piscataway, NJ 08854.

FCC

Until March, 1975, when an FCC order declared it obsolete, the FM facsimile broadcast standard adopted in 1948 to implement a "fax-in-the-home," radio-delivered newspaper service (see Chapter 1) had remained in the FCC's official *Rules and Regulations*—specifically Volume III, parts 73.266 and 73.318—as a valid broadcast service. What it had covered, in essence, was a multiplex subcarrier system (22 to 28 kHz) capable of delivering a letter-size page in about $3\frac{1}{2}$ minutes. Thanks largely to the efforts of broadcast pioneer John Porterfield, it has been replaced by a less specific and potentially more utilitarian standard adopted in February, 1975 as FCC Docket No. 20012, RM-1927.

Though not a standard per se (until the basic technical parameters are included in the amended *Rules and Regulations*), the guidelines in Docket 20012 have been officially in effect since the Docket's finalization in April, 1975. Its somewhat broad wording allows for subsidiary commercial FM broadcast services ranging from simple data transmission to fax and slow-scan TV. The special transmissions are on a subcarrier basis, separate from—but usually simultaneous with—the normal aural transmissions. All such subcarrier services rendered under the terms of the docket are regarded as *visual transmission*, which the FCC defines as "transmissions of a broadcast nature on a subcarrier modulated with a signal of such characteristics as to permit its employment, in receivers of suitable design, for visual presentation of the information so transmitted, e.g., on a viewing screen or a graphic record,"[2] It is further defined as covering the transmission of (for example) "photographs, drawings, printing and handwriting," as well as signals generated by teletypewriter or character generator terminals.

Although at this writing no specific technical standards have been applied to the new concept, there may eventually be special requirements regarding signal level, channel separation, and subcarrier bandwidth. The requirements may have to vary according to the specific type of visual transmission and whether the regular aural broadcasting is monophonic or stereo.

Implementation of a service within the scope of Docket 20012 can be as simple a matter as performing experimental transmissions, submitting the results to the FCC for approval, and receiving a license in return. Licenses for all such special broadcast services on a multiplex basis via commercial stations come under the general heading of Subsidiary Communications Authorizations—thus the term "SCA subcarrier," by which such services are usually designated.

Military

The principal standards for U.S. military fax systems are contained in Defense Department Document MIL-STD-188—specifically -118-100, "Common Long Haul & Tactical Communications System Technical Standards;" -188-260, "Equipment Technical Design Standards for Analog End Instruments and Digital Terminal Equipments;" and -188-347, "Equipment Technical Design Standards for Digital End Instruments & Ancillary Devices." Of the three, -260 is brand new and not yet published as of this writing (mid 1976), and -347 is undergoing revision to include digital fax systems. These three standards effectively supersede MIL-STD-188C and Defense Communications Agency (DCA) manual 330-175-1, the former having covered tactical communications and the latter having stressed global or long-haul systems. The new MIL-STD-188-100 covers both categories, with the present emphasis on meteorological systems, whereas -188-260 is more tactically oriented, and -188-347 concentrates on the long-haul aspect. Also covered in these standards, besides fax, are teletype, telephone, and various other types of electronic communication systems.

Responsibility for preparation of the military communications standards is shared by two U.S. Army units—the Electronics Command at Fort Monmouth, New Jersey, and the Communications-Electronics Engineering Installation Agency at Fort Huachuca, Arizona. Once finalized, all military standards are approved by the Department of Defense (DOD) as "mandatory" for use by all

STANDARDS　　　　　　　　　　　　　　　　　　　　　　　　　　　227

its departments and agencies. (In the development of new equipment, adherence to the standards is mandatory only to the extent that it does not inhibit advances in communications technology.)

Two basic classes of fax systems are covered by these standards: meteorological and general-purpose ("message") fax. As will be seen further on, except for the relative flexibility of certain parameters, the DOD specifications are generally compatible with the international standards recommendations of the World Meteorological Organization and the International Telecommunications Union. Those for general-purpose systems had been substantially in agreement with EIA Standard RS-328, but are currently undergoing extensive changes.

Following is a condensation of Subsection 5.7.3.1 of the present MIL-STD 188-100 (1975 revision), dealing specifically with meteorological systems:

Transmitter
　Copy Size, drum scanner: $18\frac{5}{8}''$ (473 mm) × 12" (305 mm)
　Copy Size, flat-bed scanner: $18\frac{5}{8}''$ (473 mm) × any length
　Scan Line Length (total): 18.85" (478.8 mm)
　Scan Direction: "normal" (corresponding to left-hand helix) [L-R]
　Dead Sector: 0.56" to 0.94" (3 to 5% of total scan)
　Scan Speeds: 60, 90, or 120 LPM, selectable by manual control
　Line Advance: $\frac{1}{96}$ inch (0.25 mm) [96 LPI]
　Index of Cooperation [CCITT]: 576 [IEEE: 1809]
　Signal Output Impedance: 600 Ω
　Carrier, AM: sinusoidal, 1800 Hz for 60 LPM; 2400 Hz for 90 & 120 LPM
　Carrier, FM: 1500 Hz for black; 2300 Hz for white
　Signal Contrast [and Signal Sense]: 20 dB, +2 dB; black *or* white transmission provided for on a selectable basis
　Signal Halftone Characteristic (voltage output/copy density relationship); nominally linear
　Synchronization: scan rate controlled by self-contained frequency standard maintained at assigned frequency (300 Hz, or integral multiple thereof) within 3 parts in 10^6
　Control Functions: (1) "start" command, (2) phasing signal, (3) "stop" command. (Phasing signal is black, interrupted by white, coincident with leading edge of dead sector.)

"Large Format" Receiver:
　Recorded Copy Size, drum recorder: $18\frac{5}{8}''$ × 12" (or integral multiples thereof)
　Recorded Copy Size, flat-bed recorder: $18\frac{5}{8}''$, 400-ft. roll
　Receiver Input Impedance: 600 Ω
　[Other specifications effectively coincide with those for the transmitter.]

"Small Format" Receiver:
  Recorder Type: continuous
  Resolution Capability: 200 LPI
  Recording Medium: stable, indefinite storage life, nonfading on exposure to light [This applies, as well, to large format receivers.]
  Recorded Copy Size: $8\frac{1}{2}''$, 400-ft. roll
  Recorded Line Length (total): 8.64" (220 mm)
  Recording Direction: "normal" [L-R] [This applies, as well, to large format receivers.]
  Scan Speeds: 60, 90, 120, or 180 LPM, selectable by manual control
  Line Advance: $\frac{1}{209.5}$ or $\frac{1}{96}$ inch, selectable by manual control [209.5 and 96 LPI, respectively]
  Index of Cooperation: 576 and 264 [CCITT] [IEEE: 1809 and 829, respectively] (264 will serve for charts transmitted at an index of 288. Dimensional distortion will be negligible.)
  Receiver Input Impedance: 600 Ω

The present thrust of new military fax standards—going beyond the current revision of existing standards—is toward the development of a single, highly flexible system that will employ one standard terminal to meet virtually all of the various graphic communication requirements of all branches of the service. Such a terminal is, in fact, currently under development as part of the so-called TRI-TAC program to standardize and improve all military tactical communications.

The prototype for the Tactical Digital Facsimile (TDF) transceiver, representing the fax aspect of TRI-TAC, is, at this writing, in the final stages of development by Litton-Datalog in compliance with a Naval Electronic Systems (NAVELEX) contract. It will use a helium-neon laser as the light source for both scanning and recording, and will employ digital redundancy reduction techniques to effect transmission speeds of less than 15 seconds per page via a choice of voiceband analog or 16 to 32 kb/s digital circuits. It will record on a dry silver medium—either paper *or film*, as required for the specific application. The production unit is expected to occupy less than three cubic feet of space and weigh less than 70 lbs.

In a related project, Litton-Datalog has developed a similar transceiver for the National Security Agency to handle the relaying of graphic intelligence data between intelligence and military command centers within the Washington, D.C., area. The several full-duplex transceivers built for the so-called *Washfax III* system employ some of the same technology as the TDF units, but are configured as consoles and produce their copies on plain, rather than dry silver, paper. They use the solid-state, charge-coupled principle for scanning (see Chapter 3), and operate in an encryption mode over a broadband digital message-switching network at a speed of about 6 seconds a page.

STANDARDS

Meanwhile, the Defense Communications Agency (DCA) is reportedly developing a digital fax standard that will apply to all government agencies. The project falls within a category of standards known as *Federal Telecommunication Standards*, for which the DCA is primarily responsible.

Weather

The effective dissemination of weather data obviously requires a measure of standardization. Accordingly, the U.S. National Weather Service has set the following basic criteria for its voice-grade fax nets:

Scan Speed: 120 LPM
Resolution: 96 LPI or 48 LPI
Index of Cooperation (CCITT): 576 for 96 LPI; 288 for 48 LPI
Signal Characteristics: double sideband AM, 2400-Hz carrier
Start Signal: carrier modulated by 800-Hz tone
Phasing Signal: Black signal interrupted by 12.5 msec white pulse twice per second prior to start of recording
Stop Signal: carrier modulated by 450-Hz tone

Note: In certain instances, speeds and resolutions higher or lower than those specified may be used. For example, where bandwidth compression is employed, the standard 120 LPM speed may be doubled to 240.

As for the reception of cloud cover pictures from weather satellites, there are no published standards per se, but there are certain basic specifications on satellite signals to which receiving equipment must obviously conform. These are as follows for the APT system, in which the signals can be received directly by anyone with suitable radio equipment:

Scan Speed: 240 and 48 LPM (the lower speed for nighttime infrared readouts)
Resolution: 100 LPI
Index of Cooperation (CCITT): 267
Input Signal Characteristics: 2400 Hz, AM
Start Signal: carrier modulated by 300-Hz tone

With the advent of new satellites whose transmissions are less widely accessible, some new standards have had to be established for transmission parameters and for terrestrial distribution of the received signals by the Weather Service. In the Geostationary Operational Environmental Satellite (GOES) program, for example, the pictures sent from the satellite are "stretched" to reduce bandwidth, and, upon reception at Wallops Island, Virginia, are relayed to Marlow Heights, Maryland, where they are automatically "gridded" and retransmitted via telephone circuits to selected cities. From there they are distributed by wire to end users (TV stations, newspapers, airports, etc.). The basic specifications for terrestrial transmission of the GOES photos are essentially the same as those previously cited for transmission of charts via phone circuits. The one exception is resolution, which normally ranges between 209 and 240 LPI, depending on the length of the recording stroke. For other satellite systems, scan speed may range as high as 400 LPM, and resolution as high as 456 LPI.

Standards for the international dissemination of weather data are controlled by the World Meteorological Organization, with headquarters in Geneva. These will be examined in the section on International Standards.

Microfilm

Although at present there are few instances where microfilm images are being transmitted by fax, the potential for "microfacsimile" is substantial, particularly in view of the steadily increasing pace with which paper documentation is being reduced to microfilm for space-saving and automated retrieval purposes. Recognizing this, a joint committee of the Electronic Industries Association and the National Micrographics Association (NMA) has prepared an industry standard spelling out some basic parameters for such systems. The standard, issued in 1972 by the NMA, is titled "Facsimile Transmission of Microfilmed Documents" (NMA Standard MS3), and covers two broad application areas: business documents and technical documents.

For the former, the parameters are such that a microfilm-input system designed to this standard will be compatible with paper fax systems complying with EIA Standard RS-328. The basic distinction is the relatively high scan resolution—90.7 scan lines per mm (2304 LPI)—required for the specified 24X linear-reduced input film images. It is, however, the equivalent of 96 LPI for the original document, and, in fact, the index of cooperation is the same (nominal 829) as that specified for the equivalent paper system in RS-328.

For technical documents, a higher scan resolution is specified: nominally 156 scan lines per mm directly on the microimage (equivalent to about 165 LPI on the original document). This naturally changes the index of cooperation—to 1426—but all other parameters are the same as those for the business document application.

Copies of the standard are available from the National Micrographics Association, 8728 Colesville Road, Silver Spring, MD 20910, and from the American National Standards Institute, 1430 Broadway, New York, NY 10018. When purchased from the latter organization, the standard bears the designation C16.45-1973.

INTERNATIONAL STANDARDS

Standards regarding international facsimile and telephotography (phototelegraphy) transmissions are chiefly the province of the International Telecommunications Union (ITU),[3] which was formed in 1932, in Madrid, as the successor to the International Telegraph Union. The latter had its origins in Paris and Berne in the late 1860s. The present ITU became a special agency of the United Nations in November 1947.

The specific organizations responsible for fax standards within the ITU are the Comité Consultatif International Télégraphique et Téléphonique (CCITT) and the Comité Consultatif International des Radio Communications (CCIR). Both are permanent organs of the parent organization, and both are, in turn, divided into study groups, one of which (for each organization) is concerned specifically with facsimile and telephotography.[4]

Another special UN agency of more recent origin, the World Meteorological Organization (WMO), is involved with international facsimile standards relating specifically to the exchange of weather data (maps, etc.). It has collaborated closely with the ITU in establishing standards for the format and speed of transmissions of weather charts by facsimile and for the remote-control signals employed in transmission and reception. The WMO was created directly as a UN agency in 1950.

The CCIR's standard for the transmission of weather maps by radio is confined to recommendations regarding frequency-shift limitations and transmission polarity in FM systems, in which respect it matches the WMO standards.

Standards activities of all three organizations—CCITT, CCIR, and WMO—consist of the studying of questions and the making of recommendations, as opposed to the establishment and enforcement of rigid standards (as is usually the case with national governments). The recommendations do, however, have the semblance of rules or directives, which, while not mandatory, are generally observed by all concerned. Copies of published ITU standards are available from the Sales Service of the ITU General Secretariat.

CCITT

The current fax standards of the Comité Consultatif International Télégraphique et Téléphonique ("International Telegraph and Telephone Consultative Committee") are contained in the committee's "Green Book," Volume VII, as "Series T" Recommendations: *Apparatus and Transmission for Facsimile Telegraphy*. Most of the basic technical standards are contained in recommendations T.1, T.2, and T.16, which are outlined below. Unless otherwise noted, the Recommendations reflect revisions made at the Fourth Plenary Assembly in Mar del Plata, Argentina, in 1968.

Recommendation T.1, "Standardization of Phototelegraph Apparatus"
1. Scanning track: Message area should be scanned in "negative" direction at transmitter; "negative" direction at receiver for "positive" reception; "positive" direction at receiver for "negative" reception. ["Negative" *direction* means right to left; "negative" *reception* means production of a photographic negative at the recorder, in which case a mirror image of the subject copy is desired for subsequent contact printing.]
2. Index of cooperation:
   Normal–352; preferred alternative–264 [These are equivalent to IEEE-defined indices of 1105 and 829, respectively. The International Index of Cooperation = drum diameter × scan lines per unit length in inches × LPI. The IEEE index is also known internationally as the *Factor* of Cooperation.]
3.1. Dimensions affecting drum scanning:
   Drum diameters–66, 70, and 88 mm [approx. 2.6, 2.75, and 3.5 inches, respectively]
   Width of "dead sector"–max. 15mm [approx. 0.6 inch]
3.2. Dimensions affecting flat-bed scanning:
   Total scan line lengths–207, 220, and 276 mm [approx. 8.15, 8.66, and 11 inches, respectively]
   Width of "dead sector"–max. 15 mm [approx. 0.6 inch]
3.3. Scanning densities:
   3, 3.77, 4, 5, and 5.3 lines/mm [approx. 76, 96, 100, 127, and 135 LPI, respectively]
5.1. Drum rotation speeds (normal):
   60 RPM [LPM] for index of cooperation 352
   90 RPM [LPM] for index of cooperation 264
7. Synchronization:
   (a) Standard frequency maintained with 5 parts in 10^6, or
   (b) a.c. sync tone of 1020 Hz transmitted to receiver [5 parts in 10^6 amounts to 0.5 parts in 100,000, as compared with 1 part in 100,000, as specified in EIA Standard RS-328.]
10.1. Modulation parameters–AM:
   Level of output greatest for white and least for black (ratio of nominal white to nominal black: 30 db)
   Carrier freq. for regular telephone lines: 1300 Hz
   Carrier freq. for conditioned telephone lines: 1900 Hz

10.2. Modulation parameters—FM:
mean [center] freq.—1900 Hz; white freq.—1500 Hz; black freq.—2300 Hz

Recommendation T.2, "Standardization of Black-and-White Facsimile Apparatus" [*Note:* Systems complying with this recommendation are loosely referred to internationally as "Group 1" systems.]
1. Scanning track: left to right; top to bottom ["positive"]
2. Index of cooperation: nominal—264 [IEEE 829]; or optional—176 [IEEE 553]
3. Dimensions:
   (a) width of "dead sector"—15 mm [approx. 0.6 inch]
   (b) total scan length for flat-bed scanners—215 mm (incl. "dead sector") [approx. 8.5 inches]
   (c) width of paper in continuous recorders—210 mm [approx. 8.25 inches]
4. Scanning density: normally 3.85 lines/mm [approx. 98 LPI]
5. Scanning-line frequency [Scan rate]: 180 LPM
6. Reproduction ratio [This item merely reaffirms the 264 index of cooperation, pointing out that no specific input/output size relationship is implied.]
7. Modulation:
   (a) AM—signal level higher for black and lower for white
       carrier freq. range: 1300-1900 Hz, depending on circuit characteristics
   (b) FM—
       mean freq. (f_o) for switched ckts.—1700 Hz
       mean freq. (f_o) for leased ckts.—1300-1900 Hz, depending on circuit characteristics
       freq. shift for black— f_o − 400 Hz
       freq. shift for white— f_o + 400 Hz
[*Note:* It was provisionally recommended in 1974 that the black and white frequencies be reversed.]

Recommendation T.16, "Facsimile Transmission of Meteorological Charts Over Radio Circuits" (CCIR, Oslo, 1966) [This is a combined CCIR and CCITT Recommendation.]
1. Center frequency— 1900 Hz
   Freq. corresponding to black—1500 Hz
   Freq. corresponding to white—2300 Hz
2. Frequency characteristics for direct FM transmission:
2.1. HF (decametric) circuits
   Center frequency (assigned freq.) − f_o

Freq. corresponding to black—   fo − 400 Hz
Freq. corresponding to white—   fo + 400 Hz
2.2. LF (kilometric) circuits
Center frequency (assigned freq.)— fo
Freq. corresponding to black—   fo − 150 Hz
Freq. corresponding to white—   fo + 150 Hz

Other recommendations in the T Series, none of which need be outlined here, are T.4, "Remote Control of Black-and-White Facsimile Apparatus" (covers control procedure—or "transmission sequence"—as explained in Chapter 4); T.10, "Black-and-White Transmissions on Telephone-Type Circuits Permanently Used for Facsimile Service (Leased Circuits)"; T.10 bis, "Black-and-White Facsimile Transmissions in the General Switched Telephone Network"; T.11, "Phototelegraph Transmissions on Telephone-Type Circuits" (Geneva, 1964); T.12, "Range of Phototelegraph Transmissions on a Telephone-Type Circuit" (Geneva, 1964); T.15, "Phototelegraph Transmission Over Combined Radio and Metallic Circuits."

Some of these repeat the same basic parameters given in the three recommendations outlined above. Others consist of details that go somewhat beyond the scope of this book. One of them, T.4, is currently undergoing a complete revision.

Recently finalized (but not published at this writing) are recommendations covering the so-called Group 2 fax system: in essence a two-speed (3 and 6 minutes per page) AM machine using a 2:1 analog bandwidth compression scheme to achieve the higher speed. Message frequencies will be specified in Recommendation T.3, and the "handshake" procedure in T.30. The recommended carrier frequency is 2100 Hz, and the message signalling scheme is to consist of a combination of AM/VSB (see Chapter 4) and phase modulation. Recommended scan rates are 180 and 360 LPM. In all other respects, the present technical standards listed above for Recommendation T.2 will apply.

Still in the offing—and probably one of the more pressing concerns of the fax Study Group at the moment—is a recommendation concerning digital data compression ("Group 3") fax systems. The first order of business is to determine what coding scheme is to be standardized upon (see Chapter 4, *Run-Length Coding*). The so-called Meyr code, developed by Switzerland's Hasler Research Labs, and a variation of the Huffman code have been among those considered. But, to the best of this writer's knowledge, as of mid 1977, no firm decision has been reached on this pivotal matter.

One remaining CCITT recommendation that bears mentioning is T.20, "Standardized Test Chart for Facsimile Transmission." In it the recommended inter-

Fig. 7.1. Second edition of CCITT facsimile test chart, per Recommendation T.20. (*Note:* This reproduction of the chart cannot be used for actual tests.) (Courtesy CCITT)

STANDARDS

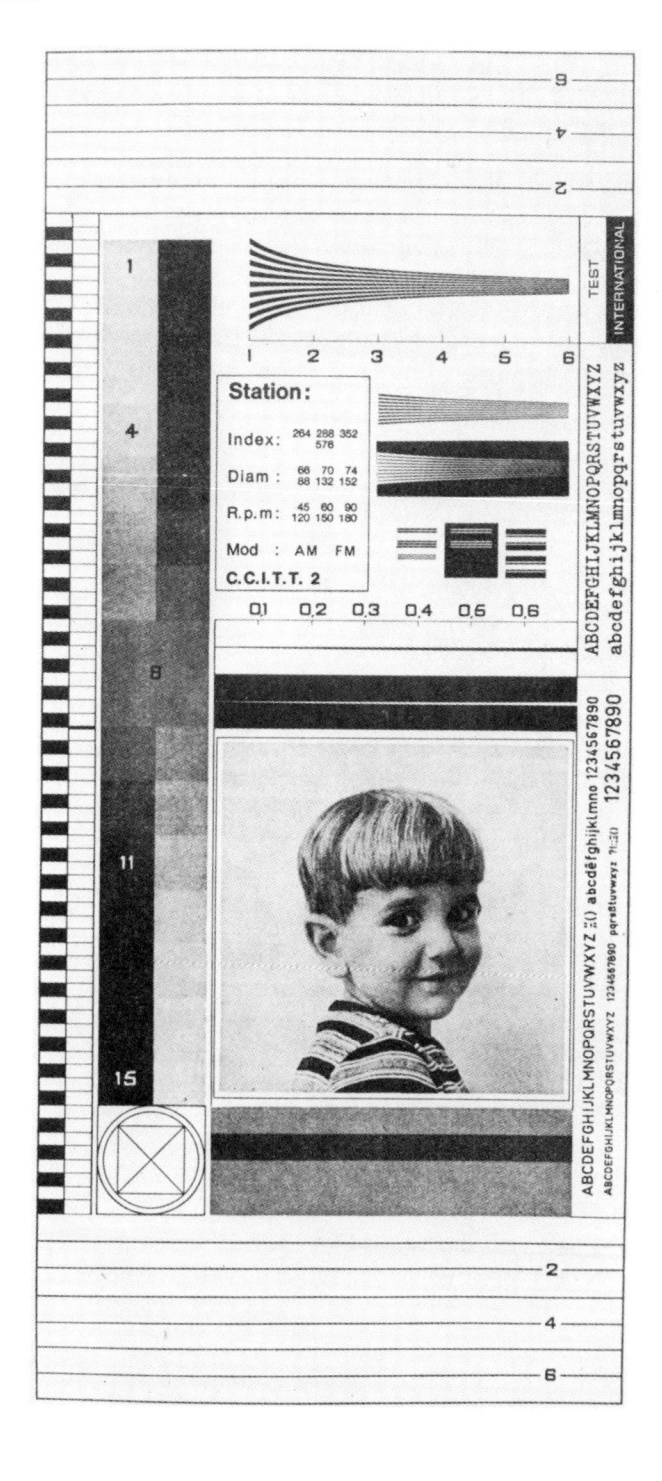

235

national test chart is described and reproduced (at a slight reduction). The actual chart measures 110 X 250 mm (approx. 4.3 X 10 inches) and contains the following test patterns, symbols, etc.:

- Two tone scales, each containing 15 density steps (ranging from black to white);
- Three wedge-type resolution patterns: black and white, gray and white, gray and black;
- Letters, digits, and punctuation marks in various styles;
- Photograph (portrait of an Argentine boy in latest edition); and
- Miscellaneous patterns.

The second (latest) edition of the chart is shown in Fig. 7.1.

CCIR

In addition to CCIR Recommendation 341-1 (corresponding to CCITT Recommendation T.16—see above), which covers meteorological chart transmission via radio, the Comité Consultatif International des Radio Communications ("International Radio Consultative Committee") has also issued Recommendation 344-2, "Standardization of Phototelegraph Systems for Use on Combined Radio and Metallic Circuits" (Plenary Assembly XIII, Geneva, 1974).

At the outset, the standard lists its purposes, among which are (in paraphrase): (a) to facilitate interworking of systems for long-distance telephotograph transmission via HF (decametric) radio circuits; (b) to standardize certain characteristics of the systems so as to make them equally suitable for use with wire circuits; and (c) to better secure the advantages of FM techniques by standardizing certain modulation parameters. The specific recommendations are

1. "that over the radio path ... the preferred method of transmission of half-tone pictures is by subcarrier frequency-modulation, of a single-sideband or independent-sideband emission with reduced carrier ...." [The specific recommendations for subcarrier FM are the same as in CCITT Recommendation T.1, item 10.2; and those for direct FM are the same as those in CCITT Recommendation T.16 and in CCIR Recommendation 343-1 (item 2.1 in both cases), but with the black and white frequencies reversed. In addition, 344.2 recommends that the lower frequency (corresponding to white) in each case also be used for phasing. It also specifies limits on frequency deviations within the various sections of a combined radio and wire circuit.]

STANDARDS

2. "that, for the present, the following alternative characteristics should be used:

	a	b
index of cooperation	352	264
speed of rotation of drum in r.p.m.	60	90/45

In due course, characteristic b will become obsolete."

3. "that frequency-modulation or amplitude-modulation may be used in the metallic portions of the combined circuit. When conversion from amplitude-modulation to frequency-modulation (or vice versa) is required, the conversion should be such that the deviation of the frequency-modulated carrier varies linearly with the amplitude-modulated carrier . . . ."

WMO

The World Meteorological Organization's recommended standards for the collection and exchange of meteorological data by facsimile are contained in **WMO** Publication No. 386, "Manual on the Global Telecommunication System," Vol. I, Part III, Sect. 7, excerpts from which follow:

7.1.1 Index of cooperation:
576, for min. black or white picture elements of 0.4 mm, [about 1.5 mils]; and 288, for min. picture elements of 0.7 mm [about 3 mils]. [These are equivalent to IEEE-defined indices of 1809 and 904, respectively.]

7.1.2 Drum speed:
60, 90, 120, 240 revolutions [or strokes] per minute

7.1.3 Diameter of drum:
152 mm [approx. 6 inches]. In the case of flat-bed scanners this will be the length of the scanning line (including the dead sector) divided by π.

7.1.4 Scanning density:

$$\text{Scanning density} = \frac{\text{Index of cooperation}}{\text{Diameter of drum}}$$

It is approximately:
4 lines per mm [about 100 **LPI**] for index 576;
2 lines per mm [about 50 **LPI**] for index 288

7.1.5 Length of the drum:
at least 55 cm [about 22 inches]

7.1.6 Direction of scanning:
at the transmitter, left to right, commencing in the left-hand corner at the top and finishing in the right-hand corner at the bottom

7.1.7 Dead sector:
4.5 percent ±0.5 percent of length of scan line. Signal transmitted during passage of "dead sector" should correspond to white, but a black pulse may be transmitted within, and not exceeding, one half the length of dead sector.

7.1.8 Synchronization:
Scanning speed should be maintained within 5 parts in 10^6 of its normal value.
[same as CCITT Recommendation T.1, item 7(a)]

7.2 Remote control signals transmitted:
[means for remotely controlling selection of index of cooperation, speed, starting of recorders, phasing, adjustment of recording levels, and stopping of recorders are specified]

7.3.1 Modulation characteristics:
(a) AM—Max. carrier amplitude should correspond to signal black. Value of carrier freq.: about 1800 Hz for 60, 90, and 120 RPM; about 2600 Hz for 240 RPM. *Note:* vestigial sideband (VSB) transmission to be used for 240 RPM.

(b) FM—Center freq.: about 1900 Hz; black freq.: 1500 Hz; white freq.: 2300 Hz

7.3.2 Levels of signals for AM:
Receiving equipment should accept any level between +5 db and −20 db, zero reference level corresponding to one milliwatt dissipated in a resistance of 600 ohms.

7.3.3 Contrast ratio: between 12 and 25 db (same for picture and control signals in any one transmission)

7.4 Halftone transmission [AM]: Spacings of 3 db should be observed [between discrete gray levels].

7.5.1 FM subcarrier transmission via radio circuits: [same as 7.3.1(b), above]

7.5.2 Frequency characteristics for direct FM—or Frequency-Shift Keying (FSK):
[same as CCITT Recommendation T.16, item 2]

In addition to these recommendations, the WMO has standardized on a meteorological facsimile test chart, the official description of which is contained in Part II of Publication No. 386, Attachment II-10. It is reproduced here as Fig. 7.2. The actual chart measures 153 × 449 mm (approximately 6 × 18 inches).

As can be seen in the illustration, the WMO chart contains a variety of resolu-

STANDARDS 239

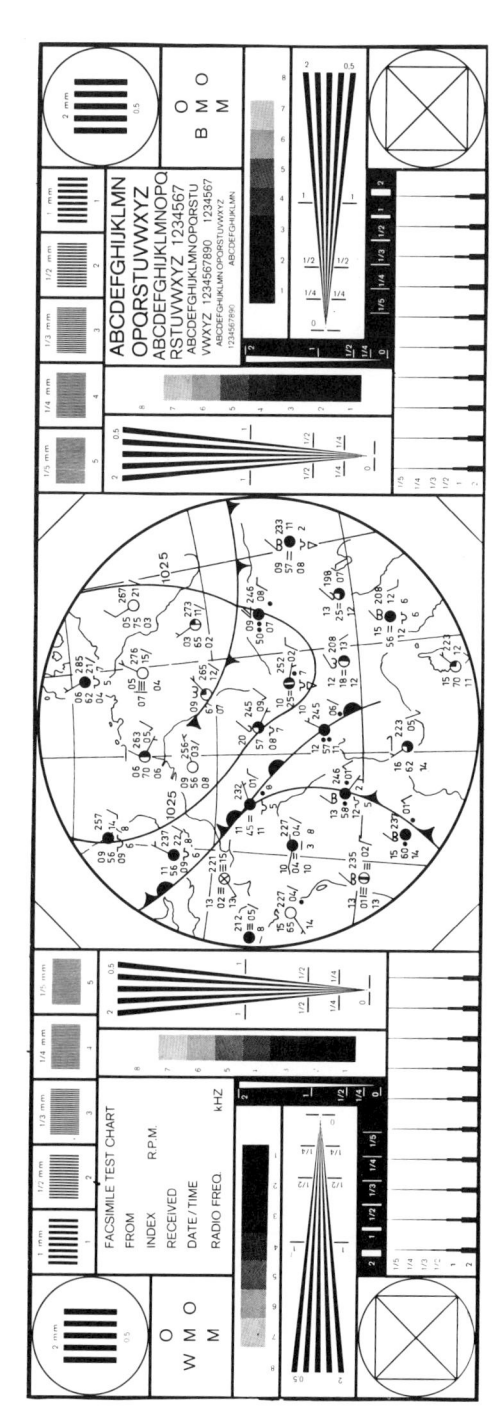

Fig. 7.2. World Meteorological Organization (WMO) Facsimile Test Chart (Courtesy WMO)

tion patterns. The range covered by the wedge patterns is 0.5 lines per mm (12.7 LPI) to 5.0 lines per mm (127 LPI). The "step-tapered" bars at the bottom of the chart range in thickness from 0.20 to 2 mm (approx. 0.008 to approx. 0.08 inch) and are used to assess the reproduction quality of separate lines. At the center of the chart is a representative segment of a typical meteorological chart.

It is likely that in the next few years, additional technical specifications will be adopted to permit more efficient transmission of weather maps through use of digital redundancy reduction techniques.

INTERSYSTEM COMPATIBILITY

In spite of all the effort that has gone into the development of domestic and international standards, certain incompatibilities have persisted, both within and across national boundaries. Within the U.S., for example, there is somewhat less than universal compatibility even among systems in the same specific category (FM analog business fax systems, for example).

Traditionally, next to the basic type of modulation (AM or FM), the most critical compatibility determinant has been scan rate: the number of strokes per minute. Signalling frequencies may differ slightly, but so long as the modulation scheme and scan rate are the same, intercommunication between terminals of different manufacture may still be possible without modification. Similarly, if the scan resolutions and/or scan-line lengths are all that differ—i.e., different indices of cooperation—there is still compatibility to the extent that one is willing to accept a geometrically imperfect copy of the transmitted original.

In recent years, however, some new causes of incompatibility have entered the picture. One is, of course, the advent of redundancy reduction systems—notably the digital systems for which the individual manufacturers have developed their own unique data compression techniques and their own proprietary compression algorithms. But even within the straight analog realm, there are differing transmission sequences, or control procedures, to contend with. As explained in Chapter 4, these are the automatic premessage and post-message exchanges of control signals by which the intercommunicating terminals "get to know one another," and, as pointed out in that chapter, there are some marked differences in procedure and in signalling frequencies between different makes of equipment.

Fortunately, there has been some inclination for compromise among the competing fax terminal vendors, and equally fortunately a design trend of recent years has been to put the electronic circuitry of a fax terminal on modular, interchangeable circuit boards. That trend has made it feasible for the vendor to make the engineering changes necessary to provide "interbrand" compatibility, at least on the basis of customer demand. Table 7.A illustrates the extent to which compatibility has been maintained, on a domestic basis, despite mounting technical pressures against it. The table is confined to analog transceivers intended for use on voiceband transmission circuits—notably the Direct Distance Dial (DDD) telephone network.

Where fax transmissions must cross national boundaries, one would naturally expect an even greater compatiblity problem. The economic problems alone in trying to achieve complete international compatibility of the many independent facsimile systems operating around the globe are formidable. In the case of the news agencies, for example, bringing a domestic system into line with international standards could mean the scrapping and replacement of millions of dollars' worth of existing equipment—a price that no private business is going to pay unless it is necessary for its survival.

But in spite of the incompatibilities that give rise to such dilemmas, there is, in fact, substantial agreement on most of the fundamental design parameters affecting international compatibility of fax systems. A review of the various standards outlined in this chapter will bear this out. Table 7.B summarizes the areas of general agreement.

Mention was just made of the high cost that would have to be incurred to bring various national newspicture networks into conformance with a single international standard. A less expensive way around that has been the design of somewhat elaborate "standards converters," through which the transmissions incoming from abroad are automatically reformatted for compatibility with the receiving country's terminal machines. The technological complexity of the process is illustrated by the block diagram of one such converter as shown in Fig. 7.3. This particular one took the brainpower of a team of MIT scientists[5] to develop.

Prior to the development of such sophisticated electronic converters, it had been necessary for the incoming pictures to be received on a machine meeting the sender's standards, and the processes picture then retransmitted via the local system. Needless to say, the procedure was costly, time-consuming, and detrimental to output quality.

At the root of the problem is the fact that there are differences in drum speed, scanning density, and direction of scanning between the U.S. and European systems. In the automatic converters (the first of which was built by the Shintron Company of Cambridge, Massachusetts, and began operation in New York City in January, 1968), these differences are resolved automatically by a combination of analog-to-digital conversion, buffering, and core storage. The electronics of such systems consist of thousands of integrated circuits.

Apart from its having simplified the interfacing of international and domestic systems—improving output quality in the process—the automatic standards con-

TABLE 7.A. Compatibility of Voiceband Analog Fax Transceivers

	DATAFAX 180 (S.W.)	DATAFAX 240 (S.W.)	DATAFAX 360 (S.W.)	GSI DEX I	GSI DEX 180, 181, 182	GSI DEX 580	GSI DEX 700	GSI DEX 4100	MAGNAFAX 850 (3M)	MAGNAFAX 854 (3M)	MAGNAFAX 856 (3M)	MAGNAFAX 860 (3M)	QWIP 1000	QWIP 1200	3M VRC 600	3M VRC 603	XEROX TC I	XEROX TC II	XEROX TC III	XEROX 400 TC	XEROX 400-I TC	XEROX TC 410	XEROX TC 200
DATAFAX 180 (S.W.)	X					6	6	6		6	6		6	6	6			6	6	6	6	6	6
DATAFAX 240 (S.W.)		X																					
DATAFAX 360 (S.W.)			X																				
GSI DEX I				X	6	6	6	6															
GSI DEX 180, 181, 182				6	X																		
GSI DEX 580	6			6					6		6	6		6	6	6	6	6	6	6	6	6	6
GSI DEX 700	6			6					6	4		6					6	6					
GSI DEX 4100	6			6					6	4		6					6	6					
MAGNAFAX 850 (3M)	6					6	6	6		6	6		6	6	6			6	6	6	6	6	6
MAGNAFAX 854 (3M)						4	4			4			4	4	4					4	4	4*	4*	4*
MAGNAFAX 856 (3M)	6					6			6	4		6					6	6					
MAGNAFAX 860 (3M)	6					6	6	6		6			6	6	6	6	6	6	6	6	6	6	6
QWIP 1000																							
QWIP 1200						6				6	4		6					6	6				
3M VRC 600						6				6	4		6					6	6			*	*
3M VRC 603						6				6	4		6					6	6			*	*
XEROX TC I	6					6	6	6		6	6		6	6	6			6	6	6	6	6	6
XEROX TC II	6					6	6	6		6	6		6	6	6			6	6	6	6	6	6
XEROX TC III	6					6				6	4		6					6	6				
XEROX 400 TC	6					6				6	4		6					6	6				
XEROX 400-I TC	6					6				6	4*	*	6			*		6	6				
XEROX TC 410	6					6		†		6	4*	*	6			*		6	6				
XEROX TC 200	6					6		†		6	4*	*	6		†			6	6				

Explanation of Symbols

X = Fully compatible.
4 = Compatible at 4-minute speed only.
6 = Compatible at 6-minute FM speed only.
⋋ = Compatible at more than one speed, but not all.
/ = Compatible (fully or partially), but only with special option.
BLANK = Not compatible.

Abbreviations

GSI = Graphic Sciences, Inc.
S.W. = Stewart Warner
TC = Telecopier

*One of the machines may require a special option for control signal compatibility under certain conditions.
†Under certain conditions, these machines may not be able to intercommunicate in a fully unattended mode.

Note: The term "compatible" implies that there will be no more than negligible geometric distortion in the received copy.

TABLE 7.B. Areas of General Agreement in International Standards

Item	General Purpose EIA	General Purpose CCITT	Meteorological DDD*	Meteorological CCITT	Meteorological WMO (&CCIR)†	Photo CCIR	Photo CCITT
Copy width (drum circum. or total scan length, in inches)	8.5–9.2	8.5	18.6/8.5	—	18.75	—	8.1/8.5/11
Scan (or recording) direction	—	L-R	L-R	—	L-R	—	L-R/R-L
Signal sense (freq. or amplitude, as applicable)	Max. black	Max. black	Max. black	Max. black	Max. black	—	Max. white
Index of cooperation IEEE: CCITT:	829 264	829/553 264/176	1809/829 576/264	— —	1809/904 576/288	1105/829 352/264	1105/829 352/264
Line feed (LPI)	96	98	96/209	—	50/100	—	76/96/100/127/135
Scan rate (LPM)	180	180	60/90 120	—	60/90 120/240	45/60 90	60/90
FM freq. swing (Hz)	—	1500–2300‡	1500–2300	1500–2300	1500–2300	1500–2300	1500–2300

*MIL-STD-188-100.
†where applicable.
‡at highest mean freq., leased circuits.

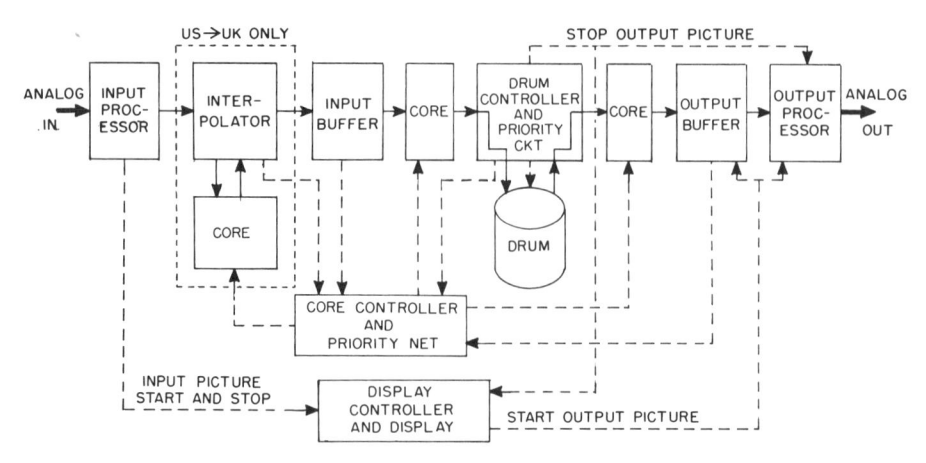

Fig. 7.3. Block diagram of an automatic standards converter for use in the international exchange of newspictures. Thin black lines denote digital signals, and dotted lines denote control signals. (Courtesy IEEE)

verter has also greatly enhanced the practicality of exchanging *color* pictures internationally. Separate transmissions must still be made for each primary color, but now only once for each, rather than twice as before. The separate recordings have to be superimposed to obtain a full-color reproduction, and in the old system of retransmission there was always the prospect of cumulative registration (superimposition) errors, which could considerably degrade reproduction quality.

The evolution of international standards satisfactory to all concerned is a long and tedious process. As noted, appreciable progress has been made, but there are still a few gaps to be closed.

FOOTNOTES

1. It was also known briefly as RTMA, Radio-Television Manufacturers' Association.
2. As published in the *Federal Register*, Vol. 40, No. 49 (March 12, 1975). Eventually to be covered in 73.310 of the *Rules and Regulations*.
3. Official name: Union Internationale des Télécommunications (UIT).
4. CCITT Study Group XIV, and CCIR Study Group 3.
5. Donald E. Troxel, William F. Schreiber, and Charles L. Seitz.

8
OTHER DOCUMENT AND GRAPHIC COMMUNICATION SYSTEMS

So far this book has dwelt upon facsimile communication (fax) with only occasional reference to other methods by which documents and graphics can be transmitted electronically. There *are*, of course, other methods, each quite effective in its own way—notably "handwriting machines," closed-circuit television (CCTV), slow-scan television (SSTV), and communicating word processors—and this chapter will briefly explore each of them. It will also provide a quick glimpse at some selected peripheral devices: plasma panels and other non-CRT electronic imaging devices, CRT printers, scan converters, video recorders, Optical Character Recognition (OCR) devices, and Computer Output Microfilming (COM) devices—all of which can be used in one way or another to help expedite the flow of documentation (of various sorts) electronically. Two other peripheral devices, image digitizers and computer output graphic recorders, will be reserved for brief discussion in Chapter 9, within the context of the merging of previously separate technologies.

Before getting down to descriptions, we must point out that we are dealing here with basically two categories of information transfer systems.

The first one, which can be dubbed *copying-type* systems, includes TV systems, fax, and handwriting machines, and is distinguished by the ability of the systems to produce faithful (or reasonably faithful) copies of the transmitted graphics. A further distinction is that regardless of the signal-processing steps involved and the transmission technology used, copying-type systems are generally *analog* by nature.

The second category might be termed *conversion-type* systems because the original material or information undergoes a conversion process that results in its being reproduced in a different form—or format—than that in which it originates. Predominant in this category are systems in which the original document is converted via a manual keyboard (e.g., a communicating word processor) or via a character recognition device (e.g., a communicating OCR system). Typically, in the first example, the actual original document—if, indeed, there is one[1]— would be a longhand draft; in the second, it would be a standard form typed in some specified font.

In both of these two basic system categories, the output may be either "soft" or "hard" copy, as required. "Soft" copy is a nontangible screen image such as the output of a TV system. "Hard" copy is tangible output, usually in the form of a sheet of paper containing the reproduced text or graphics. Addition of a hard copy capability to a TV system used to transmit document images has the effect of converting the TV system to a fax system. (The *form of output* is really the basic distinction between these two last-mentioned graphic communications media.)

But, more than faithfulness of copy or form of output, perhaps the important distinction to be made is between those systems that transmit *already existing, tangible documents*—ordinarily fax or TV systems of some sort—and those that, in effect, *create the received document from raw information*. In the latter case, no document per se need have existed prior to transmission of the affected information (hence the "paperless office" or "office of the future" we keep hearing about). Here too a fax or TV system may be used at the output end of the system to create a "document" from data transmitted from an intangible "memory bank" (located possibly at some central point in the system).

One type of system that does not exactly fit into either category in this last-mentioned method of classification is that in which handwriting machines are used as the communicating terminals. As will be explained below, these machines are unique in that although they are in the *copying* category along with fax and TV, they transmit the original document graphically *as it is being created*.

MANUAL GRAPHIC TERMINALS

The invention of the "handwriting telegraph"—or *telautograph*, as it was originally called—in the latter part of the last century is generally credited to Elisha Gray, an American inventor who had earlier come remarkably close to preceding Bell in the invention of the telephone.[2] The original system used a relatively simple arrangement of a pair of rheostats in the transmitter varying the current flow through an associated pair of electromagnets in the receiver. The rheostats were operated via mechanical linkages with a hand-held stylus. The movement of the operator's hand thus produced "x" and "y" current variations, which were communicated via separate wire loops to the associated electromagnets in the receiver. The moving armatures of the electromagnets were similarly linked to a stylus, arranged to record its movements on a piece of paper.

Servomechanical Devices

Modern handwriting machines continue to employ much the same principles as Gray's system, but with servomechanisms replacing the crude rheostats and electromagnets of the earlier units. The modern units also include solid state circuitry for amplification, modulation, and demodulation so that they can be operated via long-distance telephone circuits. The earlier devices were limited to operation over relatively short lengths of d.c. telegraph lines. Being analog by nature, the signals could not get beyond the first telegraph repeater.

Figure 8.1 depicts the basic scheme of a handwriting communication system.

These devices serve literally as an electrical extension of the operator's hand. Remarkably, a receiving unit in Tokyo can follow almost perfectly the hand movements of a person in New York *as the message is being written*. A handwriting receiver is quite fascinating to observe in action. It gives the impression of being operated locally by an invisible hand.

The fact that whole characters are being formed as they are written makes the handwriting machine a *real-time* device more nearly comparable to a telephone set than to a fax machine. Remember that whereas the telephone and handwriting machine messages are being transmitted and recorded *as* they are being spoken or written, the fax message has to have been fully prepared *prior* to its transmission.

This classification of handwriting terminals as real-time devices should perhaps be qualified. With an appropriate algorithm, it is entirely possible for such devices to perform computer input/output functions as well. For example, a person's signature could be input to a computer's memory, and a line printer could conceivably be adapted to reproduce it at the end of a computer output document.

At the present time, handwriting machines are being produced and marketed by at least four companies. The leading two are Telautograph Corporation of Los Angeles and Infolink, a Chicago firm that acquired the *Electrowriter** line from a division of Victor Comptometer in 1976. Telautograph calls its machines *Telescribers.** In addition, Talos Systems Incorporated of Scottsdale, Arizona, produces the *Telenote** system, and Feedback Instruments Ltd. of Great Britain offers a system called *Cygnet*. Figures 8.2 and 8.3 show typical examples of modern handwriting terminals.

All of these commercial systems use the same fundamental operating principles, the most significant design departure being the use of a coordinate contacting grid in some transmitters to permit "unshackling" of the writing stylus from its servomechanism linkages.

**Electrowriter, Telescriber,* and *Telenote* are registered trademarks of the producers of the machines bearing those names.

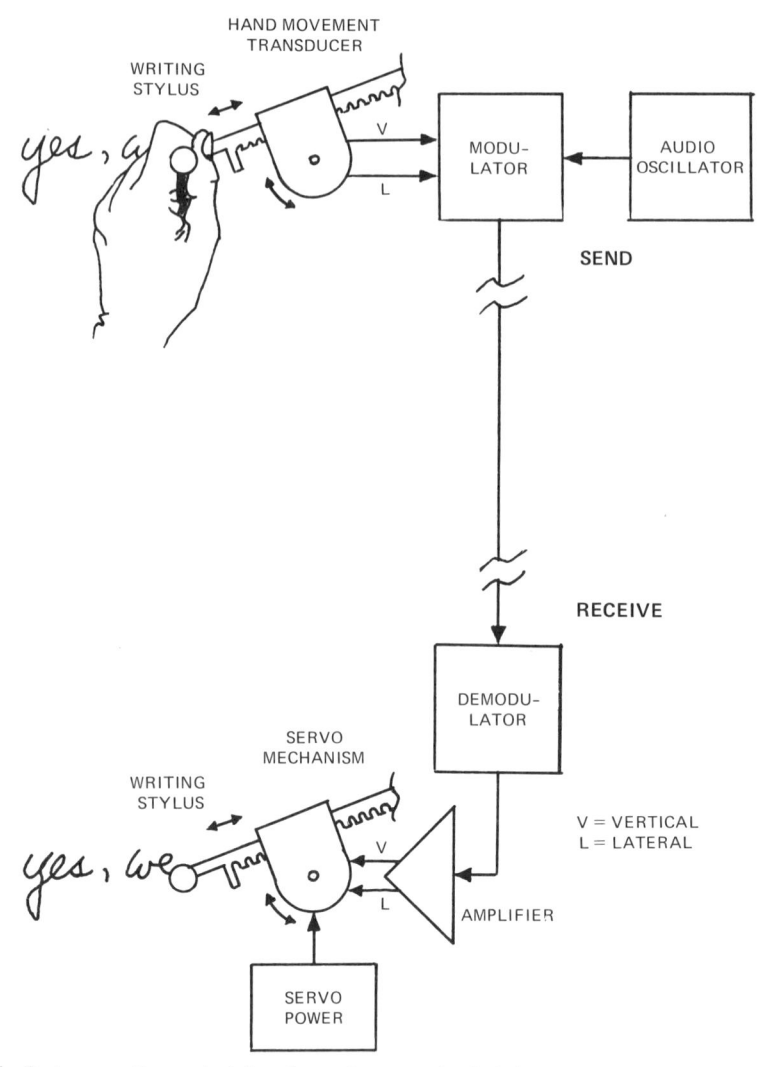

Fig. 8.1. Basic operating principle of an electromechanical handwriting communication system. The hand movement transducer in the transmitter controls the modulation of the audio carrier in such a way that the information for vertical and lateral movements can be separately transmitted. At the receiver, the incoming signal is modified and boosted as required to permit faithful reproduction of the transmitted message by the receiver's stylus-equipped servomechanism.

OTHER DOCUMENT AND GRAPHIC COMMUNICATION SYSTEMS

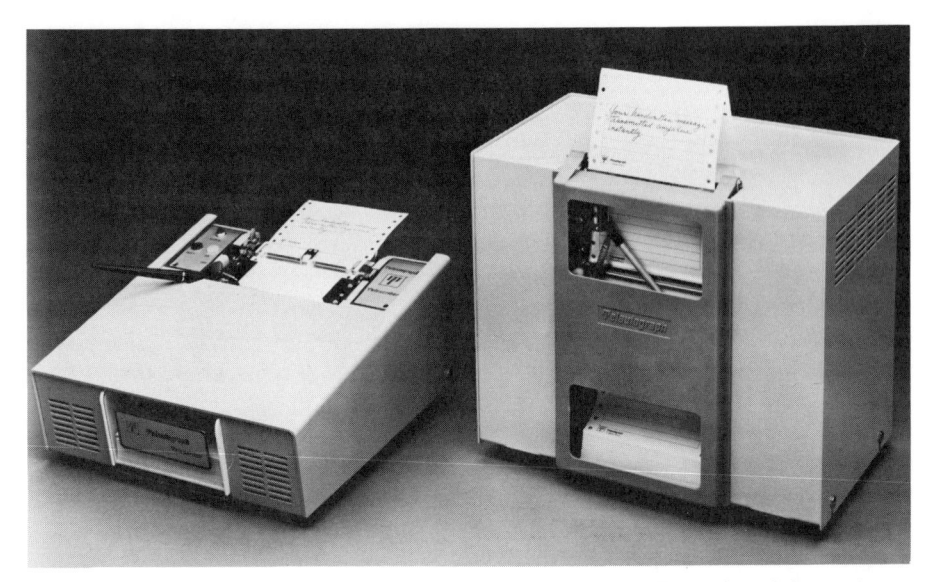

Fig. 8.2. A typical pair of commercial handwriting terminals. Transmitter left, receiver right. (Courtesy Telautograph Corp.)

Telautograph and Infolink also produce fax machines. One of them recently ran an interesting magazine advertisement showing how a handwritten request could be dispatched to a central document file via handwriting terminals, and the requested document dispatched back to the requestor via fax. In their own right, the handwriting terminals are used extensively for fire, police, and truck dispatching; coordination of air and rail terminal activities; production control (parts ordering, etc.) in factories; and various other rush dispatching applications.

A typical phone-coupled handwriting terminal costs between $1200 and $3000 to purchase, and between $35 and $70 a month to rent. This is roughly comparable to the cost of a basic fax terminal.

Electronic Systems

Close kin to the traditional handwriting terminal is a relatively new family of devices distinguished by the application of new technologies and the use of integral, *nonpaper* writing and recording media. This latter distinction would tend to remove such devices from the strict classification of document transmission. However, their similarity to handwriting machines in terms of general application and the fact that they *are* graphic communicating devices puts them sufficiently within the scope of this chapter to warrant a brief look.

Fig. 8.3. This handwriting receiver is equipped to project its received messages on a screen in a remote classroom environment. (Courtesy Talos Systems, Inc.)

Perhaps most representative of this new breed is the *Electronic Blackboard* developed by Bell Labs. The send terminal consists literally of a chalk board, but one of quite special design. The writing surface is backed by a special pressure-sensitive, coordinate-sensing element consisting of two resistive membranes. The instantaneous location of the chalk on the writing surface is thus sensed and translated into an electrical signal that is then tailored for transmission via phone lines. At the receiver, the signal undergoes a conversion that permits re-creation of the chalk movements on the face of a CRT. Erasing the received graphics is a simple matter of pressing a button at the transmitter. The writing surface is erased in the normal manner, using a conventional chalkboard eraser.

An earlier version of the Electronic Blackboard, introduced in 1971, used "strip" microphones in the blackboard to detect ultrasonic pulses generated by a small device attached to the chalk. The received graphics were recorded on special film and optically projected onto a screen. Figure 8.4 shows the Electronic Blackboard send and receive terminals as presently configured.

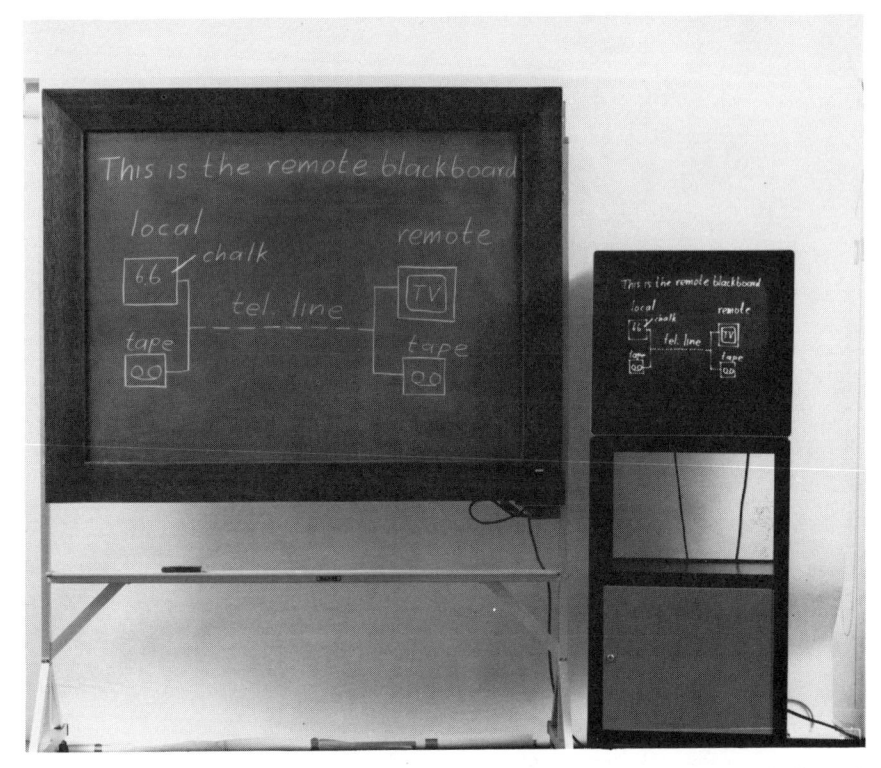

OTHER DOCUMENT AND GRAPHIC COMMUNICATION SYSTEMS

Fig. 8.4. Bell Labs "Electronic Blackboard." Easily mistaken for a conventional chalkboard. the send terminal (left) transmits its conventionally chalked "message" via phone lines to a CRT display (right), where it is reproduced in near real time. (Courtesy Bell Telephone Labs)

In a variation of the Electronic Blackboard concept, a laser beam "writing" arrangement and a special liquid crystal light valve are combined to form the image-reproducing mechanism in the receiver. The special light valve consists of a liquid crystal material sandwiched between conductors that are transparent to visible light but opaque to infrared. Wherever a laser beam impinges upon the valve, the temperature rises and the state of the crystalline material at that point is altered in a way that causes it to scatter any visible light passing through it. The images thus formed by the laser beam are reproducible by optical projection. They are subsequently erased by first raising, then lowering, the temperature of the whole valve in the presence of an intense electrical field. The scheme is illustrated in Fig. 8.5.

Similar systems have been developed using essentially the same principles, but

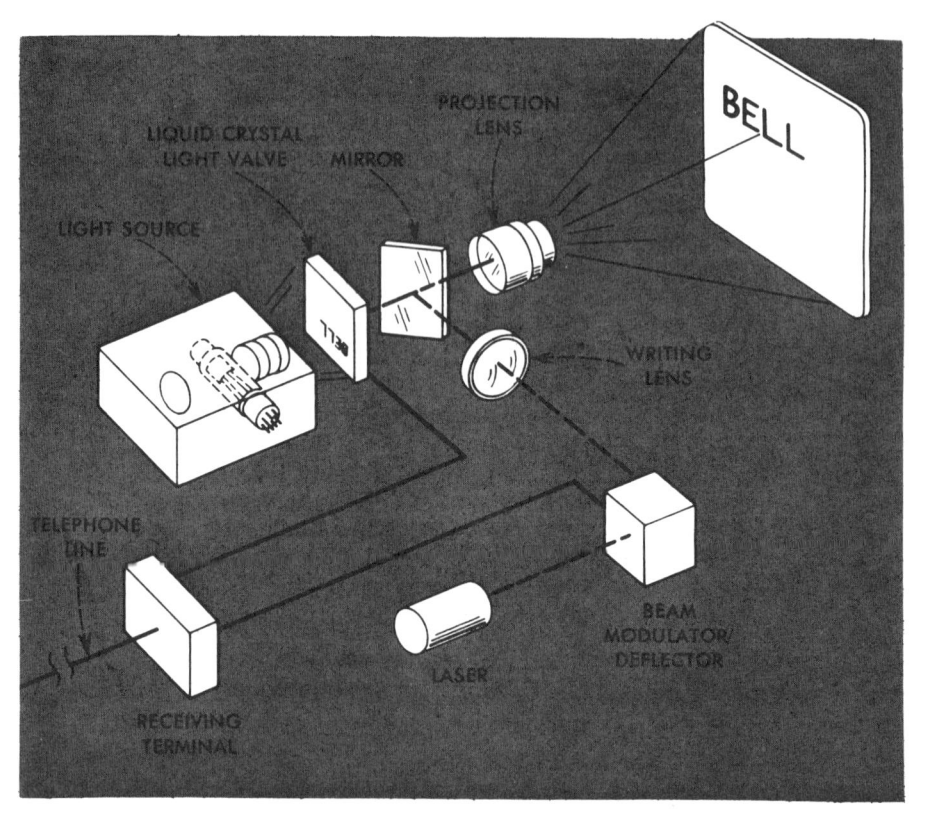

Fig. 8.5. Conversion of received graphics to a projected display via a special liquid crystal light valve. A modulated laser beam "writes" the graphic message on the crystalline material of the valve, temporarily altering its light-scattering properties so that the message can be reproduced on a screen by optical projection. (Courtesy Bell Telephone Labs)

with solid-state-imaging media. An example is *Cerampic*, a light-valve-imaging scheme developed in 1971 by Sandia Corporation of New Mexico. At the heart of the imaging device is a special ceramic material known as PLZT,[3] which, by application of a suitably strong electrical field, is able to retain an image by changing its refraction properties in proportion to variations in the intensity of the electrical potential applied across it. As in the liquid crystal device above, the PLZT ceramic is sandwiched between two transparent conducting surfaces. One of the two, however, is a photoconductor, so that when a strong electrical potential is applied across the whole device, its effect is greatest wherever light strikes the photoconducting surface, and the effect on refraction is therefore greatest at those points. The recorded image can be retained indefinitely simply by removing the potential, and can be erased by the simple expedient of reapplying the potential and flooding the ceramic material with light.

OTHER DOCUMENT AND GRAPHIC COMMUNICATION SYSTEMS

In a more recent, simpler version of the Cerampic device, a "sputtered" conductive surface is substituted for the photoconductive film surface previously used. Besides making the device easier and less expensive to fabricate, the new method facilitates reversal of image polarity by simply reversing the polarity of the applied d.c. potential.

In yet another variation of manual graphic communication, a device called a *plasma panel* is used for both transmission and reception. The device is named for an effect originated by a research group at the University of Illinois a few years ago, and it consists physically of a type of flat-screen gas discharge display unit, variations of which are commercially available from several manufacturers, notably Owens-Illinois, IBM, Burroughs, Control Data, and Fujitsu. The devices are sometimes used in place of CRTs for computer output display purposes, and are at the heart of some of the flat-screen TV sets that have been perennially promised by various vendors over the past ten years or so.

The plasma panel consists of tens of thousands of tiny neon gas cells sealed between two glass plates, each containing an array of linear, transparent conductors. The conductors in one plate are at right angles to those in the other, thus providing a field of x/y crosspoints, at each of which is one of the Ne cells. In normal operation as an image-recording device, a sustaining a.c. potential is applied across the entire unit, sufficient to hold all the cells just below the firing (ionization) threshold. Therefore, when the potential on the correct x/y pairs is momentarily increased, selected cells can be made to glow. As in a conventional neon-glow lamp, the cells thereafter continue to glow on succeeding cycles of the sustaining voltage because of the finite life of the "bridging" ions. Transmitted graphics can thus be reproduced as patterns of glowing gas cells.

The system can also be arranged so that all cells are normally glowing and are selectively extinguished to form a dark-on-light image. One way to accomplish this is by a complex phase-altering scheme through which the apparently extinguished cells are actually still blinking, but in such short pulses as to produce negligible illumination.

Used in a manual graphic transmitter, the panel can be arranged to react to the movement of a wired stylus by producing either an illuminated or dark trace of the stylus' path. At the same time, it produces an electrical output representing the series of coordinates constituting that path. This output signal is then amplified and formatted for transmission over phone lines in the usual manner.

At the receiver, an identical plasma panel reproduces the transmitted trace as described above. Figure 8.6 shows plasma panel image reproductions in each of two polarities.

VIDEO SYSTEMS

Because television was originally conceived—and is normally applied—as an images-in-motion medium, it has become indelibly stamped with that connotation.

Fig. 8.6. Negative (top) and positive (bottom) displays of the same graphic message on an a.c. plasma panel. The light portions of the images consist of selectively energized cells in which an inert gas has been made to glow by ionization. (Courtesy Bell Telephone Labs)

However, if we bear in mind that TV reproduces motion by transmitting a series of time-sequential still pictures in rapid succession, it is obvious that when the normal input is altered, the same system can be used to steadily transmit a given still image by repeating the same frame indefinitely (as with a test pattern, for example), or to transmit a rapid sequence of separate, unrelated frames not intended to re-create the illusion of motion. Moreover, the technical specifications of a TV system can be altered to permit the transmission of separate frames at relatively slow speeds consistent with available transmission bandwidth.

These various capabilities have made it possible to apply TV systems to the transmission of individual document images on either a *real-time* ("instantaneous") or *slow-scan* basis.

Real-Time Closed-Circuit TV

Instantaneous closed-circuit TV (CCTV) is used in some document retrieval systems as the means by which documents in an isolated central file can be accessed remotely. Its main drawback is its voracious appetite for transmission bandwidth. Fifteen to thirty MHz are required for a reasonably high resolution system (1000-1200 scan lines per frame). What this means in terms of transmission facilities is the need for coaxial cable (or perhaps a dedicated microwave link) with provision for signal boosting at frequent intervals along the transmission path—an expensive proposition for distances beyond a couple of miles. For that reason, the use of real-time CCTV is generally confined to very limited geographic areas, usually within a building or between buildings of a complex.

It is possible, as in fax, to reduce bandwidth requirements by application of sophisticated redundancy reduction techniques. But even with as much as a 300:1 data compression ratio (if that were possible), the bandwidth requirement would remain quite high. Bell Labs has developed special redundancy reduction techniques in connection with the Bell System's *Picturephone** visual telephone service, the ultimate objective being to reduce the present 1 MHz bandwidth to something more feasible for transmission via existing phone lines.

Another limitation of CCTV is that it normally delivers only a "soft copy" (screen image) of the transmitted document (which may, however, be adequate for reference purposes). It is possible, of course—at additional expense—to add a "hard copy" print capability to a CRT monitor.

A distinct advantage of instantaneous CCTV for remote document viewing is the practicality of including a *zoom* capability, remotely controllable from the receiver/display terminal. If portions of the original document contain especially fine detail or especially small characters, those portions can be zoomed in upon for easier reading. This capability permits the system's basic resolution requirements to be limited without fear of illegible reproduction. When character size is magnified optically at the camera (which is what zooming does) without altering the total scan lines per frame, the scan resolution is effectively increased.[4]

Similarly, if the document being viewed is too large to fit the screen—perhaps as a result of having been zoomed in upon to make it legible—a remote positioning control can permit selective viewing of a section at a time. Figure 8.7 shows a fairly typical remote-controlled CCTV receive terminal.

High-resolution TV cameras and monitors suitable for document transmission applications are available from a selection of firms specializing in TV apparatus—e.g. Cohu Incorporated of San Diego, GBC Corporation of New York City, and Doron Precision Systems of Binghamton, N.Y.—as well as from some of the industry's big names like G.E. Prices for a basic high-resolution (black and white) camera typically range between $3000 and $5000, and for a compatible passive monitor, between $500 and $1000. Provision for remote camera control can be expected to add at least $100 to the monitor price and several hundred dollars to the camera price.

Picturephone is a registered service mark of the AT&T Co.

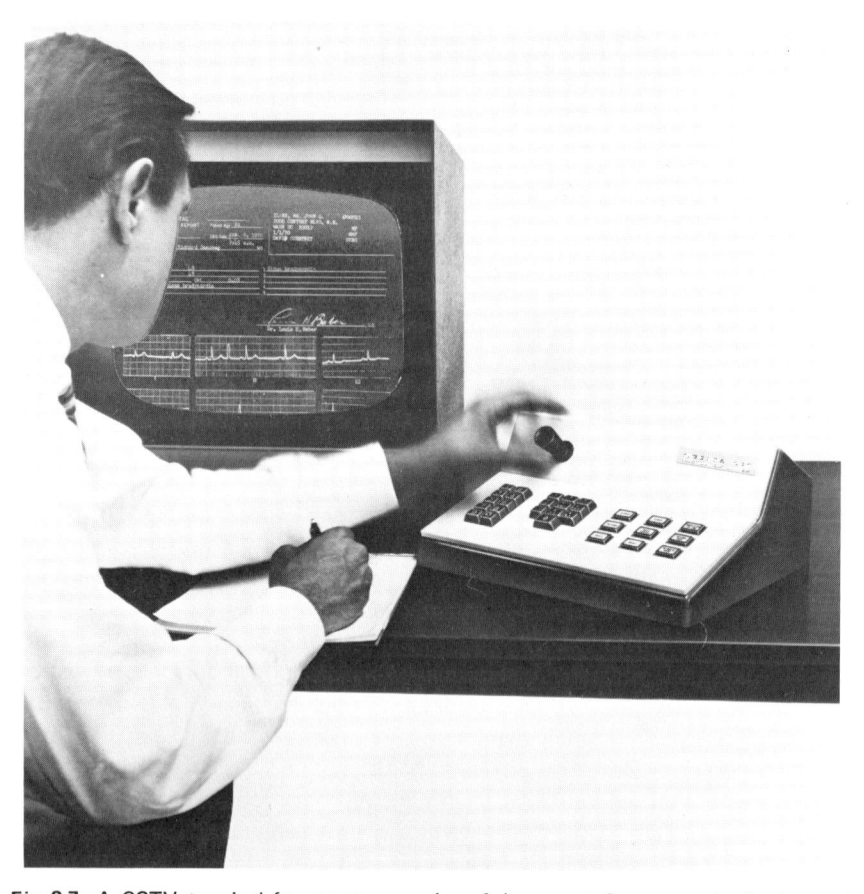

Fig. 8.7. A CCTV terminal for remote accessing of documents from a mechanized central file. Controls permit the user to focus the image, shift its position on the screen, zoom in on a particular portion, etc.

The camera price may or may not include the lens, or separate control unit or power supply where required. A lens alone may cost anywhere from twenty to several hundred dollars, depending on requirements—or as high as four figures where a special motorized unit is required for a remote zoom capability. Separate electronics can start at about $100 and go much higher, depending on what capabilities are provided and what proportion of the total electronics is already contained in the camera unit. Usually, however, the price quoted for the camera includes the control unit (if it is separate).

As already indicated, the addition of a remote-control capability to a CCTV system can increase the cost substantially. Such systems are, in fact, not very

OTHER DOCUMENT AND GRAPHIC COMMUNICATION SYSTEMS

widely available, especially in configurations suitable for document transmission. One of the few total systems currently being marketed for that purpose is the Microform Information Dissemination System (MIDS) produced by Photomatrix Corporation of Santa Monica, California. The transmit station of the system provides for semi-automated retrieval of specific pages contained on *microfiche*— 4 X 6-inch (105 X 148 mm) sheets of film, each containing as many as 98 micro-images—and transmission of the images to selectable display stations, from each of which the image-selection function of the transmit station can be remotely controlled. The system has a scan resolution of 1225 lines per frame and a direct-wire transmission range of 1000 feet (thus confining the total system within a fairly tight geographic radius). A typical configuration, consisting of four imaging units and a video distribution unit at the transmit station, plus eight remote-display stations, is currently priced at about $425,000.

Slow-Scan TV

By slowing down the system's scan rate, it is possible to send TV images over ordinary phone lines. The speed reduction has to be considerable: from the 15,750-lines-per-second rate that is standard for U.S. commercial TV down to perhaps 15 lines per second. Such slowed-down TV systems are known as *slow-scan* TV, or SSTV, and have been in use by radio amateurs on a limited basis for several years.

Besides the reduction in scan rate, an SSTV system differs from conventional TV in one other respect: The CRT in the receiver has to be able somehow to retain the slow-building image so that it can be viewed in its entirety. In practice this is done in one of three ways: (*a*) by use of a *long-persistence* CRT, such as is used in a radar monitor; (*b*) by use of a special, *storage-type* CRT; or (*c*) by somehow recording the received picture signals for slightly delayed readout, on a high-speed, continuous-recycling basis, into a conventional TV monitor.

Of the various methods of transmitting documents electronically, SSTV comes closest to fax in both technology and capability. By simply adding the means to record the CRT-reproduced images on a photosensitive medium, the SSTV system becomes, in fact, an all-electronic fax system. Unfortunately, the typical commercially available SSTV system is aimed at the amateur ("ham") radio market, and is not really designed for document transmission applications. The 128-line resolution that has been standardized by the amateur radio community provides substantially less legibility than the closest thing to a comparably priced commercial fax system. By any of the various rules cited in Chapter 5, even on something as small as a 4-inch-high monitor screen, 128 lines would be barely adequate to resolve a $\frac{1}{8}$-inch-high character.

Such SSTV systems are quite inexpensive. A basic one consisting of transmitter and receiver is obtainable for under $1000. The *Scanvision* system produced by SBE Linear Systems of Watsonville, California, and the Venus *Slo-Scan* system (Venus Scientific Incorporated, of Farmingdale, N.Y.) are typical examples. For $529[5] SBE offers a receiver complete with cassette recorder for delayed playback of the received video.

Further up the price scale is the *compressed video* system offered by Colorado Video Incorporated (CVI) of Boulder. Among other things, it permits transmission at somewhat higher resolutions over voiceband circuits at approximately a minute or less per frame, and the compression and expansion modules alone cost $8000[6] for a basic 2-point system. Even more expensive, at about $15,000, is the ED 6030B system produced by General Electrodynamics of Garland, Texas, the main feature of which is its speed and bandwidth flexibility. The frame frequency can be varied continuously from 0.02 per second (slightly less than a minute per frame) to 2 per second. The resolution is 600 lines per frame, and the system can be specially tailored to meet individual customer requirements.

"Videovoice," an SSTV service offered by RCA Global Communications at a rental of $250 a month per transceive terminal, sends a 525-line picture over voiceband circuits in 30 seconds. Like the previously mentioned CVI compressed video system, it processes the video through *scan converters* to permit use of more or less standard (real-time) CCTV cameras and monitors. (At this writing, it is rumored that the popularity of the service has not met with RCA's expectations and that the terminals are being sold at substantially reduced prices.)

Through use of scan converter tubes or solid state memory devices, the more elaborate SSTV systems—the CVI and RCA systems, for example—are able to apply electronic *frame-freezing* techniques to avoid having to slow down the camera and receiver scan rates to voiceband speeds. Such systems thus also avoid use of long-persistence CRTs with their characteristic fading away of the received images. Several companies market accessory devices that utilize frame-freezing techniques to capture images from TV monitors and CRT character displays and reproduce them as paper prints. The 4632 Video Hard Copy Unit produced by Tektronix Incorporated of Beaverton, Oregon, and the video-adaptable printer/plotters produced by Versatec Incorporated of Santa Clara, California, are typical examples of such devices. With a simple hook-up to a compatible terminal, the press of a button will print out the frozen frame in a matter of seconds.

A hybrid version of this type of device is the Frame Grabber produced (in part) and marketed by Alden Electronic & Impulse Recording Equipment Company of Westborough, Mass. A complete stand-alone terminal is pictured in Fig. 8.8. The two units constituting the Frame Grabber are a commercial scan converter (Hughes, PEP, or other) and an Alden 600 fax recorder.

Systems of this general variety—i.e., accessory video page printers—are avail-

OTHER DOCUMENT AND GRAPHIC COMMUNICATION SYSTEMS

Fig. 8.8. TV "Frame Grabber" permits a CRT image to be frozen and printed out in a matter of seconds on a fax recorder. (Courtesy Alden Electronics)

able in high-resolution configurations (upwards of 1000 lines per frame) suitable for some document reception applications, and can be expected to cost between $4000 and $5000 each.

The application of frame-freezing techniques also makes possible the sloweddown transmission of *color* images, using conventional color TV cameras and receivers. One such commercially available system is the Nippon Electric DFP-751 "Color Freeze" Picture Transmission System, which uses digital pulse code modulation (PCM) and data compression techniques to optimize transmission efficiency. Instead of a scan converter tube, a digital IC memory device is used to store the digitized video for reading out at speeds consistent with the bit rate of the transmission circuit employed. Compared with the more conventional black-and-white systems, this one is relatively expensive—between $15,000 and $19,000 per unidirectional (simplex) terminal for the special electronics alone. With a reasonably priced color camera (est. $20,000), a send-only terminal would cost in the neighborhood of $40,000.

Within the document transmission realm, a color system of this sort might find application in situations where it is necessary to transmit charts or other graphics containing significant color.

Video Storage and Remote Retrieval

A system was described above in which document images on microfilm are remotely—and semi-automatically—retrievable via CCTV. Television lends itself well to such systems because it permits the retrieved document to remain stationary and in position for refiling while being scanned for transmission.

But video technology can satisfy the *storage* as well as transmission requirements of an electronic document retrieval and delivery system. Separate individual TV frames can be recorded sequentially on videotape or video discs, and a system can be arranged to retrieve and transmit a given document image in response to a search code keyed in at a remote receive terminal. Several thousand separate document pages can be recorded on a single roll of tape, and "freeze" techniques can be applied to the display of a given retrieved frame on a CRT monitor. At least one such system—the Ampex *Videofile** system—has been commercially available for several years.

A variation of video recording in which the inputting device is some form of image digitizer—a digital fax scanner perhaps—has been applied in a handful of commercial systems, known variously as electronic records management systems or electronic filing systems (notably Trans-a-File, Precision Instrument RMS-180, and Infodetics 2000). The paper originals are fed into the digitizer, whose output is merged with a manually keyed-in identifier. The combined digital data is recorded by magnetic or other means in a secure storage "vault" that is remotely accessible via standard communication circuits. The PI RMS-180 system (Precision Instrument, Santa Clara, California) uses as its recording medium a special rhodium-coated mylar strip into which a modulated laser beam burns microscopic "holes" representing the bits of the encoded digitizer output. Documents are retrieved by manual keying of identifier codes and delivered either as CRT images or paper copies.

Conceivably, the new video disc systems developed by Philips, RCA, Telefunken and others, can be similarly applied. The discs would be divided into thousands of separate bands, each containing a different document page. The disc would lend itself to a high-speed, automated random access retrieval scheme by which a given document could be retrieved almost instantly.

---

*Videofile is a registered trademark of the Ampex Corp.

OTHER DOCUMENT AND GRAPHIC COMMUNICATION SYSTEMS

COMMUNICATING WORD PROCESSORS

Word processing (WP), one of the more recent innovations in office automation, involves a whole range of relatively new capabilities—notably automated, high-speed typing, computer-aided editing, and automatic recording and storage of typed matter for later retrieval and/or revision. It also makes feasible certain tasks that were previously impractical for office typing. One example is the automated justifying of pages of text (the adjustment of spacing to make all text lines equal in length). Some of the word-processing functions, such as automated editing and justification, had formerly been among the exclusive functions of professional typesetting operations.

Inasmuch as word processing requires the typewriter to output an encoded electrical signal for character recording, the first requirement for electronic communication between geographically separated points has been met. Consequently, with the relatively simple addition of amplification, modulation, and "handshaking" functions, a WP terminal can double as a "record" communications terminal. Because of the relative ease of achieving this added capability, several such systems have been developed commercially, and a relatively small but growing number of them are in operation.

A good current example of a communicating word-processing (CWP) system that is effectively an "electronic mail" operation in the truest sense is one offered by Wiltek Incorporated of Norwalk, Connecticut. At the heart of the system is a compact "work station" consisting of an executive electric typewriter with full editing capabilities made possible by a built-in memory module (see Fig. 8.9). The memory module also serves to store the typed documents for later retrieval.

For the communications function, the work station hooks up to a computer-controlled communication service provided separately by the same vendor. At predetermined intervals the computer automatically "polls" each work station terminal in the system, instructing it, in effect, to "dump" the contents of its memory module into the transmission stream, where they are automatically routed to the appropriate receive terminal. The receiving work station automatically types out the received "mail" at 450 words a minute. Where a given document is intended for multiple recipients, the communication system automatically distributes it to all addressed terminals, and it also automatically confirms to the send station that the documents have been properly delivered. Urgent messages can circumvent the predetermined pick-up/delivery cycle and be sent immediately, any time, while low-priority messages can be held off for lower-cost overnight transmission.

The work station can also function as a *TWX* or *Telex* terminal for more far-

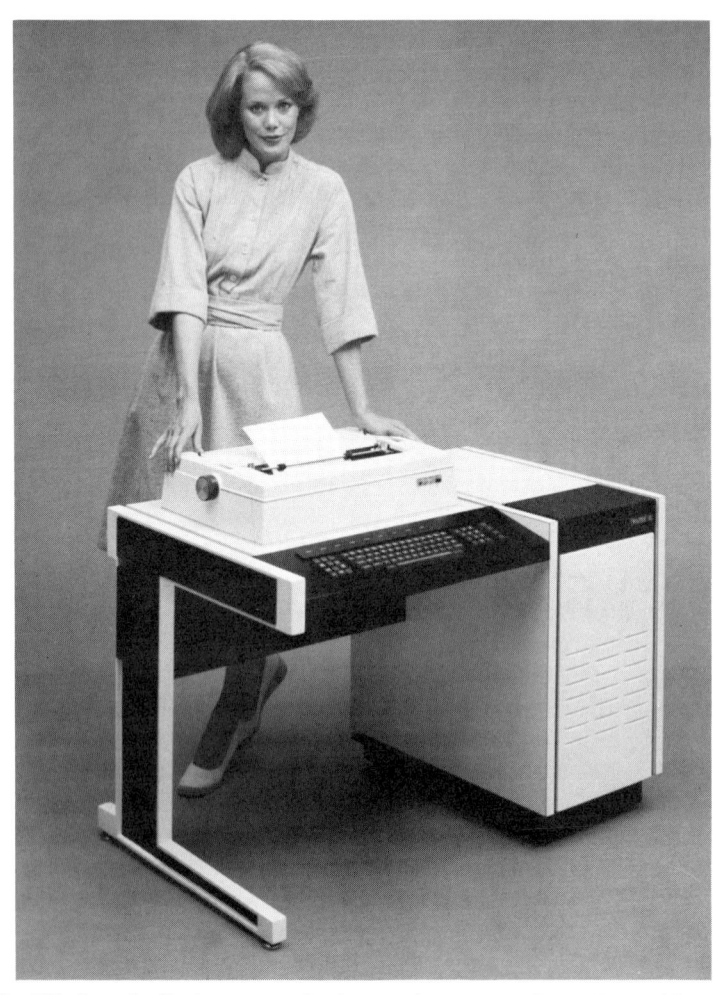

Fig. 8.9. "Work station" of a communicating word processor. It can be used for ordinary typing chores, or, through its built-in memory module (box at the right), can store the manual keystroke input for electronic modification and subsequent distribution of the finished document via automatic "electronic mail." (Courtesy Wiltek, Inc.)

reaching communication. (There are, at present, some 150,000 such terminals reachable via common-carrier circuits in North America alone.) Or it can be totally isolated, the typewriter portion being used in the conventional manner for everyday typing chores.

There is also a security aspect to the system, whereby a recipient may be required to enter an identity code in order to receive confidential messages.

OTHER DOCUMENT AND GRAPHIC COMMUNICATION SYSTEMS

Similarly, key personnel are assigned an "electronic signature" to authenticate the messages they send. The signature is a digital code produced by merely entering the sender's name. A built-in algorithm recognizes the name and makes the conversion automatically.

By nature the system is more secure than conventional message-handling procedures in that the messages need pass through fewer hands between the originating point and their ultimate destinations.

As previously pointed out, such systems differ basically from *most* copying-type systems—notably fax and TV—in that the normally separate processes of document generation and distribution are intimately enmeshed. Moreover, there is no graphic capability per se (although one could conceivably be added). The types of documentation handled are confined pretty much to text and numerics.

Each work station in a system such as that described above can be expected to cost about $300 a month on a rental basis. It is difficult to put a price tag on transmission because of the infinite variety of possible network configurations and message volumes. But assuming use of WATS (Wide Area Telephone Service) circuits in a network of some 35 stations, and a message volume of about 50 a day per terminal (each message averaging 500 characters), the cost per message can be expected to average between 30 and 40 cents.

Among the larger firms currently active in word processing are IBM, Xerox, 3M, and Burroughs. The fact that the last three happen also to be leading suppliers of fax terminals is significant. All three are inclined to put the two technologies under the same organizational umbrella in apparent recognition of an imminent interrelationship. In fact, all four of the aforementioned firms already offer WP terminals with built-in communications interfaces, as do a number of other vendors (Base, Honeywell, Vydec, and Wang, to give just a sampling).

MISCELLANEOUS SCHEMES

As was noted at the outset of this chapter, some document and graphic communication systems transmit *and reproduce* the information in a form different from that in which it originated. One such system was just described: the communicating word processor. There are various other conversion-type approaches to record communications, some of which will be reserved for discussion in the next chapter. There are two, however, that warrant a brief look here moving on to the future. One is optical character recognition (OCR), and the other is computer output microfilming (COM).

OCR

Automatic recognition of printed (and to some extent written) characters has been a topic of serious discussion within the information-processing sphere for at least 20 years. Indeed, the past decade has witnessed some encouraging technological advances. There are today OCR systems that can recognize characters in a limited number of fonts at split-second speeds and with an impressive degree of reliability. What this means to record communications is that transmission redundancy can be reduced to a practical minimum with near real-time speed and without need for manual translation of the input by a keyboard operator.

This basic advantage becomes clear when one compares OCR with the redundancy reduction techniques currently used in digital fax systems. While the latter are able to preserve every detail of the original document—signatures and pictures, for example—they do so at the expense of having to generate a separate digital code for each significant tonal transition encountered in the multi-line segmenting of a character or symbol. OCR, by contrast, confines all of that analysis data within the reading and recognition units, the combination of which outputs only one code representing the interpreted character or symbol *as a single entity*.

One example of a communicating OCR system is the *Alpha* terminal produced by Compuscan Incorporated of Teterboro, N.J. Its normal input is a form containing information typed in either of two switch-selectable fonts. The relatively cryptic information is read by the character scanner and interpreted by the system's built-in logic as an instruction to produce a complete output document of one sort or another. The output document may be produced locally or on a remote terminal (or on several remote terminals simultaneously) via communication links at rates up to 9600 bits per second. Transmission may be direct or on a store-and-forward basis. An on-line keyboard is provided on the terminal for editing or making corrections—or simply to permit use of the *Alpha* as a conventional teletypewriter terminal. A single self-contained *Alpha* terminal is priced at about $30,000.

Another example is the *OCR/FAX* system designed by Dest Data Corporation of Sunnyvale, California. As the name implies, it combines in a single terminal unit the efficiency of OCR and the flexibility of fax, either mode of operation being manually selectable according to the nature of the input. The complete input document is scanned in as little as $5\frac{1}{2}$ seconds (about 1000 characters per second) regardless of the selected mode. The remote receiver automatically switches to the proper mode, producing the OCR-processed copy in real time and the fax copy at approximately $\frac{1}{3}$ the speed (3X the duration) of the transmitter throughput. Digital data compression techniques are employed in both modes for maximum transmission efficiency. A system costs somewhere between $30,000 and $50,000 to purchase, depending on the particular configuration.

A similar system is being readied for commercial marketing by Stewart-Warner Electronics of Chicago. Besides character recognition and high-speed fax, other

OTHER DOCUMENT AND GRAPHIC COMMUNICATION SYSTEMS 265

capabilities to be offered by this newest *Datafax* system are pattern recognition, store-and-forwarding, and transmission of a page of copy in as little as two seconds via broadband circuits.

Still another example of a communicating OCR system is the *Autoreader* produced by the ECRM Company of Cambridge, Massachusetts, a primary application of which is the conversion of unformatted typed copy to a teletypesetting (TTS) tape format for transmission to a remote printing site. There the received signals control automatic typesetting equipment for the production of printing plates. Various models of the *Autoreader* (for various applications) are available at purchase prices ranging from $16,000 to about $65,000 each.

As for the current state of the OCR art in general, there are several systems commercially available that are capable of reading and interpreting hand-printed characters and mixtures of numbers, letters, and punctuation marks with a fair degree of accuracy. Most systems are arranged to read only one particular font at a time—and possibly only uppercase characters—but are "switchable" within a range of standard fonts. There are a few highly sophisticated (and correspondingly expensive) systems that can automatically read a wide variety of *intermixed* fonts. Input is normally paper containing typed or ink-printed characters, but there are a couple of systems available that accept microfilm as well.

Two of the main drawbacks of OCR in its present state are (1) the high cost resulting from its great complexity, and (2) the fact that even the most elaborate and expensive systems cannot be relied upon consistently to interpret characters correctly.

COM

Computer output microfilming, like OCR, is generally viewed as a localized and peripheral computer operation. In its normal environment, a COM device does essentially the same job as a line printer, except that it records on photographic film rather than paper and is somewhat more versatile. But more pertinent to the present discussion is the fact that like OCR, a COM device has the capability of functioning as one end of a record communications system.

The principle of COM is fairly simple: Digital data output from a computer or tape reader—or from a remote transmitter—is automatically reformatted to be reproduced as characters and lines on the face of a CRT. A microfilm camera aimed at the CRT has its shutter held open while a series of characters and symbols momentarily appear on the screen. When a "frame" of data is completed, the film is advanced to record the next frame, and so on. In a typical situation, the completed strip of film must be removed under darkroom conditions and chemically processed to develop the latent images. The developed film can then be inserted into a viewer for temporary reference, or in an enlarger-printer for reproduction of selected frames as permanent, eye-readable paper copies.

There are COM devices available capable of producing elaborate graphics as well as alphanumerics, and most devices are capable of mixing CRT images with fixed, optically imaged form designs. The advantage of the film output is that it saves both paper and space, thereby attacking the growing problem of wasted resources on two fronts simultaneously.

As the receive end of a communications system, COM fits nicely into a store-and-forward arrangement in which incoming data is recorded for delayed readout onto film whenever the COM device is idle. The devices are expensive— $30,000 to $400,000 each—and considering the number that are already installed in computation centers all over the country (some 2000 at last count, and steadily growing), such a scheme would seem to make good sense.

A cursory analysis of the comparative costs and capabilities of the systems surveyed in this chapter should serve to explain why fax, despite its age, continues to reign as the dominant medium for electronic dissemination of documents and graphics. This chapter also reveals, however, that there is sufficient potential competition from newer, more sophisticated technologies to force some kind of technological evolution—perhaps a synthesis of the best that all the technologies have to offer.

The evolution has already begun in the blending of fax and computer technologies to improve the efficiency of record communications. The next chapter will attempt to make some projections of this and other emerging trends in an effort to determine what is ahead for this increasingly important branch of the communication arts.

FOOTNOTES

1. A voice on a reusable tape from a dictation machine is another typical source.
2. Gray also founded the original Western Electric Company, which was later to become the manufacturing arm of the Bell Telephone System.
3. PLZT: Pb (lead), La (lanthium), Zr (zirconium), Ti (titanium).
4. Despite this flexibility, it is generally preferred that a TV system used for document transmission have a somewhat higher raster resolution than commercial TV (e.g., 800–1000 scan lines per frame rather than 525).
5. Price quoted as of May, 1976.
6. Price quoted as of May, 1976.

9
THE FUTURE

Having examined virtually every facet of the art of dispatching documents and graphics electronically, we can now round out the picture by looking ahead to what may be in store for it in the next decade or so. A knowledge of future possibilities is always helpful in testing the validity of present decisions.

In Chapter 2 we looked at some current trends, and in Chapter 6 did some speculating on the future of facsimile communication from an economic standpoint. This chapter will focus on technological advances and the merging of previously separate technologies. It will also look at some prospective new application areas and the possible reshaping of the way in which fax and related graphic and document communication systems are applied.

TRADITIONAL ROLE; NEW TRENDS

There is no question but that the role of facsimile communication as a business convenience will endure—even expand—for some time to come. The high probability of rapidly declining costs for basic fax terminals as new technologies begin to have their impact not only ensures increased acceptance of such devices in the business and professional communities, but also makes it increasingly likely that they will eventually find their way into the home (as will be discussed in more detail further on). Moreover, most such devices will continue to be telephone-coupled and, like the telephone instrument itself, will remain within the analog realm. Voice communication and low-cost convenience fax terminals are, in fact, likely to be among the last electronic communication devices to output or accept anything other than a digital signal.

On the other hand, fax systems for more specialized applications—fingerprint and newspaper page-proof systems, for example—are likely to change drastically in the basic technologies they employ and in the way in which the systems are applied. It has already begun to happen in large measure. The typical high-speed business fax transmitter, for example, is no longer merely an analog device that has been sped up to operate over wideband circuits. It is instead a scanner connected to a minicomputer (or, perhaps more accurately, a microprocessor), which analyzes the scanner output and converts it to an encoded digital signal for more efficient transmission over voiceband lines. Similarly, the typical modern weather chart system relegates the fax scanner to the role of a computer-inputting device, the actual on-line transmitter being a mainframe computer that merges the fax-input graphics with current data and then broadcasts the resulting fax-compatible signal to conventional fax receivers.

The time is fast approaching when it will no longer be valid to refer to fax per se. Instead we should perhaps begin thinking in terms of *graphic* versus *alphanumeric* communication—or, in a broader context, *total data communication.*

Systems capable of handling both graphic and alphanumeric materials with equal facility have already been developed by various communications equipment manufacturers, notably Alden, Dacom, Dest Data, and Stewart-Warner. Some of these were discussed in previous chapters.

As for transmission media for "closed network" applications, we have already witnessed the advent of newspaper page fax systems in which the terminals are linked via earth-orbiting communication satellites. (See chapters 2 and 4.) Any reasonable projection of that trend will point to a proliferation of such systems and to the emergence of compact "dish" antennas as commonplace fixtures on the rooftops of office buildings and factories. Moreover, a *demand assignment* scheme of some sort will undoubtedly be implemented for more efficient allocation of satellite circuits to large customer populations.

Meanwhile, terrestrial communication will be undergoing some dramatic changes of its own. Two current developments—the substitution of optical fibers for copper wire, and the emergence of switched digital networks that will be widely accessible and highly flexible in terms of bit rate—will combine to provide increases in transmission efficiency that will be little short of phenomenal. Channel bandwidth will cease to be the major obstacle it now is to the adoption of truly high-speed systems.

MERGING OF PREVIOUSLY SEPARATE TECHNOLOGIES

As noted above and in the previous chapter, we have already witnessed the intermarriage of fax scanners and computers, the scanner becoming, in effect, an *image digitizer.* In the rapidly widening world of micrographics, the term *computer input microfilming* (CIM) applies where the input image is on microfilm.

Similarly, fax recorders have been adopted as computer output devices in some instances. The recorders in the weather chart and high-speed business fax systems mentioned above have that role, even though they are, in all other respects, fairly traditional in design.

Conversely, computer output recorders—printer/plotters, for example—can,

THE FUTURE

with the aid of appropriate software, be made to function as fax recorders. Moreover, just as microfilm images can be input to a computer via a device that amounts to a fax scanner, so a computer can produce graphic (or alphanumeric) microimages via a *computer output microfilming* (COM) device, which is, in essence, a computer-controlled electronic fax recorder (see chapter 8).

Figure 9.1 shows a commercial graphics-processing system, which, in addition to the basic components of a digital fax transceive terminal, consists of a magnetic tape drive, a keyboard/display station, and—at the heart of the system—a minicomputer controller. Besides scanning and transmitting documents to other terminals in more or less normal fax fashion, the system is capable of storing the input graphics in digital form, compressing the digital data, recalling the stored graphics as either CRT images or paper reproductions, and enhancing or otherwise altering the stored graphics.

Fig. 9.1. A digital graphics processing terminal. It can transmit and receive graphics in an efficient, data-compressed mode and can hold the digitized graphics in magnetic storage for later retrieval or for computer-controlled revision. (Courtesy Broomall Industries)

With appropriate software and perhaps some modification of the scanner, the system could conceivably be arranged to offer an *optical character recognition* (OCR) capability as well, analyzing and identifying the individual characters and symbols of the input document rather than just the juxtaposition of individual picture elements. (OCR was discussed at some length in the preceding chapter.) In this manner, even greater storage and transmission efficiency could be achieved by virtue of a substantial reduction in copy redundancy.

The system's recorder is a "pin printer," basically a computer output device that feels equally at home whether functioning as a graphics or a character recorder. (See Chapter 3.) Thus, when functioning as a receive terminal in a communications network, the system can selectively respond to either fax signals or individual character codes with equal facility and speed. This capability is by no means confined to the pin printer variety of electrostatic recorder. Modulated ink-jet recorders (see Chapter 3) have been designed that can match the pin printer's speed and quiet operation, and can surpass it in versatility.

Another domain that has been invaded in recent years by electronic scanning and recording technology is that of the preparation of printing plates and duplication masters, or "mats." Currently there are perhaps a dozen producers of commercial equipment (of varying complexity) for that purpose, at least three of which have developed and successfully tested the ability to produce printing plates remotely via a precision, computer-controlled laser fax system. This opens the door to electronic document *distribution* in "cluster" fashion—i.e., the publisher preparing a paste-up and transmitting it by high-speed fax to any number of branch printing plants, where it is received as a ready-to-use plate for high-volume printing and physical distribution within a given radius of each plant. As noted in Chapter 2, several large periodical publishers (notably Dow-Jones and the Christian Science Publishing Society in this country) have already established such electronic distribution systems, but on a relatively small scale and with the added step of having to produce the printing plate separately from a photofacsimile negative at the remote location.

Getting back to the merging of graphic and alphanumeric communication techniques, we recall that one of the nonfacsimile document transmission systems discussed in the preceding chapter was the so-called *communicating word processor*. The point should be made here that while such systems may not now include a graphics capability, there is no compelling reason why they could not be adapted to do so without compromising practicality.

Let's assume, for example, that the letter being typed into the system's memory for automatic distribution later that day reads "Attached is the sales chart for the month just past." As presently configured, the typical communicating word-processing system could not automatically handle the described attachment. But why could there not be an image digitizer alongside the typewriter keyboard—in essence a slot into which the accompanying graphics could be fed for digitizing and storage to form part of an alphanumeric/graphic package for transmission as a homogeneous digital bit stream? The terminal's integral minicomputer could be programmed to perform a data compression (redundancy reduction) functional upon recognition of a graphic input.

As for the receiving end, it has already been pointed out that there are paper-printing recorders that are equally at home with graphics and with alphanumerics. All that is required is automatic selection of the proper algorithm upon recognition of one or the other type of input at a receive terminal. There is no hard-and-fast rule that the printing portion of a word-processing terminal need to be integrated with the keyboard unit, as is the case with a conventional electric typewriter.

George M. Stamps, who as an engineer with Magnavox in the 1960s was instrumental in making fax a practical adjunct to the voice telephone, sees the culmination of the present merging of fax and digital technologies as the "mature and enduring condition" of the fax medium.[1] That condition in his view will encompass not only a marriage of alphanumeric and graphic communication, but also message-switching and universal terminal compatibility as well—the prime prerequisites for the eventual realization of universal "electronic mail."

NEW TRENDS IN THE WAY DOCUMENTATION ORIGINATES

Most of us have been conditioned to assume that the creation, revision, duplication, filing, and distribution of documents are by nature separate and distinct operations. Thus we speak of inserting an existing page into a slot to digitize it and perhaps merge it with other "pages" that may so far exist only latently as magnetic bits in the memory of a word processing terminal. But the very existence of this terminal and its rapidly increasing popularity should be a clue that there are some dramatic changes in the making as regards the creation and use of documents.

As indicated in Chapter 8, the word-processing terminal is not only the point of origination of an increasing proportion of the documentation underlying modern business operations, but it also performs editing and storage functions and has the potential (already being exploited to some extent) to serve as the terminal through which the documented information is routinely disseminated. This is in sharp contrast to the customary scheme of things, in which a text draft is brought to one place for editing, another for typing, and then the typed document is brought to yet another place for duplicating, and finally to the place where it is to be stored for future reference. Meanwhile, the duplicates are taken somewhere else to be physically distributed.

In a manner similar to that in which alphanumeric documentation is originated and disseminated via word processing, graphic documentation—charts, engineering drawings, etc.—can be formed (with manual assistance) in a computer's memory, and can thereafter be stored, retrieved, altered, and disseminated by the operation of pushbuttons that tell a computer what block of data is involved and how it is to be processed.

In view of these trends, perhaps we had better begin now to alter our thinking about the need to insert existing physical documents into a slot—either the slot at the front of a fax machine or a mail slot—in the process of disseminating printed information. We may fast be approaching the day when all the documents we will ever need will already be "in the slot," waiting to be called forth and dispatched by some electronic magic.

However, there is no reason for panic. Full realization of this emerging concept of a totally "paperless office" is still several years away. And even then there will remain an abundance of applications in which already existing physical documents will still have to be transmitted from place to place via fax or other appropriate communication terminals.

MICROMINIATURIZATION: ITS ECONOMIC IMPACT

Despite the sophisticated new concepts regarding document origination, storage, and retrieval, one of the more exciting events that this writer sees coming is the introduction of a fax terminal sufficiently inexpensive to be affordable by anyone who has a telephone, a radio, or a TV. That means terminals in most *homes* as well as business offices.

The imminent arrival of such a device is virtually assured by the breakthroughs that are continuing to occur in the design and mass production of integrated electronic circuits—and the economic ramifications thereof. It is further promoted by the unrelenting march of solid state electronics into areas formerly monopolized by cathode ray devices and precision mechanisms. The logical projection of these combined trends is a gradual decline in cost for graphic scanning and recording devices in general.

To better perceive the present state of the microminiaturization art and the potential for turning out amplifiers, modems, logic circuits, scanning devices, etc., as if they were postage stamps, one need merely look at what has already been possible in the field of microphotography for many years. Consider, for example, a process called *microprint*, in which the pages of a book are optically reduced 18 diameters onto a photoengraved printing plate and then *ink*-printed onto paper in that reduced size in a more or less conventional manner. Though these printed microimages have to be optically enlarged to be read, there is no discernible loss in quality from the original pages. A similar process is used in the production of integrated circuits.

Now consider the fact that the *microprint* technology is by no means new. It was invented and first experimented with by the noted publisher Albert Boni in the 1930s, and since 1950, microprint editions of books of all sorts have been in regular production by the Readex Microprint Corporation of New York. (The invention of microphotography per se goes back to the early 1800s.) The microphotography technology in general has steadily advanced to the point where today there are commercial micrographic systems readily available in which document pages are reduced 40,000X (in area) onto film and then enlarged back to their original size without having lost even the periods and commas of the original text. And that is by no means the technological limit.

All of this suggests the highly advanced state of the optical technology underlying the production of microminiaturized electronic circuits. It can be stated with some accuracy that before the inventors of solid state electronics were even born, about half the capability to mass-produce microelectronic circuit chips already existed. That is why today, only a comparatively short time since the invention of the transistor, an electrical engineer can point to the little square in the middle of his novelty tie clasp and boast that it contains more electronic components than a typical color TV set, and that, moreover, they are fully "wired" to form a workable circuit. On a more practical level, it is why, in the face of steady economic inflation, a fully self-contained electronic calculator small enough to fit in the palm of one's hand—a device that was undreamed of a decade ago—can now be purchased for under $10.

From 1968 to 1973, the price of integrated circuits dropped by a factor of 10. It has since continued to decline, and the bottom is not yet in sight. As for the replacement of expensive precision mechanisms with potentially less expensive solid state components, consider Stewart-Warner's advertising claim that its new, compact DDX fax transmitter has *only one moving part* (a claim that is subject to interpretation, but is otherwise reasonably accurate). In today's world, there is really no compelling reason why a fax transmitter of advanced modern design need have any moving parts at all.

FAX IN THE HOME

Consideration of the technological advances just discussed, and their potential economic impact, brings us back to a point that was made in Chapter 6: The prospects of a fax terminal in the same purchase price class as, say, a moderately priced electric typewriter ($500 or less) are by no means unreal. Naturally, there will always be the need for a manufacturer to recover development and other start-up costs—which can be substantial—and this can mean a deferment in the passing along of reduced costs to the ultimate customer. But in view of the speed with which the technology is advancing and attendant economic breakthroughs are occurring, it does not seem far-fetched to this writer to expect introduction of either a home radiofacsimile receiver or a compact phone-coupled transceiver (perhaps complete with some compression capability), purchasable for less than $500, by the end of the present decade.

The fact that this was already tried in earnest some 30 years ago and that the venture ended in dismal failure (see Chapter 1) seems no longer to be a factor in the probability of a retrial along somewhat different lines. It is, in fact, an idea that has been recurring frequently since the 1950s, when commercial TV succeeded for a time in suppressing it. There have been frequent reports that more than one large Japanese electronic firm is working toward that end, attempting to develop equipment sufficiently inexpensive to make home fax economically feasible. As recently as 1977, the British government was still actively engaged in a trial program for the delivery of pages of news and various other kinds of information via a home TV accessory (the so-called *Ceefax* and *Teletext* systems), and, at this writing, is close to implementing its *Viewdata* system, by which subscribers can receive printed matter selectively via telephone. Meanwhile, the French and Japanese governments have been talking seriously about home fax systems to be operated by their respective postal telegraph authorities, the terminals for which are expected to be priced at substantially less than the equivalent of $500.

In recent years, the emphasis has shifted somewhat from the old idea of a radio-dispatched daily newspaper to a variety of specialized services. Among the possibilities are library research services by phone, the dissemination of financial and medical data to appropriate professionals via FM radio or cable TV, and a telephone news-clipping service that could be offered to subscribers on an individual-interest-profile basis.

One obstacle to the implementation of such systems is the question of how authors, publishers, libraries, etc., will be reimbursed for the materials so disseminated. How, for example, will the copyright situation be handled when such items as the pages of published books and magazines are electronically distributed by facsimile—in some cases to many homes and offices simultaneously? The potential impact of the new (effective January 1, 1978) U.S. copyright law on such systems has yet to be fully assessed.

The schemes by which such systems might be implemented are abundant and varied. One firm has come up with an interesting system in which the transmitted information is received on a TV set as a scarcely visible flickering spot in a corner of the screen. The home receive terminal hooks up to the TV set via a photodetector pick-up that is clamped over that corner of the screen. The receiver translates the detected light pulses into characters or marks on the recording paper, thus reproducing the transmitted document.

Then there is the transmission of fax signals in "stow-away" fashion in the

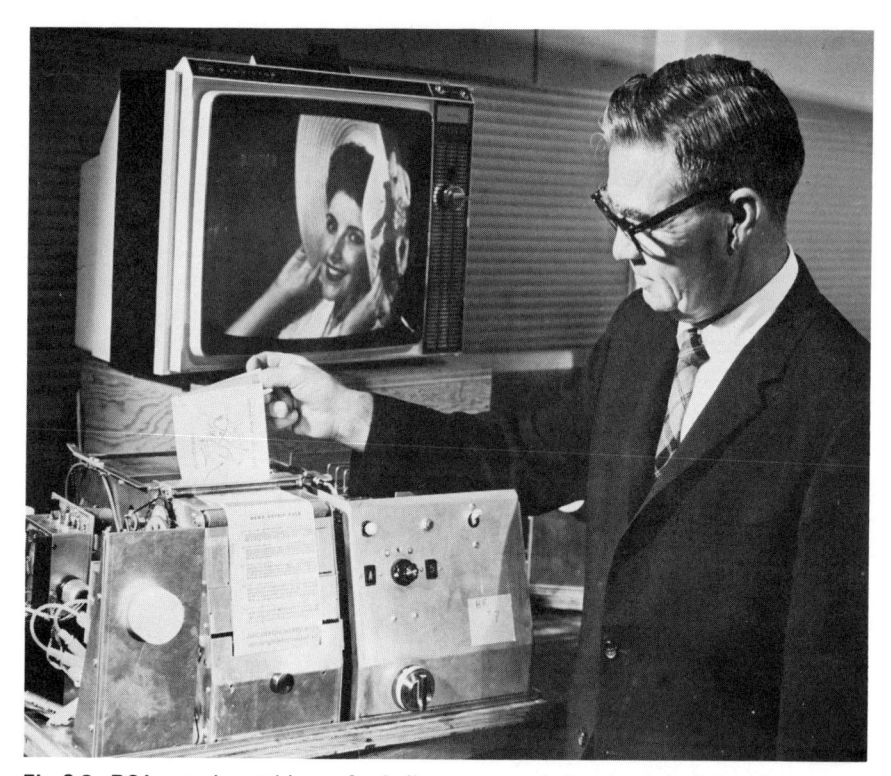

Fig. 9.2. RCA experimental home facsimile system sends fax signals in "spurts" over commercial TV channels during the brief blanking intervals that occur every 1/60th of a second. An electrostatic graphic recording emerges from the fax receiver while a TV show, broadcast simultaneously over the same channel, proceeds without interruption. (Courtesy RCA)

blanking, or "flyback," intervals of commercial TV signals (Fig. 9.2). Systems to accomplish this have been perfected by RCA in this country, and by Philips of the Netherlands and Matsushita of Japan. Britain's experimental *Teletext* system transmits its data in the first few scan lines of each TV frame, the loss of which for picture transmission goes unnoticed. All of these systems have been successfully tested and are apparently waiting until the time is ripe for their implementation.

Among the various communications media for home fax, cable TV (CATV)[2] offers perhaps the most interesting potential inasmuch as it is expected eventually to provide a supplementary communications network that will, in a sense, take up where the present commercial networks leave off. As a purchased service, CATV represents a kind of compromise between the free reception of sponsored broadcasts on the one hand, and the leasing of a private communications link on the other. As of September, 1976, there were just under 4,000 CATV systems in operation in the U.S., serving a total of 12 million subscribers (with about three times that many actual viewers). These numbers represent less than 20% of the potential audience. Authorities on information dissemination see CATV as having the potential for becoming a kind of general information "pipeline," carrying such diversified materials as mail, library books, newspapers, magazines, press releases, and information distributed according to subscriber profiles. They speculate that, by the early 1980s, such materials will be receivable in the home or office at the rate of a page a second, at costs comparable to current postal rates.

One particularly significant thing about the various schemes by which home fax will be implemented is that they all involve communications services that are already widely and economically available to home owners and businesspersons. Moreover, they all connote more efficient use of existing facilities, which should have a favorable effect on economics.

So, in view of these several promising developments, and in spite of the various obstacles, it is a safe bet that fax will eventually join radio, TV, and hi-fi as a home entertainment and information medium, and possibly even as a supplement to the telephone and the mail as a communications medium.

SPEED

One thing that can be said with absolute confidence about tomorrow's typical facsimile system is that it will be a much faster system than today's. As this is being written, there is a vigorous race among manufacturers of fax terminal equipment to capture a bigger share of the market (or to retain their present share) by offering the fastest system at the lowest cost. Some are going the "compression" route, employing special electronic techniques to reformat the transmitted signal so that more efficient use is made of channel space. Others are looking ahead to the widening availability of digital channels and are designing relatively simple terminals for direct digital transmission at various bit rates. Both are acceptable approaches, and which is preferable in any given case depends, of course, on the customer's end requirements and on how he balances desired output quality against finances.

Unfortunately, as a result of this race, we see the clashing of two equally desirable trends. On the one hand, there is the growing popularity of the "detached terminal" approach to fax communication: the desk-top transceiver coupled to an existing telephone set and utilizing the switched voice network to communicate with other, compatible terminals within a wide radius. And, on the other hand, there is the increasing speed of transmission, which—desirable as it is in itself—has resulted in the further deterioration of an already grim standardiza-

THE FUTURE

tion picture. Without standards, phone-coupled fax terminals of different makes are not likely to be compatible, and so the use of such equipment, with all of its valuable potential, is handicapped.

The situation is certain to resolve itself in time. The manufacturers themselves recognize the problem, and some have already expressed a willingness to get together and seek a solution. But there also has to be a willingness to make concessions. Standardization after the fact invariably requires sacrifices.

The development of high-speed facsimile systems dates back many years. RCA's *Ultrafax**, for example, which combines the technologies of fax and TV to achieve transmission speeds of 480 pages a minute (or better), dates back to the 1940s. Other high-speed systems have been developed before and since then, and today there are a handful—none as fast as Ultrafax—that are available commercially (Xerox's LDX,* for example, and the 3M, Dacom and Rapifax digital data compression transceivers).

The promised extension of true digital transmission facilities by common carriers should help to bring about more practical high-speed fax systems that can be afforded by anyone with a suitable application. At the same time, digital data compression techniques will have been improved—both technically and in terms of cost-effectiveness—to the point that the typical voiceband system will transmit a full page of text in a matter of seconds rather than minutes. Both of these eventualities are closer at hand than most people realize.

COLOR

The earliest example of color facsimile that has come to the author's attention is to be found in the frontispiece to an article by Dr. H. E. Ives, et al., in the April 1925 issue of *The Bell System Technical Journal*.[3] It shows a fax-recorded reproduction of a colorful painting depicting the installation of a cross-country telephone line. It is actually a composite of three separate fax recordings of the same picture, each produced with a different color filter in the scanner's optical path. The process was crude and time-consuming, but the result is impressive.

The same basic technique, automated somewhat, was later used by Finch Telecommunications in prototype fax equipment specially designed to reproduce pictures in color. England's Muirhead Company is a principal supplier of equipment for the direct production of multicolor output, and some of the newspaper page-proof systems—Litton/Datalog's Pressfax DRR system, for example—offer the precision capabilities necessary for transmission of color separations.

As for the future of color facsimile, a lot depends on the degree to which speed of transmission can be economically increased, or more efficient use can be made

---

*Ultrafax is a registered trademark of the RCA Corp.; *LDX* is a registered trademark of the Xerox Corp.

of transmission channel capacity. There are at present basically two ways of sending pictures in color by facsimile, which have in common the technique of separating colors by optical filtration (which is fundamental to all color reproduction processes) at the scanner.

One way is to send the three primary colors sequentially, which, without degrading resolution, requires three times the transmission time necessary for straight black-and-white reproduction (or four times, if black is included). The other way is to send separate signals for each color simultaneously, which, without degrading resolution or increasing transmission time, requires a broader bandwidth channel.

Thus the key to the success of color facsimile—as it is to the continued success of the fax art in general—is *increased transmission efficiency*. And the prospects of achieving that, as previously discussed, are good. If by 1985 we can realize the prediction of black-and-white copy at the rate of a page a second on a home receiver that everyone can afford, let us hope that we are not so spoiled by then as to reject a slight slowdown—to maybe three seconds a page—as an expedient for color reception.

Of course, there is also the possibility that by the end of the present decade someone will have adapted present color TV techniques to facsimile, permitting the transmission of color within the same bandwidth and at the same speed as black and white. Nippon Electric has already demonstrated the feasibility of a system along these lines with its "Color Freeze" picture transmission system, which captures selected color TV frames and transmits the static color images via phone lines (if desired) to a conventional TV receiver for reproduction. (See Chapter 8, Video Systems.)

In any event, it is reasonable to expect that both the economic and technological advances necessary to strengthen the feasibility of color facsimile are forthcoming, and that it is but a matter of time before color facsimile becomes commonplace.

THE QUESTION OF PAPER

No discussion of the future of any enterprise is complete nowadays without some thought being given to the implications of the new ecological awareness. Pollution is a growing issue, and one of the things frequently cited as an environmental pollutant is paper. Not only does its production result in the pollution of streams in the vicinity of the paper mills, but its increasing consumption has resulted in the so-called paper explosion, a side effect of which has been the increasing pollution of earth and air through disposal of paper waste.

In a typical application, a present-day fax recorder is a consumer of paper— or a generator of it if you view it from within the office environment. We can,

THE FUTURE

of course, rationalize that, as a substitute for mail, fax eliminates the need for envelopes and stamps, and thus reduces paper waste. But, being realistic about it, the probability that the war on pollution and waste will intensify with the passage of time, and that the use of paper may diminish as a result of it, raises a question of what new form, or forms, facsimile recording may take in the years ahead.

There are already very fine lines separating the facsimile process from certain other information retrieval/reproduction processes, notably computer output microfilming (COM) and video recording.

As already indicated earlier in this chapter and in the preceding chapter, COM is the outputting of manipulated, computer-stored data in the form of visual images on microfilm. The device that produces the film consists basically of a CRT, on whose face symbols and lines are reproduced in response to the computer's coded output signal, and a camera, which photographs the face of the CRT. There is, of course, also the logic necessary to translate a rapid stream of ON-OFF pulses, as outputted by the computer, into properly constructed and accurately positioned symbols on the face of the CRT.

A COM device has many of the attributes of a fax recorder, and can, in fact, be arranged to function as one. We therefore have microfilm as one possible substitute for paper in a fax system. It is a strong possibility in any event because of its steadily expanding use as an information storage and distribution medium, as was pointed out in Chapter 2.

Apart from COM, there have recently been some interesting technological developments that should greatly facilitate the direct substitution of film for paper in facsimile systems. Two or three different systems have been developed, for example, in which permanent, high-resolution microimages are producible by application of a charged toner to special film in which the latent images are electrical in nature. In a separate approach, Drs. J. S. Courtney-Pratt and H. A. Watson of Bell Labs (among others) have successfully used laser technology in conjunction with special (but inexpensive) films to produce high-quality, permanent images either directly or with minimal image processing.

Videotape recording, usually associated mainly with commercial TV broadcasting, has also proved itself an effective information storage and retrieval technique, and video discs now offer a similar capability. As pointed out in the preceding chapter, special accessory electronics can permit the capturing of a selected single frame of TV camera output to be recorded as a kind of latent image on a segment of a videotape or video disc. In essence, the TV camera has produced an electronic "snapshot," and, assuming a high-resolution system and the right kind of lens, the subject might as easily be a printed document as anything else. Many separate document pages can thus be stored—temporarily or permanently—within a relatively short length of tape or a relatively small video disc. In one system that the author has seen demonstrated, 54,000 pages can be laser-recorded onto a master film disc, which can then conceivably be mass-duplicated for a cost of less than a dollar per disc. The desk-top machine on which the discs are "played" into a TV monitor is sufficiently simple to be mass-producible for under $100.

Assuming an erasable medium, such as a magnetic disc, a temporary "memory" might function merely as a transmission *buffer*, the captured frame being re-recorded in a similar buffer at the receive terminal and retained only as long as it is needed for reference via a CRT display. (By automatic repetitive cycling, the receive buffer functions to hold the image on the CRT screen.) If an occasional hard copy is required, the receive terminal could be designed to provide it at the press of a button, using a conventional copier process. The system buffers would thereafter be released for the next transmission.

As noted in Chapter 8, video storage/retrieval systems along these lines are already commercially available from a few vendors. Though intended primarily for local electronic filing of documents, these systems offer an optional remote retrieval capability via communication links if desired.

Whether any of these techniques will influence the design of future document transmission systems is a matter for conjecture. However, there is a strong possibility that the systems produced in a few years from now will bear little resemblance, technologically, to those that are currently typical. The world is changing at an accelerating rate, and, in order to survive, people, things, and systems will have to change with it.

FOOTNOTES

1. Quoted from a letter to the author, March 16, 1976.
2. The acronym *CATV* originally stood for *Community Antenna TV*, but is usually thought of today as meaning *CAble TV*.
3. H. E. Ives, et al., "The Transmission of Pictures Over Telephone Lines," *The Bell System Technical Journal*, April, 1925, pp. 187–214.

CURRENT INFORMATION

For the reader who wants to keep current on developments in facsimile communication and in electronic text and graphic communication in general, the author suggests the following continuing information sources:

Communications News, published monthly by Harcourt Brace Jovanovich Publications, 402 W. Liberty Dr., Wheaton, IL 60186;

Data Communications, a McGraw-Hill bimonthly magazine;

Datapro Office News, published monthly by Datapro Research Corp., Delran, N.J. 08075, and covering all aspects of office automation;

Electronic Mail & Message Systems (*EMMS*), published twice monthly by International Resource Development Inc., New Canaan CT 06840;

Graphic Communications World, published biweekly by Technical Information, Inc., Lake Tahoe, CA 95705 (this publication is geared primarily to the printing and periodical publishing industries);

Telecommunications, published monthly by Horizon House, Dedham, MA 02026;

the reports of Howard Anderson's *Yankee Group*, P.O. Box 43, Cambridge, MA 02138.

In addition, the following trade publications carry applications articles that are usually well written and quite informative: *Corporate Systems*, *Data Comm User*, *Information & Records Management*, *Infosystems*, *Modern Office Procedures*, *The Office*, and *Word Processing World*.

For those desiring more in-depth information, there are, besides the already mentioned *Datapro Research Corporation*, International Resource Development and *Yankee Group*, several publishers of comprehensive reports covering all aspects of the various document and graphic communications media. Some of these publications are of a continuous-update type (loose-leaf, annual subscription), while others are of the "inside track," trend-forecasting variety, available as one-time purchases. In either case, compared with a subscription to a trade periodical, such reports tend to be expensive. Among the additional publishers are Arthur D. Little, Cambridge, MA; Auerbach Publishers, Philadelphia, PA; Creative Strategies, Inc., San Jose, CA; Frost & Sullivan, New York, NY.

APPENDIX
COMMERCIAL SYSTEM DESCRIPTIONS AND SELECTION GUIDE

This appendix provides general descriptions of specific commercial apparatus available (as of mid-1977) for use in systems for transferring documents and pictorial matter electronically via communication circuits. It also provides a "selection guide" for exercise in zeroing in on a commercial system for a specific application.

1. PRODUCT LISTINGS

Each listed product was selected as representative of a particular class (or subclass) of device, and the total list is divided into four broad categories as follows:

- A—General-purpose ("message" or "convenience") business fax systems, low to medium speed;
- B—General-purpose high-speed business fax systems;
- C—Special-purpose fax systems (including hybrid fax/data systems);
- D—Other document and graphic transmission systems (nonfacsimile) and miscellaneous devices.

Category A is confined to fax systems capable of sending a document page (nominally $8\frac{1}{2} \times 11$ inches, or 22×28 cm) in no less than two minutes via voiceband communication circuits. Such systems are usually of the analog variety (see Chapter 2, "Limitations; Relative Slowness," and Chapter 4), and are designed for use on the public switched telephone network.

Category B covers fax systems capable of sending the same size page in less than two minutes via voiceband *or wideband* circuits. As a general rule, those intended for voiceband operation are of the digital data compression variety (see Chapter 4, "Redundancy Reduction"), whereas those requiring wideband circuits are usually analog. It should be noted that, with the advent of digital data compression systems, the use of wideband analog systems for high-volume applications has declined appreciably.

Category C covers fax systems intended for special applications, e.g., the dissemination of weather charts, signature verification, the transmission of photographs and highly detailed graphics of various sorts. An important point to bear in mind with regard to the systems in this category (and in the next category) is that just because some of them are intended for specific applications that does not preclude their use for various other suitable applications as well.

Finally, Category D is devoted to graphic and document communication systems other than facsimile. It covers, for example, handwriting terminals and TV systems. Some of the systems described may prove limited in their ability to faithfully communicate the contents of typical documented messages. However, the objective here, as in the other sections, is merely to provide a glimpse of what is generally available so that the prospective user can determine the extent to which the end requirements of a planned system may be realized, within practical bounds, using available manufactured products. This segment also covers miscellaneous devices and ancillary systems.

In all four categories, unless otherwise noted, it may be assumed that the normally required communication channel for the listed system is *voiceband* (DDD telephone or the equivalent).

There is, of course, a certain degree of overlapping of system capabilities among the various categories. Some of the machines in B, for example, offer low as well as high transmission speeds, and therefore overlap the machines in A in that respect. Moreover, some of the general-purpose systems in categories A and B may be capable of achieving the same ends as some of the special-purpose systems in category C.

Although no conscious effort was made to distinguish between systems of domestic (U.S.) and foreign manufacture, it bears mentioning that, in the preparation of this material, information on domestic systems was generally easier to obtain than that on foreign products.

Two things should be kept in mind regarding cost figures given in these lists. One is that they are as of the first six months of 1977 and are merely approximations. The other is that they reflect, in large part, features and capabilities not all of which, obviously, are covered by these very general descriptions.

As for the varying degree of representation given to individual suppliers, it is important to realize that the number of separate systems shown for a given vendor has no direct correlation with that vendor's comparative standing within the industry. In keeping with the basic objective of selecting only those systems representative of particular classes (or subclasses) of devices, it was unavoidable that some vendors would receive more coverage than others. (Some suppliers offer a wider variety of distinctive systems than do others.) Also, it may appear that the degree of representation is greater for some *types* of systems than for others. However, close analysis will show that each system described in this appendix has some significant distinguishing characteristic that warrants its inclusion along with other, apparently very similar systems. The only possible

COMMERCIAL SYSTEM DESCRIPTIONS AND SELECTION GUIDE 285

exceptions are a few cases where the absence of a supplier's name from a given category might have been viewed as a glaring omission.

Any apparently unfavorable comparison of one system with another in a given category is, of course, unintentional. The reader is especially urged not to consider price alone in assessing the relative worth of the various systems described.

Immediately following the product listings are (1) a key to the various numeric codes and abbreviations used therein, (2) explanations of some of the terms, and (3) a key to the supplier identities given in the NAME and MODEL column.

A. GENERAL-PURPOSE ("MESSAGE" OR "CONVENIENCE") BUSINESS FAX SYSTEMS, LOW TO MEDIUM SPEED

Name and Model	General Description	Input Size*	Input Feed	Output Proc.	Output Feed	Speed (Secs. per full page)	Resol. (Lines per mm)	Approx. Cost Purchase	Approx. Cost Rent per mo.	Remarks
1. Alden 800	FM/analog, desk-top, separate send/receive units, manual operation	7	D	El	C	240, 360	2.5, 3.8	T: $1285 R: 1255	—	Rcvr can be wall-mounted.
2. CIT Alcatel CITEFAX 101	FM/analog, desk-top, separate send/receive units, manual operation	8	D	T	R	180, 360	3.8	2200	—	Built-in phone coupler. Model 580 is AM-FM
3. GSI dex 180 & 580	desk-top analog transceivers, manual operation	12	D	Er	D	180, 360	3.5	4000–4600	$69–96	Built-in phone coupler.
4. GSI dex 700	multi-speed, desk-top AM-FM/analog transceiver, manual operation	12	D	Er	D	120, 180, 240, 360	2.4, 3.5	4750	77–100	
5. GSI dex 3400	multi-speed, automatic desk-top AM-FM/analog receiver, unattended operation	—	—	Er	R	120, 180, 240, 360	2.4, 3.5	—	80–115	Less expensive, dual-speed model also available.
6. GSI dex 4100	multi-speed, multi-resolution AM-FM/analog, fully automatic desk-top duplex transceiver	10	F	Er	R	120, 180, 240, 360, 390, 540, 720	2.4, 3.5, 4.9, 7.0	7800	140–180	High resolution capability. Can send and receive simultaneously.
7. HELLFAX BS-137	AM-FM/analog, desk-top fax receiver, unattended operation	—	—	S	R	360	3.8	—	—	
8. HELLFAX HF 1048	AM-FM/analog, desk-top transceiver, manual operation	7	D	Er	D	120, 180	3, 3.8	—	—	
9. Infolink Scanatron	FM/analog, desk-top, separate send/receive units, unattended operation	13	F	El	R	138/276 or 180/360	4	1250–1700 per unit	53–80	Uses "Pacfax" band-compression for speed doubling.
10. Litton/Datalog Messagefax	AM-FM/analog, desk-top, separate send/receive units, unattended reception	10	D	El	R	180, 360	4	—	—	send & receive units are "stackable".
11. Mitsubishi MELFAS FA 80E	multi-speed, AM-FM/analog, desk-top transceiver, unattended reception	8	F	T	R	120, 180, 240, 360	2.6, 3.9	—	—	solid state scanning.
12. MUFAX Courier K-440	multi-speed, multi-resolution, AM-FM/analog desk-top transceiver, unattended operation	13	F	DE	R	120, 180, 240, 360	2.5, 3.7, 4.9	7000	294	Also available as separate send/receive units.

COMMERCIAL SYSTEM DESCRIPTIONS AND SELECTION GUIDE

13.	N.E.C. NEFAX-1000	analog, desk-top transceiver, manual operation	8	D	Er	D	240, 360	3.9	1800	—	solid state scanning.
14.	OKIFAX-600	analog, desk-top, separate send/receive units, unattended operation.	8	F	T	R	240, 360	3.9	—	—	1000 uses special sync shortcut to simplify electronics.
15.	QWIP 1000, 1200	FM/analog, desk-top transceivers, manual operation	7	D	Er	D	240, 360	2.5, 3.8	—	1000: $29 1200: 45	
16.	S-W Datafax "Electronic Mailbox"	FM/analog, desk-top transceiver, unattended operation.	13	F	El	R	180 240 360	3.8	3900	80–109	
17.	Telautograph Copyphone III-300	Analog, desk-top, separate send/receive units, unattended operation.	11	F	El	R	180	3.3	3000	85	
18.	3M VRC II	FM/analog, desk-top duplex transceiver, unattended operation.	10	F	DE	R	240, 360	2.5, 3.8	3395	82–99	Can send and receive simultaneously.
19.	3M VRC 603	FM/analog, portable desk-top transceiver, manual operation.	7	D	Er	D	180, 240, 360	2.5, 3.8	1645	55–64	Can be battery-operated.
20.	S.E.C.R.E. S-300 Telecopier.	AM/analog, desk-top transceiver, unattended reception.	7–8	D	Er	C	210	3.8	—	—	
21.	Xerox Telecopier III	FM/analog, desk-top transceiver, semi-auto operation.	10	F	Pp	F, R	240, 360	2.5, 3.8	895–945	59–67	Uses plain paper in conjunction with carbon paper for recording.
22.	Xerox Telecopier 200	multi-speed, FM/analog floor-standing transceiver, fully automatic	11	F	Pe	R	120, 180 240, 360	2.5, 3.0 3.8	$8500	$235 + $0.05 per transaction	laser scanning and recording, auto-dial option.
23.	Xerox Telecopier 400, 400-1	FM/analog, desk-top transceiver, manual operation.	7	D	Er	D	240, 360	2.5, 3.8	1230–1800	52–64	built-in phone coupler.
24.	Xerox Telecopier 410	FM/analog, desk-top transceiver, unattended operation.	7	F/D	Er	F/D	240, 360	2.5, 3.8	2995	94–104	built-in phone coupler.
25.	Panafax MV-1200	Multi-speed AM-FM/analog desk-top duplex transceiver, unattended operation.	11	F	DE	R	120, 180, 240, 360	2.5, 3.0, 3.8	4500	70–90 + meter charge	Auto-dial option; CCITT-compatible (groups 1 & 2)

3M VRC II

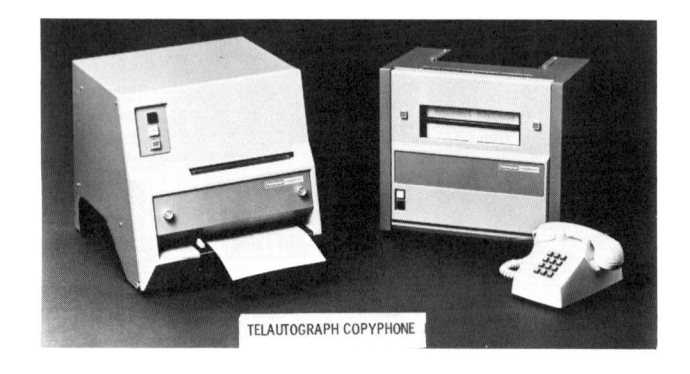

TELAUTOGRAPH COPYPHONE

COMMERCIAL SYSTEM DESCRIPTIONS AND SELECTION GUIDE

XEROX TELECOPIER 410

XEROX TELECOPIER 200

ALDEN 800

LITTON/DATALOG MESSAGEFAX

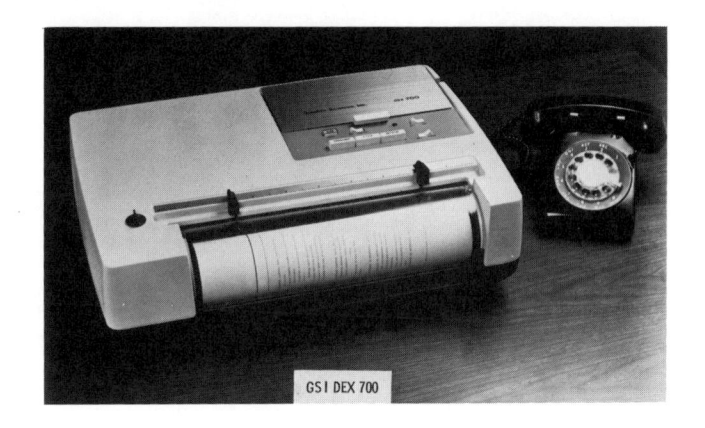

COMMERCIAL SYSTEM DESCRIPTIONS AND SELECTION GUIDE

GSI DEX 700

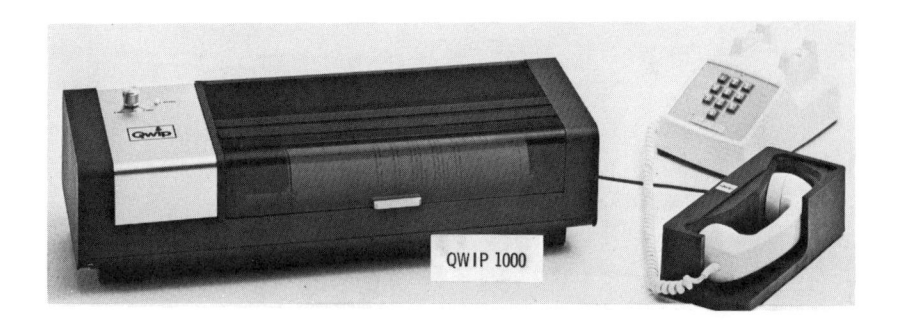

QWIP 1000

B. GENERAL-PURPOSE HIGH-SPEED BUSINESS FAX SYSTEMS

Name and Model	General Description	Input Size*	Input Feed	Output Proc.	Output Feed	Speed (Secs. per full page)	Resol. (Lines per mm)	Approx. Cost Purchase	Approx. Cost Rent per mo.	Remarks
1. Dacom 400 series	multi-speed, automatic, digital data compression transceivers.	10	F	DE	R	approx. 35 and up	2.6, 4.8	—	$350	Speed affected by am't of detail in copy: 2-level tonality (no grays).
2. Fujitsu FACOM-6556	automatic, digital data compression fax system.	14	F	DE	R	approx. 30 and up	—	$15,000	—	speed is copy-dependent 2-level tonality (no grays) auto bit-rate select.
3. Infolink Scanatron	dual-speed/mode, FM/analog, desk-top, separate send/receive units, unattended operation.	13	F	EI	R	72, 108, 144, 216	4	1220–2070	71–103	Uses band-compression technique for speed doubling. Conditioned line req'd for higher speeds.
4. Mitsubishi FA 300A	3-speed, automatic, digital data comp., separate send/receive units.	12	F	DE	R	approx. 33 and up	4, 8	—	—	solid state scanning. speed is copy-dependent. 2-level tonality (no grays).
5. N.E.C. NEFAX-6000	Automatic, digital data compression fax system, separate send/receive units	9	F	DE	R	approx. 37 and up	4, 8	—	—	solid-state scanning. speed is copy-dependent. 2-level tonality (no grays).
6. OKIFAX 7100	automatic, digital data compression fax system, separate send/receive units.	8	F	T	R	approx. 60 and up	4, 6	—	—	Solid state scanning. speed is copy-dependent. 2-level tonality (no grays).
7. Rapifax 100	3-speed, automatic, digital data compression transceiver.	10	F	DE	R	approx. 35 and up	2.6, 4.3	—	264–295 + possible meter charge	speed is copy-dependent. 2-level tonality (no grays).
8. Telautograph	Analog, desk-top, separate send/receive units, unattended operation.	11	F	EI	R	72	4	3000	95	Wideband (nom. 15 kHz) transmission ckt. req'd.
9. 3M Express 9600	totally automatic, digital data compression duplex transceiver.	11	F	DE	R	20 and up	3.8	—	295 + per-minute meter charge.	built-in auto-dialer. speed is copy-dependent. 2-level tonality (no grays). automatic bit-rate selection.
10. Toshiba COPIX-9600	automatic, digital data compression fax system.	12	F	DE	R	approx. 37, 74	4, 8	—	—	speed is copy-dependent. 2-level tonality (no grays).
11. GSI dex 5100	Fully automatic, digital data compression duplex transceiver.	10	F	DE	R	20 and up	2.6, 4, 7.8	12,000–19,000	275–450	Solid state scanning. speed is copy-dependent. 2-level tonality (no grays). many options.

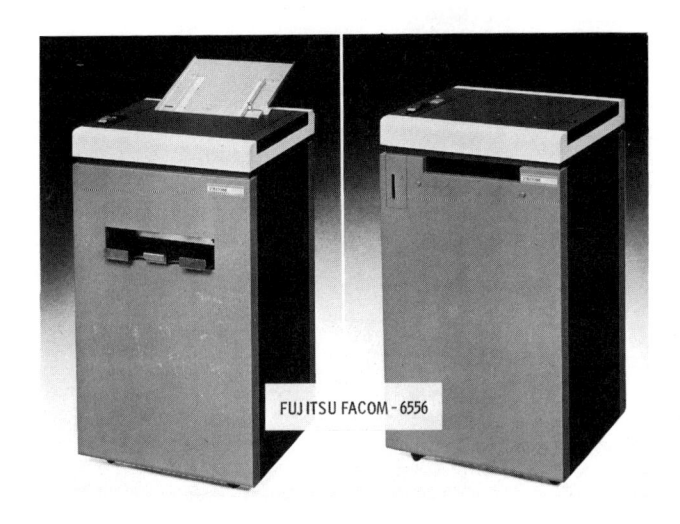

COMMERCIAL SYSTEM DESCRIPTIONS AND SELECTION GUIDE 293

N.E.C. NEFAX-6000

FUJITSU FACOM-6556

3M EXPRESS 9600

RAPIFAX 100

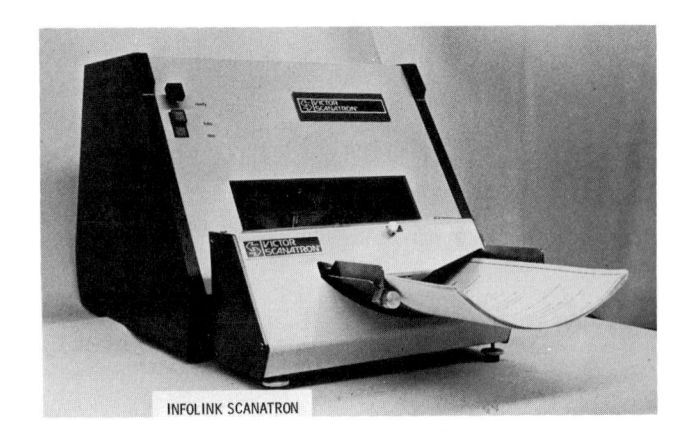

COMMERCIAL SYSTEM DESCRIPTIONS AND SELECTION GUIDE 295

INFOLINK SCANATRON

GSI DEX 5100

C. SPECIAL-PURPOSE FAX SYSTEMS (INCLUDING HYBRID FAX/DATA SYSTEMS)

Name and Model	General Description	Input Size*	Input Feed	Output Proc.	Output Feed	Speed (Secs. per full page)	Resol. (Lines per mm)	Purchase	Rent per mo.	Remarks
1. Alden 11 Marinefax	marine analog fax recorder for chart reception.	—	—	El	R	98 for 28 cm sq. chart	6.5	$2995 (incl. converter)	—	connects to HF radio receiver output
2. Alden 400 Signature	FM/analog system for signature verification, separate send/receive units.	1	F	El	R	11 per signature	3.8	—	TR: $72 REC: 39	can also send full 9 × 12 cm card in 102 sec.
3. Alden 519	floor-standing marine recorder for chart reception.	—	—	El	R	300 for 30 × 46 cm chart	3.8	6441 (incl. converter)	—	connects to HF radio receiver output.
4. Alden 600 "Push To Print"	analog recorder for relatively small input applications.	—	—	El	R	30 (wideband)	3.8	2500	—	applicable to computer output and SSTV operations.
5. Alden 1800 Auto command	basic wide-copy analog scanner, floor-standing.	19	F	—	—	150 or 300 for 28 × 43 cm	2, 3.8	12,500	—	can be equipped with band-comp. modem.
6. Alden 1800, Model 9500S	Universal, analog/digital receiver for weather chart applications.	—	—	El	R	—	3.8	—	—	16 discernible gray tones.
7. Alden 9285	floor-standing digital scanner.	19	F	—	—	(8-16 scans per seconds)	1.9, 3.8	14,500	—	intended for computer inputting and custom graphic communication applications.
8. Alden Dual Digital	combined digital fax and alphanumeric system (floor-standing scanner).	19	F	El	R	(19 cm of fax copy or 12K ASCII char. per min.)	6.5	—	—	accepts both compressed fax and alphanum. ASCII signals as received input.
9. DACOM 300 Telepress	high-speed, high-resolution newspaper page transmission system	17	D	Fc	D	180	11.8-47	100,000 ea. end	—	uses digital compression, auto. error elim.
10. Gifft Model RDR	weather chart recorder.	—	—	El	R	360 (25 × 25 cm)	4	3000	—	available in AM and FM configurations.
11. GSI dex 185/271	AM/analog, desk top, high resolution transceivers.	12	D	Er	D	360–720	5, 7	4600	80-108	Model 185 is 2-speed, uses band-compression technique for speed doubling.

COMMERCIAL SYSTEM DESCRIPTIONS AND SELECTION GUIDE 297

	System	Description									Notes
12.	GSI dex Broadcaster	AM/analog transceiver capable of selective transmission to 10 receivers simultaneously.	12	D	Er	D	180–360	3.5	—	6250–25,000	uses laser technology. Commercial version of AP Laser-photo
13.	Harris Laserfax	family of photofax systems, separate send/receive units.	14	F	Fh	R	60–900	3–11	—	—	Adaptable to "tunable" reception of signals from various sources.
14.	HELLFAX BS 114	multi-speed, dual-resolution recorder for large weather charts.	—	—	S	R	270–2160 (46 × 56 cm)	2, 3.8	—	—	multi-speed, dual-resolution recorder for large weather charts.
15.	HELLFAX BS 1035	unattended AM-FM/analog receiver.	—	—	Er	R	600–1200 (20 × 25 cm)	4.9	—	—	intended primarily for newspicture monitoring
16.	HELLFAX FA 124	AM/analog transmitter for small weather charts.	7	D	—	—	420–630	4.3	3000	—	
17.	HELLFAX FB 128	"rugedized" AM-FM/analog transceiver.	7	D	Pi	—	36–690	3.8	—	—	military style, for use in rough environments.
18.	HELLFAX TM 4005	automatic photofax receiver	—	—	Fc	C	840 (20 × 20 cm)	8.2	—	—	used by Litton/Datalog in some of its photofax systems.
19.	HELLFAX TS 1085	Portable, AM-FM photofax transmitter.	14	D	—	—	300, 600 or 1200	5	—	—	for newspicture applications. Choice of 2 dual-speed configurations.
20.	HELLFAX WF 1206	multi-speed, dual-resolution, AM-FM/analog weather chart transmitter.	18	F	—	R	540–2160	1.9, 3.8	—	—	choice of 3 carrier frequencies, linear or 2-level transmissions.
21.	HELL-PRESSFAX P912	multi-speed, multi-resolution newspaper page transceiver.	18	D	Fc	D	120–2400	4–40	—	—	requires 48 kHz to 1 MHz ckt. Uses band-comp. technique for speed-up.
22.	III HRD	high-resol. recorder for weather satellite transmissions.	—	—	Fh	R	(8000 line image in 1500 secs.)	(8000-line raster)	—	100,000–200,000	input is 99 kHz modulated subcarrier from radio receiver.
23.	JUKI Multifax 100-A	multi-speed, multi-resolution analog transceiver for small documents.	3	D	Er	D	min. 180	4–12	—	—	uses band-compression modem for speed-up or simultaneous voice/fax.
24.	Litton/Datalog Police-fax	high-resol., AM/analog system optimized for fingerprint records.	5	D	Fc	C	540–840	8	—	—	uses HELLFAX TM-4005 Receiver.

C. SPECIAL-PURPOSE FAX SYSTEMS (INCLUDING HYBRID FAX/DATA SYSTEMS) (Continued)

Name and Model	General Description	Input Size*	Input Feed	Output Proc.	Output Feed	Speed (Secs. per full page)	Resol. (Lines per mm)	Approx. Cost Purchase	Approx. Cost Rent per mo.	Remarks
25. Litton/Datalog DL-19W WEATHERFAX	AM/analog weather chart receiver.	—	—	El	R	264–1056 (46 × 56 cm)	1.9, 3.8	5000	200	uses band-compression technique for speed-up.
26. Muirhead K-500/1	Color photofax system.	6	D	Fc	C	780–1080	4	—	—	3-color reprod., available in AM or FM configurations.
27. Muirhead M-136	direct photofax recorder customized to match commercial transmitters.	—	—	DE	R	(variable)	(variable)	4000–5000	150–200	—
28. Muirhead Mercury IV	mobile fax system.	2	—	El	R	(10 cm of 11 cm-wide copy per minute)	3.5	TR: 3000 REC: 1500	—	recorder mounts beneath dashboard of vehicle.
29. S-W Datafax DD	computer-compatible fax system with various customized capabilities.	13	F	DE	R	min. 2	4, 7.5	—	—	high speed, solid state terminals, 2-sec. speed assumes broadband ckt.

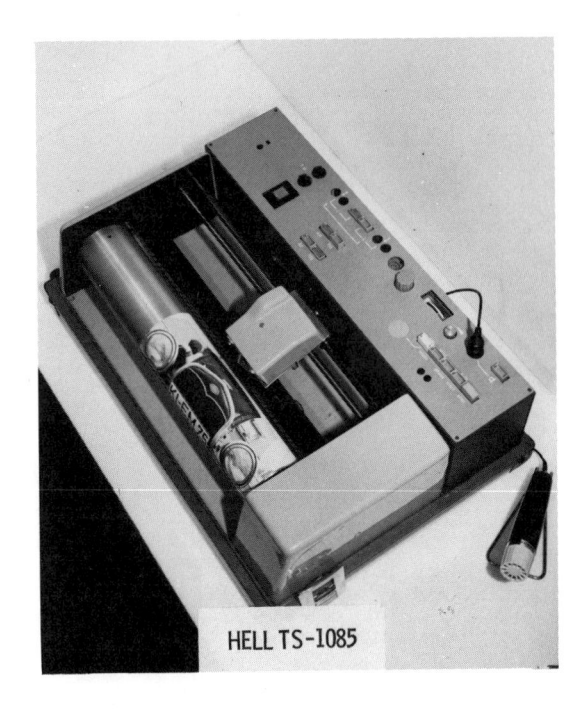

COMMERCIAL SYSTEM DESCRIPTIONS AND SELECTION GUIDE

HELL TS-1085

Litton/Datalog Weatherfax

300   APPENDIX

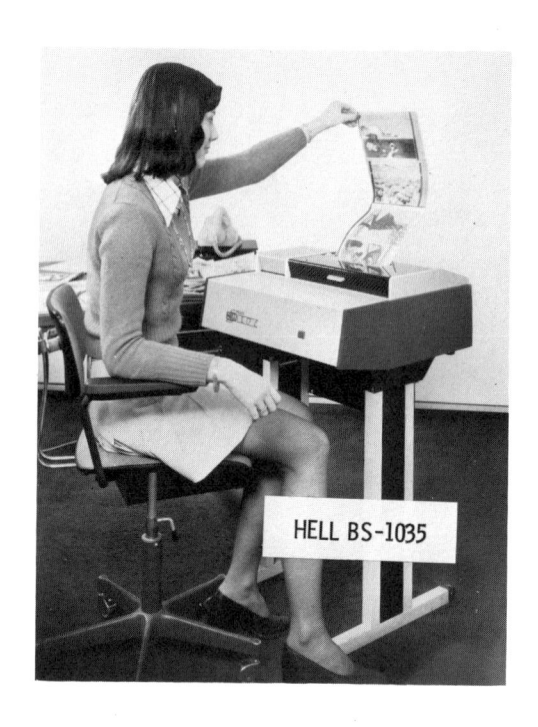

HELL BS-1035

HELL TM-4006

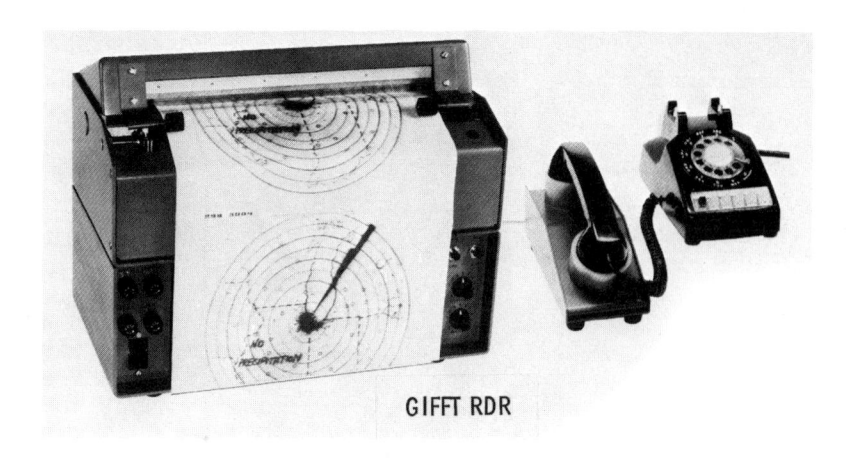

COMMERCIAL SYSTEM DESCRIPTIONS AND SELECTION GUIDE 301

GIFFT RDR

S-W DATAFAX DIMENSION DDX

ALDEN 519

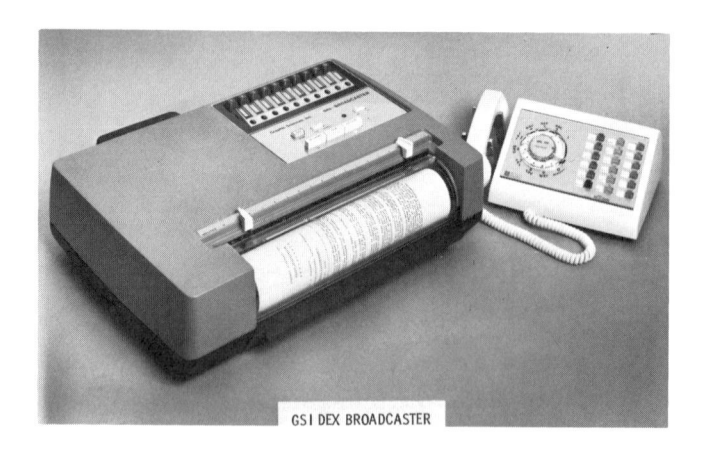

GSI DEX BROADCASTER

ALDEN SIGNAFAX

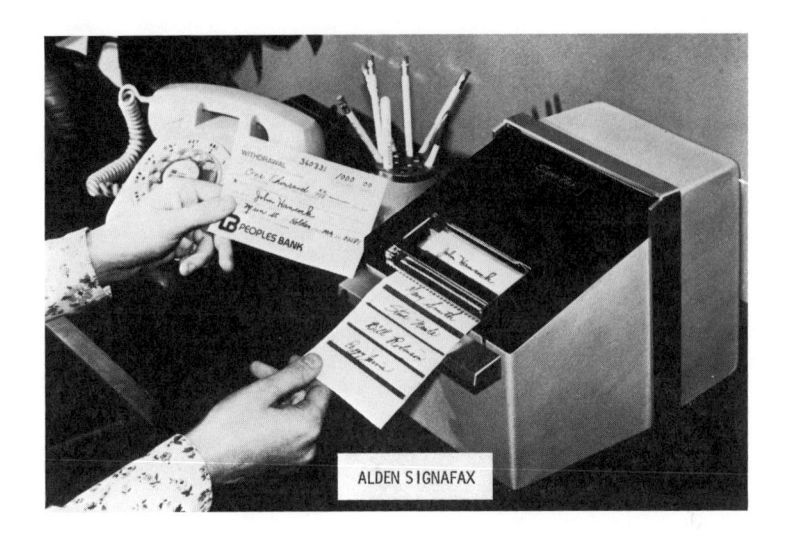

ALDEN MARINEFAX

HARRIS LASERFAX

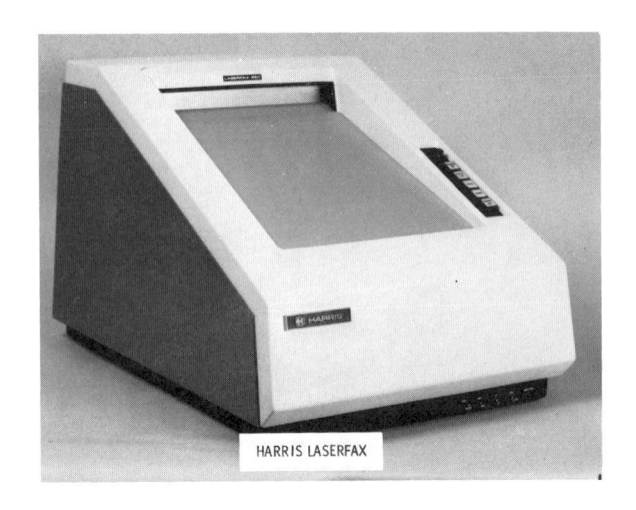

MUIRHEAD M-136

COMMERCIAL SYSTEM DESCRIPTIONS AND SELECTION GUIDE 305

DACOM TELEPRESS

HELL TM-4005

JUKI MULTIFAX

D. OTHER DOCUMENT AND GRAPHIC TRANSMISSION SYSTEMS (NONFACSIMILE) AND MISC. DEVICES

Name and Model	General Description	Input Size*	Input Feed	Output Proc.	Output Feed	Speed (Secs. per full page)	Resol. (Lines per mm)	Approx. Cost Purchase	Approx. Cost Rent per mo.	Remarks
1. ANACOMM-TLC	telecommunications line controller available with digital fax interface	—	—	—	—	—	—	$5500 and up	—	permits digital fax terminal access to existing data networks.
2. Broomall Visicon	digital graphics processing system.	15	D	DE	R	under 60	4	50,000–100,000	—	modular design, can be configured as a graphic communic. sys.
3. COHU 7000 Series	high-resol. TV camera, various line rates.	DLF	—	(video)	—	0.03 or 0.04	(525–1225 lines/fr.)	4400	—	requires 8-32 MHz channel
4. CVI Narrow Band	SSTV system of "frame-freeze" variety.	DLF	—	CRT	—	78	(525–625 lines/fr.)	9000	—	consists of camera, comp./expander, monitor.
5. Compuscan ALPHA	optical character Recognition system applicable to use as a data communic. terminal.	7	F	TP	R	(100 characters/sec.)	—	33,000–55,000	—	automatically converts printed text to digital data code.
6. Datotek DF-300	encryption system for fax signal scrambling (accessory to existing fax system).	—	—	—	—	(no change from norm.)	2.1	3950–5000	—	capable of 32 trillion code combinations.
7. Dest Data OCR/FAX	combined OCR and data-compressed fax system, selectable mode.	7	F	printer not included	—	FAX: 15 OCR: 1000 Char./sec.	5.7	37,600 39,100	—	variable transmission bit rate reqts (4.8 Kbs to 1.55 Mbs). Price is for send terminal + receiver electronics.
8. Doron 1000-C	high-resol. TV camera, various line rates.	DLF	—	(video)	—	0.03 or 0.04	(525–1229 lines/fr.)	5000	—	requires 15–34 MHz channel
9. G.E. Model TE-21	high-resol. TV camera, various line rates.	DLF	—	(video)	—	0.03 or 0.04	(525–1023 lines/fr.)	3500	—	requires 14 MHz channel
10. G.E. Model TE-31	high-resol. TV camera, various line rates.	DLF	—	(video)	—	0.03	(875, 1023 or 1225 lines/fr.)	4500	—	requires 30 MHz channel

D. OTHER DOCUMENT AND GRAPHIC TRANSMISSION SYSTEMS (NONFACSIMILE) AND MISC. DEVICES (Continued)

11. G.E.C. ED 6030B	variable speed SSTV system.	DLF	—	CRT	—	0.5–50	(600 lines/fr.)	20,000	—	transmission bandwidth variable: voiceband to 1 MHz.
12. GBC Model CTC-6008	med. high-resolution TV camera, self-contained	DLF	—	(video)	—	0.03	(850 lines/fr.)	546	—	price excludes lens.
13. Info-Link Electro-writer 2000	handwriting transceiver, real-time	—	—	SP	R	(real-time)	—	TR: 1170 REC: 1260 TRCR: 1980	36 36 61	available as transceiver or separate send/receive units.
14. N.E.C. DFP-751 "Color Freeze"	digital/PCM color SSTV system of the "frame-freeze" variety.	DLF	—	CRT	—	1–600	(525 or 625 lines/fr.)	27,500 (basic 1-way sys., excl. camera and monitor).	—	consists of electronic processor only. Uses standard NTSC camera and monitor.
15. Photomatrix MIDS, Series 700	micro-image transmission system, semi-automated microfiche retrieval.	(24×-reduced page in a multi-image form)	—	CRT	—	0.03	(1225 lines/fr)	400,000–500,000	—	price is for a complete 8-sta. system, local zoom control.
16. Talos Telenote	handwriting system, real-time	—	—	SP	R	(real-time)	—	TR: 1095 REC: 1995 TRCR: 2295 2695	—	available as transceiver or separate send/receive units.
17. Tektronix Video Hard Copy Unit	graphic recorder for use as receive terminal of "frame freeze" video system.	—	—	Fh	R	8–18	(525 lines/fr.)	—	—	accepts any analog or digital video signal.
18. Telautograph Telememo	handwriting system, real-time, separate send/receive units.	—	—	SP	R	(real-time)	—	—	38 (T or R)	—
19. Telautograph Telescriber	handwriting system, real-time, separate send/receive units.	—	—	Sp	R	(real-time)	—	—	35 (T or R)	mechanically-integrated pen for transmission.
20. TEC Datapatcher-900	phone line broadcast unit for use with voiceband data and fax terminals	—	—	—	—	—	—	445	—	permits broadcasting of data or fax to 9 receivers simultaneously via telephone dial net.

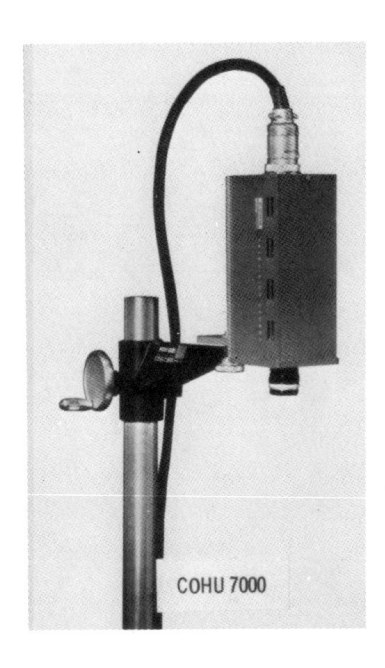

COMMERCIAL SYSTEM DESCRIPTIONS AND SELECTION GUIDE

COHU 7000

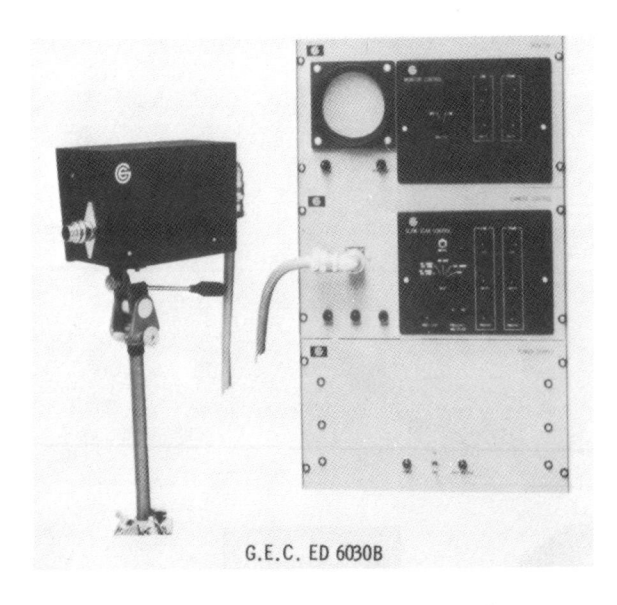

G.E.C. ED 6030B

GBC CTC-6008

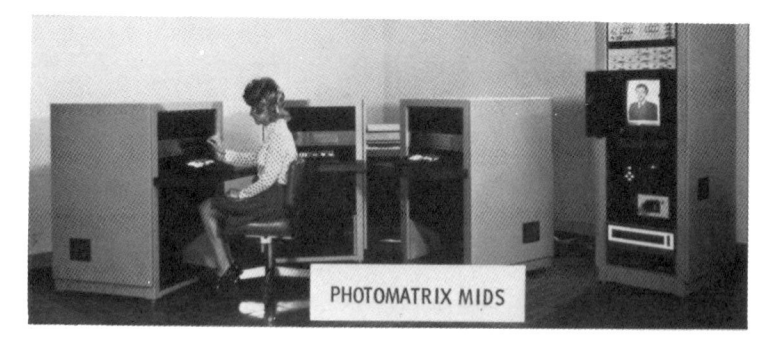

PHOTOMATRIX MIDS

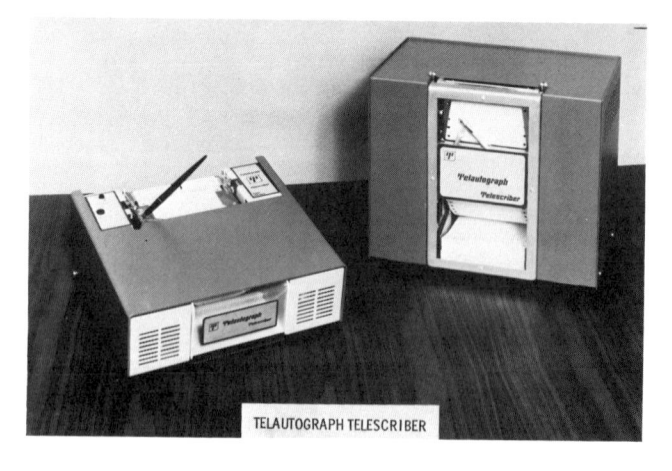

TELAUTOGRAPH TELESCRIBER

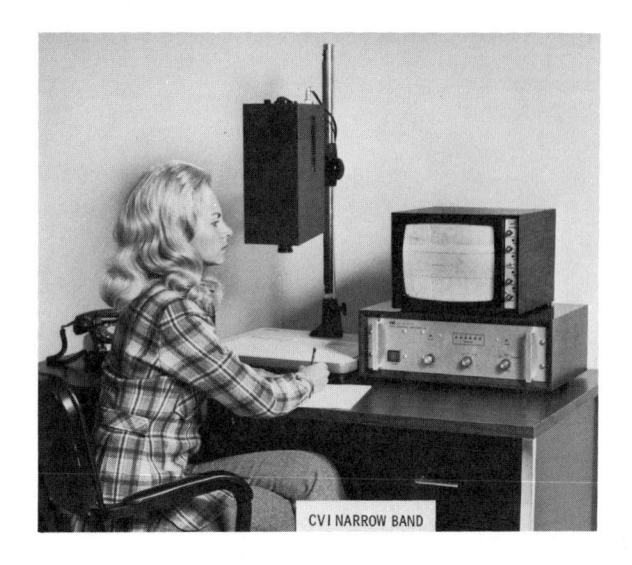

COMMERCIAL SYSTEM DESCRIPTIONS AND SELECTION GUIDE

CVI NARROW BAND

TEKTRONIX VIDEO PRINTER

312 APPENDIX

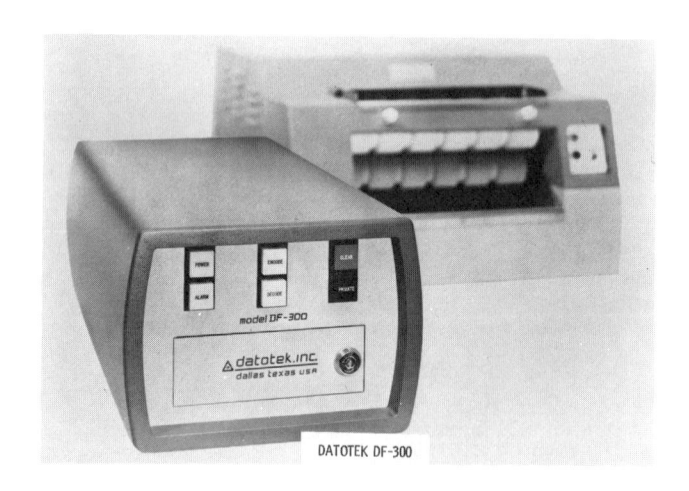

BROOMALL VISICON

DATOTEK DF-300

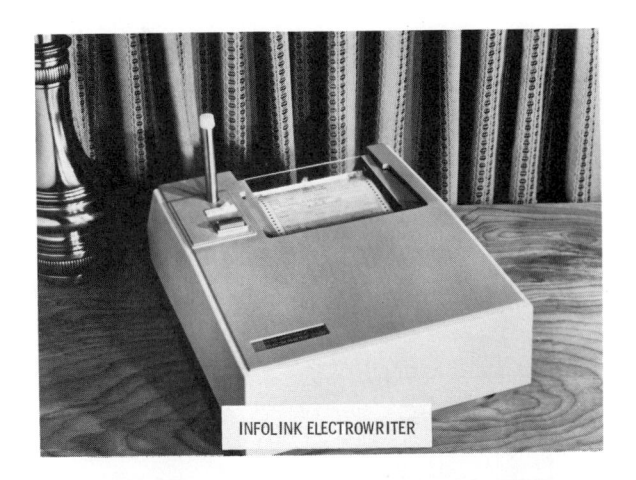

COMMERCIAL SYSTEM DESCRIPTIONS AND SELECTION GUIDE 313

INFOLINK ELECTROWRITER

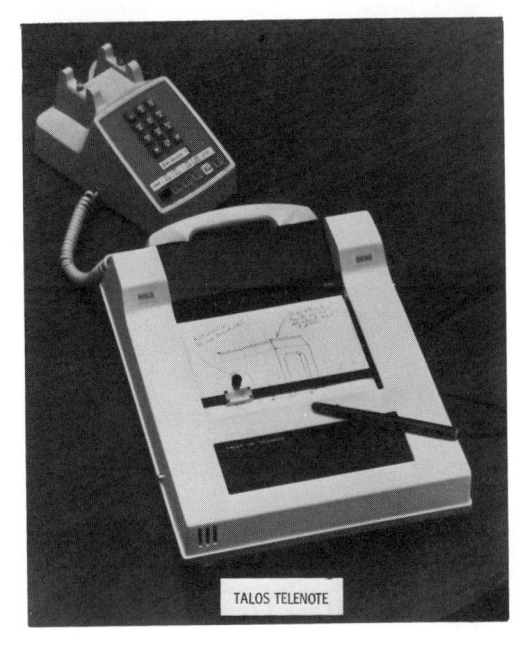

TALOS TELENOTE

Explanation of the Numeric Codes and Abbreviations Used in the Foregoing Product Listings

INPUT SIZE

1. $3\frac{1}{2}'' \times \frac{1}{2}''$ or 6'', or 9 × 1.3 or 15 cm
2. $4\frac{1}{2}''$ (or 11 cm) × virtually any length
3. Int'l "B6": approx. 5'' × 7'', or 13 × 18 cm
4. Int'l "B5": approx. 7'' × 10'', or 18 × 26 cm
5. 8'' (or 20 cm) square
6. Photograph with caption: approx. 8'' × 12'' or 20 × 30 cm
7. "Letter size," U.S.: $8\frac{1}{2}'' \times 11''$ or 22 × 28 cm
8. Int'l "A4" (slightly larger than U.S. letter size)
9. Legal size, U.S.: $8\frac{1}{2}'' \times 13''$ or 22 × 33 cm
10. $8\frac{1}{2}'' \times 14''$, or 22 × 36 cm
11. $8\frac{1}{2}''$ (or 22 cm) × virtually any length
12. 9'' × 14'', or 23 × 36 cm
13. 9'' (or 23 cm) × virtually any length
14. 10'' × 14'', or 26 × 36 cm
15. Engineering drawing size: 11'' × 17'', or 28 × 43 cm
16. 11'' (or 28 cm) × virtually any length
17. Newspaper page size: approx. 16'' × 23'', or 41 × 59 cm
18. Large-format weather chart: approx. 18'' × 22'', or 46 × 56 cm
19. 18'' (or 46 cm) × virtually any length

DLF = depends on lens field

(Note that sizes are numbered in ascending order according to width.)

FEED

C = cassette, D = drum (or cylinder), F = flat, R = roll

OUTPUT PROCESS

DE = direct electrostatic,
El = electrolytic,
Er = electroresistive ("burn-off"),
Fc = photographic/chemical process,
Fh = photographic/heat process,
Pe = plain paper/electrostatic,
Pi = plain paper/ink,
Pp = plain paper/percussive,
S = special,
T = thermal,
CRT = cathode ray tube ("soft copy"),

TP = teleprinter,
SP = servo pen

Explanation of Terms

AUTOMATIC—same as FULLY AUTOMATIC (see below) except that the auto-send capability may be optional. (In some cases, this term is used where insufficient technical information was available.)

CASSETTE (or CARTRIDGE)—a packet of precut sheets of paper, arranged to be fed into the recorder automatically, a sheet at a time.

FLAT FEED—copy is automatically fed through the machine via belts, rollers, etc., as opposed to being wrapped on a drum and spun. It does not necessarily remain flat in its progress through the machine. In many cases, the copy must be sufficiently flexible to make a turn of more than 90 degrees so that it can be ejected at an accessible location.

FULLY AUTOMATIC—automatic feeding is provided as an integral feature at both the input and output ends of the system, and the receiver automatically answers and terminates the call. (Human attendant must place the call, but an auto-dialing accessory may be available.)

MANUAL OPERATION—both the original document and the precut sheets of recording paper must be individually inserted by hand. A drum type transceiver is usually implied.

OCR—Optical Character Recognition. (See Chapter 8). The scanner performs a "feature analysis" of each encountered character, individually, and converts the character to a digital code for efficient transmission.

REAL TIME—used here only in connection with handwriting systems to convey the idea that, as with voice communication, the message is transmitted directly as it is being generated. Technically, on a dot-for-dot basis, fax and TV systems may also be regarded as real-time. In fact, conventional TV is generally regarded as a real-time medium by virtue of the ultra high speed at which the serial elements are transmitted and reconstituted as a unified whole.

RESOLUTION—as used here, implies only the scan resolution of the system, i.e., the number of scan lines within a given linear dimension along an axis perpendicular to the scan axis (usually along the vertical axis of the copy as normally viewed).

SEMI-AUTOMATED OPERATION—unattended SEND *or* unattended RECEIVE, but only with a suitable accessory.

TOTALLY AUTOMATIC—same as FULLY AUTOMATIC, but including an integral auto-dialer.

UNATTENDED OPERATION—implies (1) an integral unattended reception capability, but automatic feeding of input documents requires an accessory device, or (2) a transceiver in which a common auto-feed mechanism serves both the SEND and RECEIVE modes (operator decision required). The auto-answer capability may also require an accessory device.

Supplier Identities (Name and Model Column)

Alden Electronic and Impulse Recording Equipment Co., Inc., Alden Research Center, Westborough, MA 01581
ANACOMM, 7655 Old Springhouse Rd., McLean, VA 22101
Broomall Industries, Inc., 700 Abbot Dr., Broomall, PA 19008
CIT-Alcatel, Department Transmission, Div. Energie–Scientific-Electric, 27 rue Godillot, 93406 St. Ouen, France
COHU, Inc., Electronics Div., 5725 Kearny Villa Rd., P.O. Box 623, San Diego, CA 92112
Compuscan, Inc., 900 Huyler St., Teterboro, N.J. 07608
CVI: Colorado Video, Inc., Box 928, Boulder, CO, 80302
Dacom, Rapifax: Rapicom Corp., 6 Kingsbridge Rd., Fairfield, N.J. 07006
Datotek, Inc., 13740 Midway Rd., Dallas, TX 75240
Dest Data Corp., 1285 Forgewood Ave., Sunnyvale, CA 94086
Doron Precision Systems, Inc., P.O. Box 400, Binghamton, N.Y. 13902
Fujitsu Ltd., 6-1 Marunouchi 2-Chome, Chiyoda-ku, Tokyo, Japan (or Mitsui & Co.–U.S.A.–Inc., 200 Park Ave., New York, N.Y. 10017)
GBC closed-circuit TV Corp., 74 5th Ave., New York, N.Y. 10011
GE: General Electric Co., Imaging Systems Operation, Rm. 301, Bldg. 7, Electronics Park, Syracuse, N.Y. 13201
GEC: General Electrodynamics Corp., Electronics Div., P.O. Box 798, Garland, Texas
Gifft: T. H. Gifft Co., Inc., 1141 Fountain Way, Anaheim, CA 92806
GSI dex: Graphic Sciences, Inc. (Subsidiary Burroughs Corp.) Commerce Park, Corporate Dr., Danbury, CT 06810
Harris Laserfax: Harris Corporation, Electro-Optics Operations, P.O. Box 37, Melbourne, FL 32901
HELL, HELLFAX: Dr.-Ing. Rudolf Hell GmbH, Grenzstrasse 1-5, Postfach: 6229, D2300 Kiel 14, W. Germany
III: (See MUFAX, Muirhead)
Infolink Corp., 3900 N. Rockwell St., Chicago, IL 60618
JUKI: Tokyo Juki Industrial Co., Ltd., 23, Kabuki-cho, Shinjuku-ku, Tokyo, Japan.
Litton/Datalog, 1770 Walt Whitman Rd., Melville, N.Y. 11746
Matsushita Graphic Communic. Systems, Inc., Shimo-Meguro, Meguro-ku, Tokyo, Japan (or Visual Sciences, Inc., 900 Walt Whitman Rd., Huntington Sta., N.Y. 11746)

Mitsubishi Electric Corp., 2-12 Marunouchi, Chyoda-ku, Tokyo, Japan
MUFAX, Muirhead: Muirhead Ltd., Beckenham Kent, England (or Muirhead, Inc., 1101 Bristol Rd. Mountainside, N.J. 07092)
N.E.C., NEFAX: Nippon Electric Co., Ltd., 33-1 Shiba Gochome Minatoku, Tokyo 108, Japan (or NEC America, Inc., 277 Park Ave., New York, N.Y. 10017)
OKIFAX: OKI Electric, 10 Shiba Kotochira-cho, Minato-ku, Tokyo 105, Japan
Panafax Corp., 185 Froehlich Farm Blvd., Woodbury, N.Y. 11797
Photomatrix Corp., 2225 Colorado Ave., Santa Monica, CA 90404
QWIP Systems (Div. Exxon Enterprises, Inc.), 1270 Ave. of Americas, New York, N.Y. 10020
Rapifax: (See Dacom)
S.E.C.R.E., 214–216 Rue du Fb., St.-Martin, 75010 Paris, France
S-W Datafax: Stewart-Warner Datafax Corp., 1300 N. Kostner Ave., Chicago, IL 60651
Talos Systems, Inc., 7311 E. Evans Rd., Scottsdale, AZ
Tektronix, Inc., P.O. Box 500, Beaverton, OR 97077
Telautograph Corp., 8700 Bellanca Ave., Los Angeles, CA 90045
TEC: Telephonic Equipment Corp., 17401 Armstrong Ave., Irvine, CA 92714
3M Facsimile Products Dept., 3M Center, St. Paul, MN 55101
Toshiba: Tokyo Shibaura Electric Co., Ltd., 1-6 Uchisaiwai-cho, 1 chome, chuo-ku, Tokyo 104, Japan (or Toshiba International, 200 Park Ave., New York, N.Y. 10017)
Xerox Corp., 1341 W. Mockingbird Lane, Dallas, TX 75247

2. SELECTION GUIDE

Assuming that the commercial systems in the foregoing lists have remained available since the time of writing, and that prices and other specifications have not changed significantly, the following tables are provided as a general exercise in selecting a system for a specific application. The way to use the tables is to start with the primary broad requirement for the application you have in mind, find the heading that matches it, and make note of the alphanumeric codes listed thereunder. The codes pertain to the system category and specific listed system therein. Having made note of the applicable list, find the heading that represents the next-most-important requirement and see which, if any, of the same systems are listed there. Repeat this procedure for all applicable headings.

Say, for example, that your primary requirement is that each system terminal not cost more than $100 a month to rent. Locate the applicable list. Your next requirement may be that the time required to send each page not exceed a nominal three minutes. Find the applicable heading and note which codes from the first list have survived . . . and so on. The list can thus be narrowed to one or a few commercial systems that will presumably satisfy all (or most) of your system requirements. The final determination should, of course, be based on an actual demonstration of each applicable system, which can usually be arranged by contacting the suppliers.

Some systems may appear under two or more different subheads in a given table, which may seem contradictory. The reason is that these systems are available (or may be arranged) in two or more configurations, or they may consist of terminals of differing sizes, prices, etc. In some cases, a system will not appear at all in a given table, and that is because either insufficient information was available or the particular subhead simply does not apply.

IMPORTANT: Bear in mind that the list of systems is not complete. It is the responsibility of the individual user of these charts to keep abreast of changes in the information presented and of the availability of additional commercial systems. See the listing of suggested continuing information sources under the heading CURRENT INFORMATION, preceding the Appendix.

TERMINAL CONFIGURATION

$\frac{1}{2}$-Duplex Transceiver	Duplex Transceiver	Separate Send/Receive Units (Incl. Send Only & Receive Only)
A-3, A-4, A-8, A-11, A-13, A-15, A-16, A-19 thru A-25, B-1, B-7, C-11, C-12, C-17, C-23, D-5, D-13, D-16	A-6, A-10, A-12, A-18, A-25, B-2, B-9	A-1, A-2, A-5, A-10, A-14, A-17, B-2 thru B-6, B-8, C1 thru C-10, C-13 thru C-16, C-18, C-19, C-20, C-22, C-24 thru C-29, D-2, D-3, D-4, D-7 thru D-19

(*Note:* Systems that are available in two or more configurations will appear in all applicable columns.)

COMMERCIAL SYSTEM DESCRIPTIONS AND SELECTION GUIDE

TERMINAL SIZE—

Desk-Top, Small (Or Low Profile)	Desk-Top, Large (Or High Profile)	Floor-Standing
A-1, A-3, A-4, A-10, A-13, A-14, A-15, A-19, A-23, C-1, C-11, C-12, C-19, C-23, C-24, C-29, D-4, D-6, D-16, D-18, D-19, D-20	A-2, A-5, A-6, A-7 thru A-12, A-16, A-17, A-18, A-20, A-21, A-24, A-25, B-3, B-6, B-8, C-2, C-4, C-6, C-10, C-13, C-14, C-15, C-16, C-17, C-20, C-24, C-25, C-26, C-27, C-29, D-1, D-2, D-5, D-7, D-13, D-14, D-17, D-19	A-22, B-1, B-2, B-5, B-7, B-9, B-11, C-3, C-5, C-7, C-8, C-9, C-15, C-18, C-21, C-22, C-24, C-26, C-29, D-2, D-11, D-15

(*Note:* Some systems will be found in two or more columns because of differing size of send and receive terminals, or because the system is available in various configurations.)

NONFACSIMILE SYSTEMS, GENERAL CLASS—

Video	Handwriting/Real Time	Other
D-3, D-4, D-8, D-9, D-10, D-11, D-12, D-14, D-15, D-17	D-13, D-16, D-18, D-19	C-29, D-2, D-5, D-7, D-20

GENERAL DOCUMENT CATEGORY OR SIZE—

Smaller than Letter-Size	Letter-, Legal-Size	Photographs	Newspaper Page	Weather Charts, Eng'g Dwgs
C-1, C-2, C-4, C-23, C-24, C-28, D-13, D-16, D-18, D-19	A-1 thru A-25, B-1 thru B-11, C-11, C-12, C-17, C-27, C-29, D-5, D-7, D-16	C-13, C-15, C-18, C-19, C-24, C-26, C-27, D-3, D-4, D-8 thru D-12, D-14, D-15, D-16	C-9, C-21	B-2, C-1, C-3, C-5 thru C-8, C-10, C-14, C-16, C-17, C-20, C-22, C-25, D-2

APPENDIX

BEST SPEED PER LETTER-SIZE ($8\frac{1}{2}''\times 11''$ or 21.6 × 28 cm) PAGE—

Low Speed (Greater than 4 minutes)	Med. Speed (2-4 Minutes)	High Speed (Less Than 2 Minutes)
A-7, A-16, C-11, C-15 thru C-19	A-1 thru A-6, A-8 thru A-25, C-13	B-1 thru B-11, C-13, C-29, D-3, D-4, D-5, D-7 thru D-12, D-14, D-15, D-17

(*Note:* The above table excludes systems for other than letter-size pages.)*

AUTOMATION, DEGREE OF—

Unattended Receive	Unatt. Send & Rec.	Fully Unattended (Incl. Auto-Dial)
A-5, A-7, A-10, A-11, A-20, A-21, A-25, B-4, B-10, C-1 thru C-4, C-6, C-8, C-10, C-13, C-14, C-15, C-18, C-24 thru C-29, D-2, D-5, D-11, D-13, D-14, D-16 thru D-19	A-6, A-9, A-12, A-14, A-16, A-17, A-18, A-24, B-1, B-2, B-3, B-5 thru B-8, B-11, D-4, D-15	A-22, B-9, B-11

(*Note:* The above table indicates capabilities with or without available accessories.)

SPECIAL CHANNEL REQUIREMENTS (See Chapt. 4.)—

Conditioned Voiceband	Wider than Voiceband
B-3, B-8, C-4, C-7, C-10, C-13, C-14, C-15, C-16, C-18, C-19, C-20, C-24, C-25, C-26, D-7	C-1, C-3, C-4, C-6, C-7, C-9, C-21, C-22, C-25, C-28, C-29, D-3, D-7 thru D-12, D-14, D-15

(*Note:* Some of the above systems may not require special channels exclusively. Some may also operate via ordinary DDD telephone circuits.)

*The term "letter-size" is used loosely here to mean anything from $8\frac{1}{2}''\times 11''$ (21.6 × 28 cm) to $10''\times 14''$ (26 × 36 cm).

COMMERCIAL SYSTEM DESCRIPTIONS AND SELECTION GUIDE 321

COPY PROCESS (See Chapt. 3.)—

Direct Marking (Special Paper)	Separate Pigment (Coated Paper)	Separate Pigment (Plain Paper)	Photo-chem. (Coated Paper)*	Dry Photo (Coated Paper)*	CRT
A-1 thru A-11, A-13 thru A-17, A-19, A-20, A-23, A-24, B-3, B-6, B-8, C-1 thru C-4, C-6, C-8, C-10, C-11, C-12, C-14, C-15, C-23, C-25, C-28, D-5, D-18, D-19	A-12, A-18, A-25, B-1, B-2, B-4, B-5, B-6, B-8, B-9, B-11, C-27, C-29, D-2	A-21, A-22, C-17, D-5, D-13, D-16, D-18, D-19	C-9, C-18, C-21, C-24, C-26	C-13, C-22, D-17	D-4, D-11, D-14, D-15

OUTPUT RESOLUTION: SCAN LINES PER LINEAR INCH OF COPY (See Chapt. 5.)—

Low (Less than 80)	Medium (80 to 100)	Med.-High (101 to 135)	High (Above 135)
A-1, A-4, A-5, A-6, A-8, A-11, A-12, A-15, A-18, A-19, A-21 thru A-25, B-1, B-7, B-11, C-5, C-7, C-13, C-14, C-20, C-25, C-27, D-6	A-1 thru A-5, A-7 thru A-25, B-1, B-3, B-4 thru B-11, C-2 thru C-7, C-10, C-12, C-13, C-14, C-17, C-20, C-21, C-23, C-25 thru C-29, D-2	A-6, A-12, C-11, C-13, C-15, C-16, C-19, C-21, C-23, C-27	A-6, B-1, B-4 thru B-7, B-10, B-11, C-1, C-8, C-9, C-11, C-13, C-18, C-21, C-22, C-23, C-24, C-27, C-29, D-7

(*Note:* Multi-resolution systems are listed for *all* the resolutions of which they are capable. Video/CRT systems are excluded.)

*or film

TONAL CHARACTERISTICS OF COPY (See Chapt. 5.)—

2-Level (B–W)	Limited Grays	Photo Quality
A-6, A-9, A-22, A-25, B-1 thru B-7, B-9 thru B-11, C-6 thru C-9, C-20, C-21, C-22, C-25, C-29, D-2, D-5, D-6, D-7, D-13, D-16, D-18, D-19	A-1 thru A-11, A-13 thru A-25, B-8, C-1 thru C-6, C-10, C-11, C-12, C-14 thru C-17, C-20, C-23, C-25, C-27, C-28	A-12, C-13, C-18, C-19, C-22, C-24, C-26, C-27, D-3, D-4, D-8 thru D-12, D-14, D-15, D-17

(*Note:* Multi-mode systems are listed for *all* modes of which they are capable. Gray capability does not include screened halftones.)

TERMINAL COSTS (PER UNIT)—

Under $5000 or Under $100/Mo.	$5000–$20,000 or $101–$500/Mo.	Greater Than $20,000
A-1 thru A-5, A-9, A-13, A-15 thru A-19, A-21, A-23 thru A-25, B-3, B-8, C-1, C-2, C-4, C-10, C-11, C-16, C-27, C-28, D-3, D-6, D-9, D-10, D-12, D-13, D-16, D-18, D-19, D-20	A-5, A-6, A-12, A-16, A-22, A-24, B-1 thru B-11, C-3, C-5, C-6, C-7, C-11, C-13, C-25, C-27, D-1, D-4, D-8, D-11	C-9, C-13, C-22, D-2, D-5, D-7, D-14, D-15

INDEX

INDEX

Page numbers in *italics* indicate picture only.

(*Note:* except for three general references, this index does not cover the Appendix.)

A. B. Dick Co., 17
Accessories, fax, *46*, 132, *133*, 156-157, 201
Acme Newspictures, 8, 15
Acme (Telectronix), 15
Acoustic coupling. *See* Couplers
Acuity, 177
Adaptive encoding 145
Ader, Clement 3
Addressograph-Multigraph (A-M) Corp., 20
Aetna Casualty and Surety (Div., Aetna Life), 131
Airlines, use of fax by 26-27. *See also* Freight expediting
Alden (Electronic & Impulse Recording Equipment Co.) 16, 21(fn), *30*, 59, 62, 79, *90*, *91*, 165, 192, 258, *259*, 268
Alexanderson, Dr. E. F. W., 6
Alfax paper, 79
Algorithm, 197-198, 240, 247, 263, 271. *See also* Software
Aliasing, 106, 123
Allen D. Cardwell Electronics Production Corp., 16
Alpha (Compuscan), 264
"Alphagraphics," 62
Alphanumeric Communication, 41-42, 44, 48, 58, 62, 128, 261-266, 268
AM. *See* Amplitude Modulation
American National Standards Institute (ANSI), 224, 230
American Telephone and Telegraph. *See* AT&T
Amortization (equipment purchase), 209

Ampex (Corp.), 260
Amplification, 72, *73*, 74, 88-91, 112-113, 123, 126, 178
   direct coupling, *73*, 74, 90, 112-113
Amplitude Modulation (AM) 51, 106, 108, 109-111, 112-113, 116, *117*, 119, *120*, 130, 138, 154, 172, 179, 202, 229, 232, 233, 234, 237, 238
   double sideband, *107*, *108*, 124, 229
   quadrature, 138
   single sideband, 108
   vestigial sideband, 108, 172, 234, 238
Amstutz (Noah), 5
AM versus FM 49, 51, 64(fn), 106, 109-111, 119, *120*, 202, 236-237
Analog facsimile. *See* Facsimile
Analog signalling, 5, 47, 48, 57, 65, 75, 84, 103, 116, 117, 147, 179, 180, 223, 245, 247
Analog-to-digital conversion, 57, 75, 150-151, 179-183, 219, 224, 241
"AND" gate, *94*
ANPA Research Institute, 64(fn)
Antenna, "dish" (rooftop), 268
Aperture, scanning, 66-68, 70, 165-167, 171
   shape of 165-167, 171
Aperture distortion, 166-167
Application areas, 22-41, 59-60, 147, 192-193, 225-226, 249, *250*, 254-255, *256*, 260, 261-263, 265, 269, 270, 272, 274
Archival (quality), 79

325

INDEX

Army, U.S. *See* Facsimile, military; Standards, military; U.S. Army Electronics Command
Artzt, Maurice, 10
Asahi Shimbun, 33
ASCII code, 62
Associated Press (AP), 7, 8, 16, 19, 22–23, *24*, 54, *72*, 77, 87, *89*, 130, 157, 194
Asynchronous operation, digital, 97–98, 136–137, 139
AT&T (Co., Inc.), 5, 6, 7, 9, 107, 130, 131
Atmospheric disturbances, 116, 128
Attenuation, 119, 121
Audio systems, comparison of fax to, 66, 88, 103–104, 106, 141
Automatic answer, 45, 58, 132, 134, 157, 201, 223
Automatic background control, 90–91
Automatic circuit testing, 58, 121, 152
Automatic dialing, 157–158
Automatic document distribution, 158, 261, *262*, 270
Automatic equalization. *See* Equalization
Automatic feed. *See* Scanner, auto-feed
Automatic gain control. *See* Background control
Automatic loading, 69
Automatic photo recorder, 87, *88*
Automatic picture transmission (APT), 130, 229
Automatic polling. *See* Polling
Automatic retrieval. *See* Information retrieval
Automatic switching, 62, 109, 142, 148, 150, 152, 264, 271
  (of bit rate) 121, 152
Automatic transmission. *See* Polling
Automation. *See* Facsimile, "hands-free" aspect of; Facsimile automation; Office automation
Autoreader (ECRM) 265

Background control, automatic, 90–91
Bain, Alexander, 2, 3, 21(fn), 78, 92
Bain system, 2–3, 92
Baird, John, 1, 14
Bakewell, Frederick, 3, *6*
Balancing network, 119
Bandcom, 18
Bandpass; bandwidth. *See* Transmission
Bandwidth compression, 18, 27, 51, 139, 140–141, 149–150, 217, 229, 234, 255. *See also* Data compression; Redundancy reduction

Banks, use of fax by. *See* Financial transactions
Bartholomew and Macfarlane, 5
Bartlane system, 5
Base (Inc.), 263
Baseband, 89, *104*, 105–108, *110*, 111, 112, 113, 114, *115*, 135, 141, 149–150, 155, 172, 173, 183, 202
Becker, F. K., 198(fn)
Belin, Edouard, 3, 5
Bell (Alexander), 246
Bell System Technical Journal, The, 21(fn), 198(fn), 277, 280(fn)
Bell Telephone Laboratories (Bell Labs), 6, 9, 15, 74, 78, *89*, *129*, *134*, *135*, 140–141, *142*, 147, 148, 151, 178, 192, *196*, *197*, 250, *251*, *252*, *254*, 255, 279
Bell (Telephone) System, 9, 16, 61, 120, 126, 127, *129*, 134, 135, 203, 204, 205–207, 210, 214, 215, 216, 218, 255, 266(fn)
Bidwell, Shelford, 3
Binary code, 43–44, 47–48, 97, 144, 179. *See also* Digital Coding
Binary fax. *See* Facsimile, binary
Bipolar techniques, 149–150
Bit (data) rate, 62, 117, 121, 123, 126, 127, 138, *139*, 148, 151, 152, 201, 202, 208, 217, 259, 264, 268, 276
  automatic
    selection of, 121, 152
Bits (binary digits)
  vs cycles, 164, 183
  vs image elements, 164
Blackboard, Electronic. *See* Electronic Blackboard
Blades, printing, 78, 202
"Blanking" period (TV), 16, 173, 274–275
"Blob" (patch or feature) encoding, 147, 148. *See also* Pattern recognition
Boni, Albert, 273
Brake, phasing, 92
Britain 3, 5, 61, 274, 275. *See also* England
Broadband channels. *See* Transmission
Broadband Exchange service (WU), 125, 202
Broadcast; broadcasting, 10–14, 92, 128, 157, 268
  closed network, 61, 268
  commercial radio. *See* Radio
  sequential, 157
Broomall Industries, *269*
Buffer; buffering, 97, *98*, 138, 141, 144, 150, 151, 241, 280

INDEX

"Burn-off" process. *See* Recording, electroresistive
Burroughs Corp., 19, 20, 55, 253, 263
Business Week, 34

Cable and Wireless (Inc.), 130, 218
Cable, coaxial, 255
Cable, transoceanic, 5, 23
Cable TV. *See* CATV
California State University, 38
Camera
  microfilm, 265, 279
  TV, 255-256, 258, 259, 279
Campus (limited range) systems, 38-39, 255
Canada, 131
Carrier frequency, 90, 105-111, 113, 154, 172, 183, 227, 232, 233, 234, 238
Carriers, common. *See* Common carriers
Carriers, record. *See* "Record" communications; Transmission, international
Carrier swing 111, 135, 183, 233-234, 237, 238, 243
Carrier systems, 126, 128, 137, 158(fn)
  explained, 158(fn)
Carson, John R., 107-108
Caselli, Giovanni, 3
Catechol paper, 79
Cathode ray tubes (CRTs), 53, 54, 71, 74, 82-83, *84*, *85*, 142-143, 167, 250, 253, 257, 258, 265, 269, 272, 279-280
CATV, 62, 274, 275-276, 280(fn)
CCIR. *See* Standards, international
CCITT. *See* Standards, international
Ceefax, 274
Cerampic (Sandia), 252-253
Certification. *See* Registration program
Chalk board, electronic. *See* Electronic blackboard
Change orders, transmission of, 29, 34
Character printers. *See* Printer; Teleprinter; Teletypewriters
Character recognition. *See* Optical character recognition
Character recording, 261
Character size, 185-191, 255
Charge-coupled device (CCD), 15, 54, 74-75, 228
Chart. *See* Test charts; Weather maps
Chemical processing (of images). *See* Recording, photographic
Chicago Tribune, the, 12
"Chopping," 112
Christian Science Monitor, 34
Christian Science Publishing Society, 270

Clark, John H., 19, 149
Clock, clocked (digital), 97, 136, 137-138, 144, 150, 155, 170, 181-183
Closed-circuit TV. *See* Television
Cloud cover pictures. *See* Weather satellites
Coaxial cable, 255
Code words, 97, 138, *139*, 145, *146*, 196
Codex, 139
Coding
  address (location), 63, 144
  digital, 27, 42, 55, 56, 61, 76, 96-98, 138, *139*, 140, 141, 144-148, 151-152, 159(fn), 196-198, 234, 264
  identity, 260, 262
  optical, 58
  retrieval, 58, 260
  See also Run-length coding
Cohu (Inc.), 255
Coincidence detector, *94*
Colorado Video, Inc. (CVI), 258
Color discrimination, 175, 180, 194. *See also* Spectral response
Color facsimile. *See* Facsimile in color
"Color freeze" (NEC), 259, 278
Color response, scan system. *See* Spectral response
Color separation, 277-278
Color TV. *See* Television
Columbia Broadcasting System (CBS), 20
Combined alphanumeric/graphic systems, 20-21, 62, 264, 265-266, 270-271
Comfax-Computerpix, 20, 142-143
Comité Consultatif International des Radio Communications. *See* CCIR
Comité Consultatif International Télégraphique et Téléphonique. *See* CCITT
Commercial broadcasting. *See* Radio; Television
Commercial equipment. *See specific product name*; *See Appendix*; *See also* Facsimile makers
Common carrier fax service, 49-50, 61, 127, 130, 207, 218
Common carriers, communications
  general, 128, 129, 131-132, 139, 202-207, 208, 218, 221-222, 277
  "record," international. *See* "Record" communications; transmission, international
  specialized, 127, 128-130, 218
  telephone, 49, 52, 57, 120, 121, 123-130, 131-132, 133, 153, 217
Communicating word processors (CWP). *See* Word processing

Communication centers, 60
Communications Satellite Corp. (COMSAT), 131
Communication, facsimile. *See* Facsimile
Communications, graphic. *See* Facsimile; Graphic terminals; Handwriting machines; Television; Video systems
Commutator, *75*, 82, 97
Compatibility, 50, 52, 61, 98-101, 127, 128, 132, 152, 154, 171, 179, 210, 219, 230, 240-244, 271, 276
Competition, effects of, 55, 217-218, 220(fn), 276-277
Compressed video (CVI), 258
Compression, bandwidth. *See* Bandwidth compression
Compression, data. *See* Data compression
Compression efficiency, 142, 143, 144, 147
Compression of image details, 100-101
Compression, spectral, 150
Compuscan, Inc., 264
Computer-aided functions. *See* Editing; Retrieval; Revision
Computer-controlled networks 61. *See also* Packet switching
Computer input microfilm (CIM), 56, 265, 268-269
Computer input/output, 42, 55-56, 162, 247, 268-269
Computer output microfilming (COM), 56, 245, 263, 265-266, 269, 279
Computerpix system, 20, 142, *143*
Computer plotters, 26
Computer processing (of data). *See* Data processing
Computer processing (of pictures), 23, 27, 147, 151-152, 268, 272
Computer processing (of signals), 20, 141
Computer-produced graphics, 26, 245, 270, 272. *See also* Computer output microfilm.
Computer storage, 23, 48
Comsat, 131
Conditioning, line (circuit), 109, 121, 123, 207
Conditioning, signal, 52
Confidentiality. *See* Security
Contouring, *196*, 198
Contrast, 174-179, 180, *181*, 190, 197-198, 227, 238
  effect on resolution 177-178, 190
  effect on thresholding 197-198
Control Data (Corp.), 253
Control logic, 142, 151, 264, 279

Control procedure. *See* Transmission sequence
Control signalling, automatic, 50, 238, 240, 244
Convenience fax. *See* Facsimile
Conversion
  data, 29, 48, *56*
  image, 56, 65-66
  signal, 42, 75-76, 134-135, 137, 250
Conversion type systems, 245, 261-266
Converter, scan, 241, 244
Converter, signal, 134-135. *See also* Data sets; Modems
Cooley, Austin, 10, 15
Coordinate grid, coordinate sensing, 247, 250, 253
Copier, office, 23, 42, 280
Copying type systems, 245-260, 263, 264-265
Copyphone (Telautograph), 200
Copyright situation, 274
Costs. *See* Mail costs; system economics
Council on Library Resources, 189, 198(fn)
Couplers, 116, 131-134, 189, 205
  hard-wired (direct), 52, 205
  phone (indirect), 51-52, 132, *133,* 189, 201, 205, 207, 209, 210, 276
Courier, human, use of, 31
Courtney-Pratt, Dr. J. S., 279
CR index, 177-178
Crater tube (glow tube), 87, *89*, 178
Crosley Radio Corp., 12, *13*
Crosstalk, 109, 115-116, 128
CRT. *See* Cathode ray tubes
CRT display. *See* Display
CRT printers, 245, 255, 258
Crystal, liquid. *See* Liquid crystal
Crystal sync. *See* synchronization
Custom engineering, 124, 125
Cygnet (Feedback), 247
"Cylinder-and-screw," 3, *6*
Cylinder type machine, 3, *6*, 67-69, 84, *89*, 232

Dacom 400 fax system, 145
Dacom (Inc.), 20, 35, 55, *56*, 62, 145, 268, 277
d'Arlincourt. Ludovic, 3
Data access arrangement (DAA), 132, 133-134, 205, 218. *See also* Couplers, hard-wired
Data bank, central, 60. *See also* Information retrieval; Storage, data
Data compression, 20, 27, 31, 34, 43, 46,

INDEX

48, 54-55, *56*, 116, 130, 138, 140-152, 191, 195, 200, 205, 208, 217, 218, 223, 225, 234, 240, 255, 259, 264, 269, 271, 276-277. *See also* Bandwidth compression; Redundancy reduction; Run length coding; Two-dimensional compression
Data conversion. *See* Conversion
Datafax (S-W), 16, *17*, 216, 242, 265. *See also* Stewart-Warner Corp.
Datalog. *See* Litton/Datalog
Data rate. *See* Bit rate
Datapatcher (TEC), 157
Dataphone Digital Service (DDS), 126, 127, 204, 205, 206, 207, 214, 215, 216
Dataphone, international, 130
Datapost (TDX), 61
Data processing, 26, 61, 265, 268
Data sets, 134-137, 218
Data transmission, 22, 41-42, 43, 44, 46, 48, 61, 62, 63, 98, 126, 128, 134, 139, 140, 223-224, 225, 274. *See also* Pulse code modulation; Transmission, digital
Datotek encryption system, 156-157
d.c. component, *73*, 74, *90*, 103-104, 112
DDX (S-W), 273
"Dead sector," 104, 174, 227, 232, 233, 237, 238
Defense Communication Agency (DCA), 226, 229
Definitions, IEEE, 102(fn), 224-225
de Hondt (Francis), 5
Delay distortion, 121, *122*
Demand assignment, satellite circuits, 268
Demodulation (detection), 90, 114-115, 135
Densitometer, 175-176
Density
 copy detail, 141-143, 144, 149, 152, 194
 reflection, 175-176
 scan. *See* Resolution, scan; transmission (optical), 175-177
Department of Defense, U.S. (DOD), 226-228, 243
Desk-Fax, 7, *8*, 17, 37, 219
Dest Data Corp., 264, 268
"Detached" terminal. *See* Couplers, phone
Detection. *See* Demodulation
Development, image, 23, 77, 78, 82, *85*, 87, *88*, 202
dex (GSI), 19, 69, 150, 154, 216, 242
Dial network, telephone, 16, 23, 50, 51-52, 109, 115-123, 128, 130, 132, 134,

142, 152, 153, 157, 167, 172, 189, 202-207, 209-215, 217, 218, 222-223, 234, 241, 276. *See also* "Switched network"; Transmission, voice-grade
Dictaphone Corp.; Dictafax, 16
Dielectric paper. *See* Paper, recording
Differential coordinate coding, 145
Differential phase shift keying (DPSK), 138
Digital coding. *See* Coding, digital
Digital data compression. *See* Data Compression
Digital facsimile. *See* Facsimile
Digital transmission. *See* Transmission; Data transmission; Pulse code modulation
"Digitizing," image, 43, 179-183, 259, 260, 268, 270-271
Direct recording. *See* Recording, direct
Discoloration, copy, 81
Discrimination, light value. *See* Kell factor; Thresholding
"Dish" antennas (rooftop), 268
Display
 computer output 253
 CRT, 23, 48, 250, 255, 257, 258, 260, 269
 gas discharge, 253, *254*
 optical, 251, *252*
Distance, transmission (effect on cost). *See* System economics
Distortion
 aperture, 166-167
 delay, 121, *122*
 geometric, 99-101, 240, 242
 harmonic, 123, 128
 signal, 108, 121, 126, 127
 See also Envelope delay distortion; Kendall effect; Skew; Transmission impairments
Distribution, electronic, 270, 274
Dither, 196-198
Dividers, frequency, 94, *95*
Documentation routines, 271-272
Document size, 186, 207, 255
Domsat. *See* Satellite, domestic
Doron Precision systems, 255
Dow-Jones (& Co.), 131, 270. *See also* Wall Street Journal
Drafting standards, 186
Drawings, transmission of, 22, *33*, 34, 37, 39, 47, 58, 147, 152, 186-187, 191, 221
Drive motors. *See* Motor

Drum. *See* Cylinder type machine
Drum speed. *See* Scan rate
Dry processing, 23, 77, 87
Dry Silver recording, 23, 77, 87, 228
Dual-line coding, 145-146
Dudley, Homer, 141
Duobinary, 150
Duplex; full-duplex, 200, 201, 205, 228
Dynodes. *See* Photomultiplier

Early Bird satellite, 131
Echo, 119-120, 121
 suppression of, 119-120
Ecology, 266, 278
Economics (general), 199-220, 272-273. *See also* System Economics
ECRM (Co.), 265
Edge gradient, 171, 172
Editing, computer-aided, 151, 152, 261, 271
ED 6030B TV system (GEC), 258
Efficiency. *See* Laser efficiency; Transmission efficiency
EG&G (Inc.), 18
Electrodes, printer, 78, 202
Electrofax, RCA, 15
Electrolyte, 78
Electrolytic paper. *See* Paper, recording
Electronic Associates, Inc. (EAI), 20, 55, 142
Electronic Blackboard (Bell Labs), 250-251
Electronic Filing systems, 260, 280
Electronic Industries Association (EIA), 51, 158(fn), 190-191, 222-224, 227, 230, 232, 243
"Electronic mail," 19, 21, 62, 63, 261, *262*, 271, 276
Electroresistive recording. *See* Recording
Electrowriter (Infolink), 247
Emergency situations, use of fax in, 39, 249
Encryption ("scrambling"), 154-157, 228
Engelke, Hans, 39
End-to-end service, common carrier, 49
Energy, signal 111, 116. *See also* Signal strength
Engineering, use of fax in, 34-35
England, 3, 6, 14, 20, 35, 277
Envelope, carrier, 114, 121
Envelope delay distortion (EDD), 121, *122*
Equalization, 58, 121, 123
 automatic, 58, 121
Equipment costs. *See* System economics
Erdmann, R. L., 198(fn)
Error control, 138
Error correction, 123, 201

Error detection, 44, 123
Error immunity, 42-44
Error masking, 148
Error reduction, 29, 31, 39
Error, risk of, 29, 31, 39, 55
Error (rate), transmission, 22, 123, 148
Errors, quantizing, 180-183
Excise tax, Federal, 205, 210
Exclusion Key, telephone, 134, 152
Existing documents, transmission of, 246, 247, 272
Express Mail Service (USPS), 61
"Express 9600," 3M, 20, 64(fn), 145
Exxon Enterprises, 19, 219. *See also* Qwip

Facsimile
 analog, 19, 43, 47-48, 50, 51, 55, 92, 97, 98, 116, 117, 121, 123, 132, 134, 136, 140, 143, 147, 149-150, 154, 170, 171-172, 178, 181, 219-220, 222-223, 240, 242, 267
 binary, 104, 136, 149-150, *156*
 color. *See* Facsimile in color
 convenience, 16, 37, 51, 200, 267, 272
 definition of, 1-2
 desk-top, 7, *8, 9, 12, 13,* 16, 17, 18, 19, *29, 30, 46, 52, 53,* 219, 276
 digital, 19, 20, 27, 31, 43, 44, 46, 47-48, 50, *56, 57,* 63, 66, 75, 96-98, 104, 116, 117, 121, 127, 130, 136-139, 140, 144-148, 150, 151-152, 170, 171, 173, 177, 179-183, 195, 197-198, 201, 217, 220, 223, 228, 229, 240, 264
 flexibility of, 39-42, 46-47
 future of, 217-220
 general-purpose, 124, 272
 "hands-free" aspect of, 39, 44-45. *See also* Facsimile automation
 high-resolution, 34, 35, 47, 165, 187, 192-193, 194, 195, 200, 230, 279
 high-speed, 14, 27, 29, 38, 48, 54, 87, 88, 123, 124, 125, 130, *136,* 138, 141-152, 200, 218, 228, 264, 265, 267-268, 270
 inexpensive, 219-220, 272, 273-274, 278
 international. *See* Transmission, international
 invention of, 2-7. *See also* Facsimile history
 message, 1, 2, 6-7, 35, 37, 51, 105, 107, 124, 125, 135, 150, 161, 179, 183, 189, 193, 200, 221, 222-223, 227, 243
  definitions of, 1, 2, 35-36, 221, 222-223

INDEX

microfilm. *See* Microfacsimile
military, 41, 95, 156, 226-229
mobile, 35, 95
multi-speed, 58, 150, 153-154
"no-moving-parts," 273
phone-coupled. *See* Couplers; Fax by phone; Phone coupling
portable, 31, 52, 53
radio. *See* Radio facsimile
TV, 14, 54, 130-131, 277
versatility of. *See* flexibility (above)
Facsimile applications, 3-14, 22-41, 59-60, 192-193, 270, 274
  miscellaneous, 39-41, 270, 274
Facsimile automation, 14, 27, 44-46, 58, 132, *133*, 152, 157, 158, 223
Facsimile broadcasting, 10-14, 16, 225-226
  commercial radio, 10-14, 225-226
Facsimile economics. *See* System economics
Facsimile equipment, commercial. *See specific product name; See Appendix.*
Facsimile history, 1-21, 30, 33, 54, 67, 71, 78, 81, 130, 134, 144, 149, 174-175, 192, 217, 222-223, 224, 225, 274, 277. *See also* History, miscellaneous
Facsimile in banking. *See* Financial transactions
Facsimile in business. *See specific application area*
Facsimile in color, 14, 244, 277-278
Facsimile in engineering, 34-35
Facsimile in libraries, 37-39, 60, 64(fn), 189-190, 274
Facsimile in the home, 10-14, 225, 267, 272, 273-276, 278
Facsimile limitations, 47-51
Facsimile makers, current (first mention), 15-20, 59, 143, 264
Facsimile networks, 35, 37, 50, 60-62, 130
Facsimile newspaper. *See* Radio newspaper
Facsimile pioneers
  (companies and institutions), 1, 4-21, 33, 54, 81, 130, 134, 143, 149, 192, 222, 224, 264, 275, 277
  (individuals), 1-7, 9, 10, 12, 14, 15, 19, 21, 78, 149, 225, 271, 277
Facsimile principles (basic), 2-5
Facsimile process, description of, 65
Facsimile quality. *See* Recording quality
Facsimile recorders. *See* Recorders; recording, fax
Facsimile scanners. *See* Scanner; scanning
Facsimile signalling, redundancy of, 47-48. *See also* Redundancy reduction

Facsimile speed, 48, 51, 63, 76, 82, 87, 123, 142-152, 161, 195, 207-208, 222-223, 276-277. *See also* Facsimile, high-speed; Transmission speed
Facsimile standards. *See* Standards
Facsimile test chart. *See* Test charts
Facsimile transmission. *See* Transmission
Facsimile trends, 51-64, 217-220
Facsimile users. *See* Application areas
Facsimile versus mail. *See* Mail
Facsimile virtues, 41-47
Facsimile weather networks. *See* Weather networks
[*Note*: for other *Facsimile* listings, see *Fax*.]
Fading
  copy, 81
  radio, 110, 118, 128, 130
Fairchild Camera and Instrument Corp., 15, 18
Fano, R. M., 159(fn)
Farnsworth, P. T., 14
"Fax" (first mention), 1. *See also* Facsimile
Fax by phone, 16, 18, 19, 31, 50, 51-53, *59*, 62-63, 111, 128, 131-135, 202-207, 219-220, 267, 271, 274, 276-277, 278. *See also* Couplers; Data access arrangement; Dataphone, international; Modems; Transmission, voice-grade; Voice communication
Fax/data networks, 61, 127
FAXGRAM (Graphnet), 61
"Faximile" (Hogan), 15
Fax mechanisms. *See* Mechanisms, fax
Fax Net, Inc., 128
Faxon Communications Corp., 143
Fax I, EAI, 20, 55, 142
Fax-Pak (ITT), 61
FAXpaper (Hogan), 16
Fax services, 127
FCC *Rules & Regulations,* 225, 244(fn)
Feature-encoding 147. *See also* Pattern recognition
Federal Communications Commission (FCC), 10, 12, 128, 129, 130, 131-134, 205, 218, 225-226
Federal Register, 244(fn)
Feedback Instruments, Ltd., 247
Fiber optic communication, 268
Fiber optics, 71, 82, *85*, 87
Fidelity, reproduction, 46, 58, 106, 107, 108, 165-167, 171, 175. *See also* Quality, output; Resolution
Filing systems, electronic, 260, 280
Film images, 87, 228, 265. *See also* Micro-

Film images (*continued*)
    facsimile; Microfiche; Microfilm printback; Microminiaturization; Recording, fax, film
Film loops, cloud cover, 27
Filtering, filtration
    electrical, 94, 108, 114-115, 135
    optical, 278
Financial transactions, use of fax for, 30, 31, 192
Finch, (Capt.) W. G. H., 10, 12, 14, 15
Finch Telecommunications, 9, 12, *13*, 15 16, 277
Fingerprint transmission, 35, 192-193, 195, 267
Flat-screen display, 253
Flexibility, 39-42, 46-47
"Flyback" intervals, use of, 274-275
Flying spot scanner, 14, 71, 74, 173
Flynn (Peter H.), 5
FM. *See* Frequency modulation
FM vs AM. *See* AM vs FM
Fonts, type (printing), 186, 245, 264, 265
Frame-freeze (TV), 258, 260
"Frame Grabber" (Alden), 258, *259*
France, 3, 20, 274
Frank, A. J., 147
Freight expediting, 27-29, 249
Frequencies, signal, 74, 88-90, 94, 103-111, 132, 141, 149, 153-154, 172-174, 240
Frequency divider, 94, *95*
Frequency discrimination, 138
Frequency modulation (FM), 11, 12, 14, 49, 51, 64(fn), 106, 109-111, 113-115, *120*, 128, 134, 135, 154, 179, 225-226, 231, 233, 236, 237, 238, 243
    versus AM. *See* AM versus FM
Frequency response, 104, 105, 121
Frequency shift keying (FSK), 130, 137-139, 140, 150, 179, 231, 233-234, 238
Frequency standard, 93-94, 136
Frey, H. C., 178
Fujitsu (Ltd.), 20, 61, 145, 148, 152, 253
Full-duplex. *See* Duplex
Fultograph, 10
Fulton, Capt. Otho, 10, 21(fn)
Future of facsimile. *See* Facsimile, future of
Future, the, 217-220, 267-280

Gas discharge lamp. *See* Crater tube; Lamp
Gate circuit, *94*
GBC Corp., 255

General Electric (G.E.), 10, 16, 255
General Electrodynamics (GEC), 258
General purpose facsimile systems. *See* Facsimile, message
Geostationary Operational Environment Satellite (GOES), 27, 229
General Telephone and Electronics (GT & E), 131
Germany, 3, 4, 5, 14, 20, 35, 83, 102(fn)
"Ghosts," *120*, 121
Glow Modulator tube. *See* Crater tube
Government regulations; restrictions, 49, 62. *See also* Federal Communications Commission; Regulatory bodies
Graphic communications. *See* Drawings; Facsimile; Graphics; Graphic terminals; Television; Video systems
Graphic processing, 55, 147, 151-152, 266, 269-270
Graphic Sciences, Inc. (GSI), 17, 19, 20, 55, 69, 150, 154, 157, 216, 242
Graphics, miscellaneous, transmission of, 34, 35, 37, 42, 48, 58, 268
Graphic terminals, manual, 249-253, *254*. *See also* Handwriting machines
Graphic Transmission Systems, Inc., 18
Graphnet system (Graphnet, Inc.), 61, 127, 218
Gray, Elisha, 246, 247, 266(fn)
Gray, F., 198(fn)
Gray levels; gray scale; gray tone, 66, 76, 83, 143, 149-150, 162, 175, 177, 179, 180, 192-193, 194, 196-198, 223, 238
Great Britain. *See* Britain
"Grouped" channels, 125-126, 135, 195, 203-206, 214, 215, 216
Group 1 recommendations (CCITT), 233
Group 2 recommendations (CCITT), 51, 150, 234
Group 3 recommendations (CCITT), 223, 234

Half-duplex, 200, 205
Halftone, 6, 152, 162, 165, 179, 195-198, 227, 236, 238
Handset couplers. *See* Couplers; Phone coupling
"Handshake," 153-154, 234, 261
Handwriting legibility, 191-192
Handwriting machines, 19, 87, 245, 246-249, *250*. *See also* Graphic terminals
Hansell, C. W., 14
"Hard copy", definition of, 246
Harmonic distortion, 123, 128
Harmonics, 103, 106-107

INDEX

333

Harris Corp. (Harris Intertype), 19, 77, 87, 150
Hasler Research Laboratories, 148, 234
Heat development (of images), 23, 77, 82, 85-87, 202
HELAC Electronics Co., 149
Helix-and-blade (recording method), 79, *80*
Hell, Dr.-Ing. Rudolf, GmbH, 20, 83-84, *86*
Hellfax, 77, 84, *86*
"Herringbone," 116, *117*, 123
Hertz (definition), 162
Hertz, Heinrich, 102(fn)
Hertz, Helmuth, 77
History
   facsimile. *See* Facsimile history
   miscellaneous, 1, 54, 107-108, 126, 131, 133, 139, 140-141, 159(fn), 198(fn), 222, 231, 246, 250, 263-264, 273
Hitachi (Ltd.), 20
"Hits," 116, 123
Hochman, D., 15
Hogan "Faximile," 15
Hogan, John V. L., 9, 10
Hogan Laboratories, 15, 16
Home facsimile. *See* Facsimile in the home
Honeywell (Information Systems), 263
Huffman code, 159(fn), 234. *See also* Adaptive encoding
Huffman, D. A., 159(fn)
Hummel (Ernest), 5

IBM. *See* International Business Machines
ICC/Milgo, 139
Identfax (Dacom), 35
Illumination. *See* Lamp; Light source
Image digitizer, 245, 268
Imaging. *See* Laser "writing"; Optical imaging; Recorders
Impact recording, 77, 81
Impairments, transmission. *See* Transmission impairments
Impedance matching, 119-120, 227, 228
Incompatibilities. *See* Compatibility
Index of cooperation, 98-101, 123, 222, 227, 228, 229, 230, 232, 233, 237, 238, 240, 243
   CCITT, 102(fn), 232, 233, 237, 243
   CCITT-IEEE conversion, 102(fn), 232, 233, 237, 243

"Inflections" (baseband peaks), 150
Infodetics (Corp.), 260
Info-Fax 100 (W. U.), 60
Infolink (Corp.), 19, *95*, 149, 247, 249

Information retrieval, 35, 40, 41, 48, 58-59, 230, 254-255, 257, 260, 269, 279-280
   microfilm, 58-59, 60, 230
   See also Coding, retrieval; Remote access
Information theory, 47
Infrared (sensitivity), 251
Ink, electrostatic. *See* Toner
Ink jet recording. *See* Recording
Installation charges, 216
Institute of Electrical and Electronic Engineers (IEEE), 102(fn), 162-164, 176, 224-225
Institute of Radio Engineers (IRE), 224
Integrated circuits (ICs), 54, *56-57*, 272-273
Intelsat, 131
Interception (of messages), 128, 154
"Interconnection," 52, 132, 217-218, 220(fn)
Interfaces, 131-139, 156, 157, 223-224, 241
Interference, external, 44, 115-116, 127-128, 132, 205
Interlibrary loans, 37-39, 60
International Business Machines (IBM), 19, 131, 253, 263
International News Service, 15
International "record" (and graphic) communications. *See* "Record" communications; Transmission international
International Scanatron (Systems Corp.), 18
International Telecommunications Union (ITU), 223, 227, 231, 244(fn). *See also* CCIR, CCITT
International Telegraph Union, 231
Intertel, 139
IRE Transactions, 159(fn)
ITT (International Telephone and Telegraph Corp.), 61, 127
ITT WorldCom, 130, 218
Ives, Dr. Herbert, 6, 21(fn), 277, 280(fn)
Ives, Frederic, 6

Japan; Japanese, 17, 19, 20, 33, 37, 61, 71, 87, 145, 147, 152, 274, 275
Japan Radio, 20
Jenkins, C. Francis, 7, 14
Jitter, 121-213, 181-183
   phase, 121-123
Journalism, uses of fax in. *See* Newspictures; Periodical publishing
Juki (Tokyo Juki Industrial Co., Ltd.), 20

INDEX

Kell factor, 168–170, 171–172, 189
Kell, R. D., 198(fn)
Kendall effect, 106, 107, 123
Kerr cell, 87
Keyboard/display, *256*, 269
Keyboard type equipment, 42, 44, 245, 260, 261–263, 264, 270–271. *See also* Keyboard/display; Teleprinter; Teletype(writer); Word processing
Keystroke storage, 261, *262*
Korn, Dr. Arthur, 3–4, 21(fn)
Kretzmer (E. R.), 15
KSTP (St. Paul), 10

Labor considerations, 44–45, 199. *See also* Operator training
Lamp
  gas discharge, 54, 87, 201
  incandescent, 54
  projection, 87
Large scale integration (LSI), 54
Laser, laser beam, 54, 71, 228
Laser efficiency, 54, 64(fn)
Laserfax (Harris), 19, 87
Laserphoto (AP), 19, 54, 77, 87
Laser scanning. *See* Scanning, laser
Laser recording. *See* Recording
Laser "writing," 251, *252*, 260
Latent image, 23, 78, 82, *84*, *85*, 87, 265 279
"Lathe-style" scanning, *6*, 67, *68*
Law enforcement, 35, 60, 192–193, 195, 249, 267. *See also* Fingerprint transmission; "Mug shot" transmission
LDX (Xerox), 18, 82, *83*, 125, 277
Leased lines. *See* Transmission
Leasing. *See* Rental
Legibility, 171, 177–178, 183–191, 198(fn), 255, 257
  character vs word, 183–187, *188*, 189, 191
  library applications, 189–190
  microfilm printback criteria, 187–189, 190–191
  NMA/EIA Committee findings, 190–191
  scan lines per character, 185–187, 255, 257
Lens, TV camera, 255, 256, 279
Level fluctuations, 76, 110, 118–119
Level, output, 66, 72, 137
Level, signal. *See* Signal level
Library systems. *See* Facsimile in libraries
Light-emitting diode (LED), 97
"Light pipe" (fiber optic), 82, *83*, *85*, 87

Light source, scanning, 66–71. *See also* Lamp; Laser
Light valve (acousto-optic modulator), 3, 72, 87, 251, 252
  electromechanical, 87
  liquid crystal, 251, *252*
  solid state, 252–253
Limiter (FM), 115, 135
Line advance (line feed), 96–97, 98, 227, 228, 243
Line length, scan, 98, 99, 172, 173–174, 222, 227–228, 232, 237, 240, 243
Liquid crystal light valve, 251, *252*
Litcom (Div., Litton), 15
Litton/Datalog, 15, 35, 69, 145, 152, 228, 277
Litton industries, 15
Logic. *See* Control logic
Loss
  fixed, 119
  net, 119
  quality, 46, 47, 115–123, 132, 166, 168–170, 179–183, 207–208
  scanning. *See* Kell factor

Magnafax, 17, 81, 220, 242. *See also* Magnavox Co.; 3M Co.
Magnavox Co., 16, 17, *18*, 20, 78, 81, 198(fn), 219–220, 271
Magnetic recording, 158, 260, 261, 269, 271. *See also* Tape, magnetic; Video recording
Mail (general), 22, 31, 41, 62, 272, 279
Mail costs, 22, 63–64, 199
Mail, electronic. *See* Electronic mail
Mail-gram (W. U.), 61
Maintenance costs. *See* System economics
Manual graphic terminals, 246–253
Manufacturers. *See* Facsimile makers. *See also entries by specific manufacturer name; See Appendix*
Manufacturing, use of facsimile in, 34–35
Map transmission, 42, 50, 193–194. *See also* Weather maps.
Marconi Co., 5
Massachusetts Institute of Technology (MIT) 15, 19, 241
Mastergroup, 125
Matrixed format, 62, 172, 196–198
Mats, duplication, 40, 270
Matsushita (Graphic Communication Systems), 19, 20, 145, 275
MCI Telecommunications, 129

INDEX

Mechanisms
    fax, 3, *4*, *5*, *6*, 10, *13*, *45*, *68*, *69*, *70*, *72*, 79, *80*, *81*, *83*, 84, *85*, *86*, *89*, 92, *98*
    servo-, 247-249
    spring-driven, 3, 10
"Memory bank," 246
Memory device (memory module), 74, 158, 258, 259, 261, *262*, 271, 280
Merging of technologies, 266, 267, 268-271
Mertz, P., 198(fn)
Message facsimile. *See* Facsimile
Messagefax (Litton), 69
Message network (W. U.), 35, 37
Message switching, 228, 271
Meteorological fax systems. *See* National Weather Service; Weather maps; Weather networks
Meter plan, 209
Mexican government (fax network), 60
Meyr code, 234
Miami Herald, The, 12
Michigan State Library, 38
Microcopy test target, NBS, 162-164, 189, *190*
Microdensitometer, *176*
Microelectronics, 272-273
Microfacsimile, 55-56, 58-60, 66, 74, 78, 142, 162, 165, 167, 187-189, 190-191, 224, 230
Microfiche, 58, 257
Microfilm quality, 165, 177-178, 187-189, *190*
Microfilm facsimile. *See* Microfacsimile
Microfilm printback, 187-189, *190*
Microfilm Technology, 198(fn)
Microform Information Dissemination System (MIDS), 257
Microminiaturization, 272-273
Microphotography, 162, 272-273
Microprocessor, *56-57*, 97, 268
Microprint, 272-273
Microwave (radio), 29, 33, 128, 129, 131, 202, 218, 224, 255
Military facsimile. *See* Facsimile
Mills (Thos.), 5
Miniaturization, 52, 53
Minicomputer, 268, 269, 271
Mirror, optical, *70*, *71*, *72*, *89*
Missing persons report, fax-dispatched, 35, *36*
Mitsubishi (Industries, Ltd.), 20
Mobile systems. *See* Facsimile
Modems, 126, 127, 131-132, 134-135, 137-139, 140, 143, *145*, 151, 201, 218, 219, 223-224, 272

Mode selection, 62, *90*
Modular design, *136*, 240
Modulation
    amplitude. *See* Amplitude modulation
    definition, of, 90
    frequency. *See* Frequency modulation
    See also AM vs FM.
    laser, light beam. *See* Light valve; Modulators
    need for, 89-90, 105
    negative vs positive, 179
    See also Polarity; signal sense
    pressure, 77, 81
Modulation index, 111
Modulation schemes, *72*, 111-114, 240
Modulator-demodulator. *See* Modem
Modulators
    acousto-optic, *72*, *89*. *See also* Light valve
    amplitude, 112-113
    frequency, 11, 113-114, 135
    laser beam, *72*, *89*, 251, *252*
    multivibrator, 113, 114
    ring, *112*, 113
Moiré patterns, 162, 195, 198(fn)
Monetary transactions, 30, 31, 192
Monitor, TV, 255, *256*, 257, 258, 260, 280
Morehouse report, 198(fn)
"Morgue" file, newspaper, 60
Morse code, 159(fn)
Motor, drive, 92, 93, 94, 95, 182, 183
"Mugshot" transmission, 35, 195
Muirhead (& Co. Ltd. and/or Inc.), *8-9*, 20, *32*, 33, 35, *88*, 277
Multi-frequency techniques, 123
Multi-level signaling, 138, *139*, 140, 149-150, 179
Multi-line coding, 145-146
Multiplex; multiplexing, 14, 128, 138, 158(fn), 225-226

Nagano (Ltd.), 20
National Bureau of Standards (NBS), 162-164, 187, 189
National Meteorological Center (NMC), 24, 26, 27
National Micrographics Association (NMA), 190-191, 224, 230
National Security Agency, 228
National Weather Service, 24-27, 55, 149, 157, *194*, 221
Naval Electronic Systems (NAVELEX), 228.
    See also Facsimile, military; Dept. of Defense

NEA-Acme, 15
Neal, A. S., 198(fn)
Nelson, Dr. C. E., 177-178, 198(fn)
Neon-gas devices, 253
Netherlands, 260, 275
Network protection, 116, 133-134
Networks. *See* Balancing network; Dial network; Facsimile networks; Fax/data networks; Weather networks
News wire services. *See* Associated Press; United Press International
Newspaper page facsimile system, 20, 32-34, 44, 55, 145, 152, 178, 179, 195, 200, 267, 270, 277
Newspaper, radio- or TV-dispatched, 10-14, 62
Newspictures, 1, 7-9, 18, 19, 20, 22-24, 41, 50, 55, 77, 83, 87, 119, 124, 130, 157, 194-195, 201, 241, 244
New York *Sun*, 7
New York Times, the, 12, 34
New York *World*, 4
Nippon Electric Co. (NEC), 20, 71, 145, 259, 278
Nippon Telegraph and Telephone Corp., 37
Noise
  acoustic, 82, 119, 132
  electrical (general), 66, 76, 109, 115-120, 123, 126, 135
  impulse, 116
  internal, 116-117
  line, 49, 57, 109, 135, 175, 179
  quantizing (errors), 180-183
  random (broadband, gaussian), 109, 117, 118
  thresholding, 175
  See also Distortion; Transmission impairments.
Noise level, 119
Non-common carrier devices, 52, 205, 217. *See also* Interconnection
"Normal" scanning, 168

Obsolescence, 51, 57, 200
OCR/FAX (Dest Data), 264
Odors, recording, 50, 80, 87
Office automation, 261-263
Office-of-the-future, 246
Oil drilling, use of fax in, 35-36
Oki okifax, 20, 87
One-minute toll charge, 218
One-time charges, 207, 216
One-way communication, 200
Operator training, 22, 28, 30, 44, 152

Optical character recognition (OCR), 42, 151, 245, 246, 263-265, 270
Optical coding. *See* Coding
Optical fibers. *See* Fiber optics
Optical imaging, 78, 250, 251, *252*, 266, 272-273
Optics
  recording, *83, 89*
  scanning, 66, *67, 68, 70, 71, 72*, 74
Order entry, 29, 157
Order dispatching, 29, 30
Orient, use of fax in, 42
Output level, fax scanners, 66, 72, 137
Output quality. *See* Quality; Recording quality
Owens-Illinois (Inc.), 253

Pacfax, 55, 149
Packet switching, 61, 127, 218
Palmer (Herbert), 5
Panafax (Corp.), 19, 145
Panasonic, 19
Paper as a pollutant, 278-279
Paper, conservation of, 266-279
Paper "explosion," 278
"Paperless office," 246, 272
Paper, recording, 2, 3, 7, 201-202
  Alfax, 79
  carbon, 10, *11*, 81
  chemistry of, 78, 80, 82, 83, 85, 87
  cost of, 86, 201-202
  dielectric, 83
  dry silver, 23, 77, 87, 202, 228
  electrolytic (wet), 10, 78-79, *80*, 83, 201, 202
  electroresistive ("burn-off"), 79-81, 201
  electrosensitive (first mention), 2
  electrostatic, 77, 82-83, *84*, 201, 202
  plain, 10, 77, 81, 82, 202, 228
  photographic, 77, 83, *85*, 87, *89*, 201, 202
  Teledeltos, 7, 80
  thermal, 85-86, 201
Paper, replacements for, 279-280
Paste-up, transmission of, 270
Patch encoding, 147
Pattern recognition, 265. *See also* Feature encoding
Pels. *See* Picture elements
Periodical publishing, 20, 31-34, 270
  See also Newspaper . . . ; Newspictures
Permanence, image, 79
Persistence, CRT, 257
Phase jitter, 121-123

INDEX

337

Phase modulation, 138, 234
Phasing, 5, 51, 58, 91-92, 94, 96, 123, 153-154, 222, 227, 229, 236, 238
  coincident pulse, 92, 94
  stop-start, 92
Philadelphia *Inquirer*, the, 12
Philips of the Netherlands, 260, 275
Phone-coupling, acoustic/inductive, 31, 51-52. *See also* Couplers; Data access arrangement
Photocell, 65, 72, 73
Photoconductor, 252-253
  strip, 54
Photodiode, 72, 74, 75
  pin, 74
  strip, or line, 74
Photoelectric scanning. *See* Scanning
Photoelectric transducer. *See* Photocell; Photoconductor; Photodiode; Photomultiplier; Scanning
Photofacsimile. *See* Newspictures; Phototransmission; Picture transmission; Recorders, Photographic; Telephoto; Telephotography
Photofax (AP), 9
Photographs, transmission of. *See* Picture transmission
Photomatrix Corp., 257
Photomultiplier, 14, 53, 65, 72, 73, 74
Photophone Inc., 60
Phototelegraphy. *See* Telephotography
Phototransmission, 6, 207, 243, 270. *See also* Newspictures; Picture transmission; Telephoto; Telephotography
Pictorial quality criteria
  fingerprints, 192-193
  handwriting, 191-192
  maps, 193-194
  photographs, 194-198
Picture elements (pels, pixels), 5, 164, 165, 196-198
Picturephone visual telephone (Bell), 255
Picture transmission, 1, 3, 4-9, *11*, 21(fn), 22-24, 40, 42, 87, *88*, 124-125, 152, 175, 179, 191-198, 207, 225-226, 232-233, 234, 259, 264, 278, 280(fn). *See also* Newspictures; Phototransmission; Recording, photographic; Telephoto; Telephotography
Pigment transfer recording, 77, 81
Pin printer, 83, 270
Pitney Bowes (Inc.), 19
Pixels. *See* Picture elements
Plasma panel, 245, 253, *254*
Platen, scanning, 71

Plessey (Co., Ltd.), 20
PLZT ceramic, 252-253, 266(fn)
Points (type face), 186
Polarity
  signal, 149-150, 158, 232, 233
  tonal (image), 149-150, 158, 178-179, 232, 253, *254*
  transmission, 103, 178, 179, 231
  See also Signal sense
Polaroid film, 87
Polaroid (photos), 35, 41
Police facsimile. *See* Law enforcement
Policefax (Litton/Datalog), 35
Polling, automatic, 58, 157, 158, 261
Portability, 31, 52, 53
Porterfield, John, 225
"Positive" scanning, 168
Postal costs. *See* Mail costs
Postal Service, U.S. (USPS), 19, 61, 63
Postal systems, foreign, 61, 274
Power, commercial, 93, 95, 136, 201
Power line induction, 109, 115-116. *See also* Single frequency interference
PRC Corporation, 59
Precision Instrument (Co.), 260
Pressfax DRR (Litton/Datalog), 277
Printer
  character, 14, 21, 62, 87, 137, 270
  CRT. *See* CRT printers
  See also Teleprinter
Printer/plotters, 258, 268-269
Printing plates, production of, 265, 270, 272
Proceedings of the IRE, 21(fn), 198(fn)
Processing
  chemical. *See* Recording, photographic
  dry, 23, 77, 87
  signal. *See* Signal processing
Projection
  flood; spot. *See* Scanning
  image, 78, 250, 251, *252*
Projector, motion picture, 14, 66, 88
Protection, network, 116, 133-134
Protocol, 153
Public telephone network. *See* Dial network
Publishing. *See* Periodical publishing
Pulse code modulation (PCM). *See also* Digital transmission.
Pulse-comparing circuit, 92, *94*
Pulse generation, 135. *See also* Transmission, digital
Pulse
  phasing, 92, *93*, 104, 173, 221, 223, 227, 229, 238

Pulse (*continued*)
  reference, 91
  sync, 95, 136, 173
Pulse transmission, 57, 126, 135–137, 150, 155. *See also* Transmission, digital

Qix, 18
Quality, circuit. *See* Transmission quality
Quality, output, 46, 47, 49, 64(fn), 106, 107, 109, 111, 115–123, 124, 132, 160–198, 207–208, 220(fn), 241, 272, 276. *See also* Contrast; Recording quality; Resolution; Transmission quality
Quantizing
  amplitude, 49, 76, 137, 177, 179–180, 182–183, 196, 198
  time, 181–183
Quantizing errors, 180–183
Queuing, 38
QWIP systems (Exxon), 19, 219
QWIP transceiver, 19, 216, 219, 242

RACE, 150–151
Radar display, weather, 27
Radiation, Inc., 19, 150
Radio amateurs, 257
Radio, commercial, 10–14, 49, 128, 225–226, 272, 274, 276
Radio Corporation of America. *See* RCA
Radio facilities, terrestrial, 23, 24, 123, 127–130, 131, 233. *See also* Satellites
Radio facsimile, 1, 7, 10–14, 24, 26, 27, 30, 34, 35, 116, 127, 128, 130, 225–226, 229, 231, 233, 236–237, 274
Radio Inventions, Inc., 9, 15, 16
Radio newspaper, 10–14, 225–226, 274
"Radio Pen," 10
Radiophoto, 6, 130
Radio reception, 4, 10–14, *24*, *36*, 110, 118
Railroads, use of fax by, 41. *See also* Freight expediting
Random access, 260
Randomness of fax signal, 103
Ranger, Capt. Richard H., 6
Rapifax (Corp.), 20, 145, 157, 216, 277
Rapifax 100 transceiver, 55, 145, 157
Raster scan, 151. *See also* Cathode ray tubes; Television
Rates, transmission. *See* System economics
"Ray Photo," 10
RCA, 1, 5, 6, 7, 9, 10, *11*, 12, 14, 15, 16, 131, 198(fn), 260, 275, 277

RCA Glōbcom, 130, 207, 218, 258
Reader
  microfilm (viewer), *59*, 266
  tape, 61, 137, 265
Readex Microprint Corp., 273
"Reado" (Crosley), *13*
Real-time, 247, 254, 264
Receiver, fax. *See* Recorders
"Record" communications, 130, 207, 218, 261, 264, 265, 266
Recorders; recording, fax
  arcing. *See* electroresistive
  "burn-off." *See* electroresistive
  cathode ray, 82, *83*, *84*, *85*, 142–143, 257
  color, 244, 277–278
  cylinder, 76, 84
  direct, 7, 23, 78, 84, *168*, 179
  electrolytic, 77, 78–79, *80*, 201, 202
  electromechanical, 76–77
  electropercussive (impact), 77, 81
  electroresistive ("burn-off"), 77, 79–81, 201
  electrostatic, 15, 23, 77, 78, 82–83, 201, 202, 270, *275*
  film, 32, 56, 59, 78, 87, 178, 228, 232, 244, 265–266, 269, 277, 279
  flat bed, 76, 84
  ink jet, 14, 77–78, 270
  laser, 78, 82, 87–88, *89*, 228, 270, 279
  optical fiber ("light pipe"), 82, *83*, *85*, 87
  photographic (wet and dry), 3, 23, 27, 78, 87–88, *89*, 167, *168*, 178, 179, 228, 265
  roll feed, 45, 58, 76, 84
  thermal (electrothermal), 85–87, 201
  wet ink offset, 77, 83–84, *86*
Recorders
  mechanics of, 76–88, *89*
  video. *See* Video recording
Recording quality, 46, 47, 49, 79, 80, 83, 87, 106, 107, 109, 115–123, 165, 167, 180–183, 244. *See also* Contrast; Quality, output; Resolution
Record retrieval, remote. *See* Remote access
Records, criminal. *See* Law enforcement
Records management system, electronic, 260
Redundancy reduction, 15, 27, 34, 47–48, 54–55, 97, 139–152, 208, 228, 240, 255, 264, 270, 271. *See also* Bandwidth compression; Data compression
Redundancy, transmission, 47–48

INDEX 339

Reflectance, copy background, 175-176, 177
Reflectometer, 175
Reformatting (of system input), 245
Regeneration, digital, 57, 126, 136
Registration program (interconnect), 132, 133, 134, 205
Regulatory bodies, 129. *See also* Federal Communications Commission; Government regulations, restrictions
Reliability, 220, 264, 265
Remote access, 48, 60, 254-257, 260, 265, 280. *See also* Information retrieval
Rental, equipment, 49, 199-201, 209, 220
   vs. purchase, 200, 209
Repeater, d.c. telegraph, 126, 247
Reproduction, image. *See* Laser "writing", Optical imaging; Recorders, recording
Reproduction quality. *See* Quality, output; Recording quality
Resolution, 160-174, 185-195
   "cross-grain," 169, 170, 171, 172
   definition of, 161
   degradation, 51, 278
   effect of contrast on, 177-178
   effect of "gaps on, 173-174
   effect on speed and efficiency, 147, 148, 208, 264, 278
   elemental, 43, 97, 164, 169, 170, 171, 172, 173
   lines, "line pairs," elements, cycles, 161-164, 187
   scan (density), 35, 43, 98-101, 116, 147, 152, 161, 164-165, 167, 185-191, 222, 228, 229, 230, 232, 233, 237, 240, 241, 243, 255, 257, 258
   "square" (equal), 105, 164, 171-173
   TV, 161, 195, 255, 257, 258
   vertical vs. horizontal, 161, 167-173
Resolution determinants, 164-173
Resolution test charts. *See* Test charts
Response curve. *See* Frequency response
Response speed, 54, 168
Reticon (Corp.), 75
Retransmission, 191, 241, 244
Retrieval, automatic, of filed material. *See* Information retrieval
Retrieval, computer-aided, 261, 269
Revision, computer-aided, 261, *269*
Ricoh, Ltd., 20
Ridings, G. H., 14, 37, 64(fn)
Ringing signal (telephone), 132, *133*, 153
Rixon (Inc.), 139
Rockwell International, 139

Rooftop antennas, 268
Root mean square (RMS), 170
RS-232-C standard (EIA), 223-224
RS-328 standard (EIA), 222-223, 224, 227, 230, 232
Run length coding (RLC), 144-148, 234
   multi-line, 145-146

St. Louis *Globe Democrat*, 34
St. Louis *Post-Dispatch*, 34
Sampling (digital), 76, 96, 97, *98*, 144, 150-151, 155, 171, 181-183
Sandia Corp., 252
Satellite Business Systems (SBS), 131
Satellite, domestic (Domsat), 131, 203, 204, 206, 214, 215, 216, 218, 268
Satellite network (Intelsat), 131
Satellites, communication, geostationary, 1, 32-33, 131, 218, 268
Satellites, weather. *See* Weather satellites
Savin (Business Machines), 20
Savings, cost. *See* System economics
Sawyer (Wm.), 5
SBE Linear Systems, 258
Scan-a-fax (Fairchild), 18
Scanatron, 18, 19, 55, 149. *See also* Victor; Infolink
Scan axis, *100*, 101, 161-163, 166, 167-168, 170, 172, 173, 181
Scan converter, 245, 258, 259
Scan density. *See* Resolution
Scan direction, 167-168, 227, 232, 233, 238, 241, 243
Scan head, 66-69
Scan lines. *See* Line length; Resolution
Scanner; scanning, 26, 37, 65-76
   auto-feed, 45, *46*, 58, 158, 201
   cathode-ray, 14, 54, 71, 74, 142-143, 167
   character recognition (OCR), 42, 264
   coin-operated, 37
   contact, 2, 3, 6, 71
   cylinder (drum), 67-69, 232
   digital, 55, 75-76, 170, 260, 268
   electromechanical, 54, 65
   electron beam, 165
   electronic (CRT and solid state), 14, 15, 54, 65-66, 71, 74-76, 142-143, 167, 173, 174, 228, 272
   electronics of, 71-76
   flat bed, 69-71, 168, 232, 233, 237
   flood, 66-*68*
   "flying spot," 14, 71, 74, 173
   high resolution, 165
   laser, 15, 20, 71, *72*, 165, 167, 228

Scanner (*continued*)
  mechanics of, 65-66, 67-71, 72
  microfilm, 41, 55-56, 58-59, 66, 74, 142, 167, 257
  optical fiber, 71
  photoelectric, 3-5, 65-76, 158(fn), 166, 174-175
  rotary, 71, 97
  solid state, 15, 54, 65-66, 74-76, 272
  spot, 66-67
Scanning aperture. *See* Aperture
Scanning optics. *See* Optics
Scan rate, 51, 99-101, 105, 111, 130, 141-143, 171, 172, 173-174, 181, 222, 227, 228, 229, 230, 232, 233, 234, 237, 238, 240, 241, 243, 257, 258
Scan speed. *See* Scan rate
Scan spot, 65, 71, *85*, 142, 146, 151, 164-167, 173
  size of, 164-167
Scanvision (SBE), 258
Schreiber, Wm. F., 244(fn)
Scrambling. *See* Encryption
Security, message, 154-157, 262-263
Seitz, Chas. L., 244(fn)
Selenium, 3, 82
Self-testing, 49
Separation of frequencies, 94
Separations, color, 40
Sequence, transmission. *See* Transmission sequence
Service centers, 62-63
Servicing difficulties, 49-50
Servicing, remote, 62-63
Servo-mechanism, 247, *248*
Shaft angle encoding, 97, *98*, 183
Shannon, C. E., 159(fn)
Shintron Co., the, 18, 241
Shipping of goods. *See* Freight expediting
Shrinkage (of recorded images), 79
Sidebands, 106-108, 110-111
  suppression of, 107-108, 139, 140
Siemens (AG), 20
Signal formatting, 61, 217. *See also* signal processing
Signal level, 110, 111, 115, 116, 131, 132, 137, 238
Signal magazine, 64(fn)
Signal processing, 55, 62, 134-152, 178, 245. *See also* Signal formatting
Signal sense, 222, 227, 243
Signal strength, 114, 116, *180*
Signal, sync. *See* Pulse, sync; Synchronism
Signal-to-noise ratio, 117, 175

Signature, electronic, 247, 263
Signatures, transmission of (signature verification), 22, 30-31, 191-192, 264
Single frequency interference (SFI), 116, 117, 123
Singer (Simulation Products), 15, 147, 152, 191
"Singing," 117
602 Data Set, 134-135, 137
Size, relative (fax copy), *99*, 100, 227-228
Sketches, transmission of. *See* Drawings
Skew, skewing, 94, 95
"Slope" filtering, 114-*115*
Slo-scan (Venus), 258
Slow-scan TV (SSTV). *See* Television
Smear, fax copy, 121
Smokelessness (recording paper), 80
"Soft copy", definition of, 246
Software, 269, 270. *See also* Algorithm
Solid state electronics, 15, 53-54, *73*, 74-75, *90*, *91*, *94*, 95, *112*, 113-114, 126, 219, 228, 247, 252-253, 258, 272-273
Solid wire (metallic) circuit, 89-90, 207, 234, 236-237
Sound track, motion picture, 66, 88
Southern Pacific Communications (SPC), 61, 127
"Sparking paper." *See* Paper; Recorders, electroresistive; Teledeltos
Spatial cycles; spatial frequency, 105, 161-162, 172, 173, 175, 187
Spectral compression, 150
Spectral response; spectral sensitivity, 66, 175
Speed. *See* Facsimile, high-speed; Facsimile speed; Transmission speed
Speedfax (SPC), 61
Spot size, scan. *See* Scan spot
Spreading effect (image details), 169-170
Stamps, Geo. M., 271
Standards
  character and symbol, 42
  compatibility, 240-244
  digital signaling, 223, 234
  domestic, 51, 154, 194, 195, 222-230
    ANSI, 224, 230
    EIA, 51, 158(fn), 222-224, 230, 232, 243
    FCC, 225-226
    IEEE, 162-164, 224-225
    military, 226-229, 243
    weather, 50, 229-230

INDEX

drafting, 186
Federal Telecommunication, 229
international, 51, 150, 154, 195, 230, 231-240, 241, 244
  CCIR, 231, 236-237, 243, 244(fn)
  CCITT, 51, 150, 223, 231-236, 238, 243, 244(fn)
  WMO, 230, 231, 237-240, 243
  lack of, 50-51, 174, 179, 276-277
  microfilm-facsimile, 190-191, 230
Standards converters, 241, 244
Statistical coding, 159(fn)
Stencil cutter, electronic (stencil maker), 40
"Stepladder" effect, *170*
Stepping motor, 96
Stewart-Warner Corp., 16, 17, 20, *46*, 62, 216, 242, 264, 268, 273
Storage, computer, 23, 48
Storage, data, 48, 260, 271
Storage/retrieval, images and records. *See* Information retrieval
Store-and-forward, 61, 127, 158, 264, 265, 266
Stylii; stylus, 2, 3, *6*, 10, 77, 78, 79, 81, 82-83, 85-86, 91, 92, 202, 246, 247, *248*, 253
  multiple, 82-83
Subaudio frequencies, 103-104
Subcarrier, 110, 127, 130, 225-226, 236, 238
Subject copy (first mention), 65
Subscriber profiles, 274, 276
Subsidiary Communications Authorization (SCA), 128, 226
Supergroup, 125, 135
Suppressor, echo, 119, 120
Switched ("dial-up") broadband, 52, 124-125, 202
"Switched network," limitation of, 109, 110, 121. *See also* Dial network
Switzerland, 148, 234
Symbol recognition. *See* Optical character recognition
Synchronism; synchronization, 2, 3, 5, 51, 78, 91, 92-98, 99, 123, 137, 142, 183, 222, 227, 232, 238
  analog, 92-96, 222, 227
  crystal, 93-94, *95*, 96, 222, 232
  digital, 96-98, 137, 138, 142, 183
  Power grid, 93, 136, 201, 222
  stop-start, 95, 98
Synchronization methods, 3, 92-98
Synchronizing "clocks." *See* Clock (digital)

Synchronizing pulses. *See* Pulse
Synchronous operation, digital, 97-98, 136-137, 138
System applications. *See* Application areas
System economics, 22, 38-39, 42, 44, 55, 59, 86, 124, 152, 157, 199-220, 257-259, 263, 264, 265, 266, 267, 273-274, 276, 277, 280
  budgeting, financing, 209
  comparative costs, 22, 39, 42, 63-64, 72, 124, 200-201, 203, 204, 206, 208, 209-216, 266
  cost-per-copy, 39, 63, 209-213, 263
  cost savings, 31, 143
  development costs, 220
  distance, effect of, 203, 204, 205-207, 208, 210, 213, 217
  fabrication costs, 220
  labor costs, 44
  "line charge" (group channel), 125
  maintenance costs, 63, 128, 199, 200, 209
  material costs, 80, 86, 201-202, 209, 210
  "meter" charges, 209
  one-time charges, 207, 216
  terminal costs, 48, 55, 63, 86, 143, 199-201, 208, 210, 217, 219-220, 249, 255-256, 257-258, 263, 264, 265, 266, 274
  termination costs (station charges), 124, 125, 127, 131, 137, 205
  trade-offs, 46-47, 147, 173, 194, 207-208
  transmission costs, 22, 27, 42, 45, 60, 63, 64(fn), 124, 125, 127, 129, 131, 133, 137, 139, 158, 202-207, 218, 263
  trends, 217-220
  See also Appendix.
Systems planning, 160, 172-174, 208
Szabo, Dr. N. S., 152

Tactical Digital Facsimile (TDF), 228
Taiyo Musen, 20
Talos Systems, Inc., 247, *250*
Tape
  keystroke generated, 44
  magnetic, 48, 56, 266, 269
  paper, 5, 61
  video, 260, 279
Tape transmission, 44, 48, 265
Tariffs, common carrier. *See* System economics

TDX Systems (Inc.), 61
Technological advances, latest, 20-21, 27, 32-33, 54-60, 61-63, *72*, 74-76, 77-78, *84*, *85*, 87, *89*, 96-97, 126-127, 128-129, 138, 139-140, 142-152, 154-157, 158, 196-198, 241, 244, 249-253, 258-266, 267-277, 278-280
Tektronix (Inc.), 258
Telautograph (device), 246
Telautograph (Corp.), 15, 19, 200, 247, 249
Telecopier, 17, *18*, 54, 69, 81, 82, 198(fn), 216, 242. *See also* Xerox Corp.
Teledeltos, 7, 80. *See also* Paper, recording, electroresistive
"Telefacsimile," 38-39, 64(fn), 198(fn)
Telefunken (GmbH), 5, 260
Telegraph
  automatic facsimile (W.U.), 7, 35, 37, *38*
  d.c., 103, 126, 247
  handwriting, 246
  Morse, 159(fn)
  pictures sent by, 3, 5
  printing, 2, 3, 102(fn)
Telenet, 218
Telenote (Talos), 247
Telephone companies. *See* Common carriers
Telephone bandwidth. *See* Transmission, voice-grade
Telephone circuits (lines). *See* Transmission, voice-grade
Telephone couplers. *See* Couplers
Telephone, invention of, 246
Telephone network. *See* Dial network; "Switched network"
Telephone rates. *See* System economics, transmission costs
Telephonic Equipment Corp., 157
"Telephoto" (AT&T), 6
Telephoto service, 124-125, 205, 206
Telephoto, UPI, 15
Telephotography 22-24, 64(fn), 221, 231, 232, 236. *See also* Newspictures; Phototransmission; Picture transmission; Recording, photographic; Telephoto
"Telepix" (W.U.), 6
Teleprinter, 20, 39, 42, 48, 61, 62, 78, 128. *See also* Printer, character
Telescriber (Telautograph), 247
Teletext, 274, 275
Teletypewriters, 16, 28, 29, 30, 61, 226, 264
Teletypesetting (TTS), 265

Teletypewriter exchange service (TWX), 39, 261-262
Television
  cable. *See* CATV
  closed-circuit, 22, 221, 224, 245, 246, 254-257, 258, 260, 266(fn), 279
  color, 259-260, 273, 278
  commercial, 1, 14, 16, 27, 49, 161, 165, 195, 221, 230, 257, 266(fn), 272, 274-275, 276, 278, 279
  early, 1, 6, 14
  flat-screen, 253
  general, 46, 74, 95, *96*, 136, 161, 165, 168, 198(fn), 253-254, 266(fn)
  high-resolution, 224, 255, 257, 258, 259, 266(fn), 279
  low-resolution, 257
  mechanical, 1, 14
  principle of, 74
  slow-scan, 22, 128, 225, 245, 246, 254, 257-260. *See also* Video systems
Telex, 261-262
Telikon (A-M), 20
Terminal equipment. *See* Couplers; Data sets; Facsimile; Handwriting machines; Modems; Television
Test Charts, 161-164, 176, 189, *190*, 191
  CCITT, 234-236
  IEEE, 162-164, 176, 191, 224
  NBS microcopy, 162-164, 189, *190*
  WMO, 238-240
Test procedures (IEEE), 224-225
"Theory of scanning, A," 198(fn)
Thermal recording. *See* Recording
Three-level "binary," 149-150
303-Data Set, 135-137
3M Co., 17, 19, 20, *45*, *52*, 55, 64(fn), 77, 81, 87, 145, 157-158, 220, 242, 263, 277
"3000 type" channels. *See* Transmission, leased line
Threshold; thresholding, 76, 143, 149, 170, 175, 177, 179, 191, 196-198
  "floating," 76, 180
  temperature response, 85-87
Times Facsimile Corp., 8-9, 10, 15
Time zones, crossing of, 45, 158
Tonal latitude, 87, 119, 167, 177, 196-198
Tonal transitions, *120*, 141, 144-148, 161, 166, 171, 264
Toner, 23, 82-83, *84*, 279
Toshiba (Tokyo Shibaura, Ltd.), 20
"Total data" communication, 62, 268
Trade-offs. *See* System economics

INDEX

Training, operator. *See* Operator training
Transaction volume, 202–207, 208, 210–215, 217, 263
Trans-a-File, 260
Transceiver, fax, 7, *8*, 16, 17, 18, 19, 37, 49, 52, 53, 81, 132, 200–201, 216, 219–220, 228, 241, 242, 269, 274, 276, 277
Transducer
  hand movement, *248*
  photoelectric. *See* Scanner
  recording. *See* Recorders
Transistor. *See* Solid state electronics
Transmission
  analog. *See* Analog signaling
  broadband; wideband, 14, 18, 34, 52, 123–131, 135–137, 158(fn), 192, 195, 202–204, 208, 228, 255, 265, 267, 278
  closed network, 22–23, 24–27, *28*, 29, 32–34, 38–39, 61, 123–124, 128, 254–255, 257, 268
  dial-up versus leased line, 119–120, 121, 124–125, 202–205, 212–215
  digital, 42, 57–58, 61, 63, 76, 121, 123, 124, 126–127, 135–139, 179–183, 202, 218–219, 228, 259, 276, 277
  international, 60, 95, 130, 204, 205, 207, 214, 215, 218, 228
  leased line (private wire), 55, 109, 119, 121, 123, 124–131, 153, 157, 202–207, 209–216, 234
  limited distance, 89–90, 255, 257
  radio. *See* Radio
  satellite. *See* Satellite
  switched digital, 126, 202, 268
  voice-grade (voiceband), 27, 38, 41, 46, 49, 51, 55, 62, 103–111, 115–123, 124, 127, 131–135, 137–139, 140, 143, 148–150, 151, 192, 203–207, 209–215, 217, 222–228, 229–230, 242, 246–251, 253, 257, 258, 268, 277, 278. *See also* Fax by phone; Voice communication
Transmission costs. *See* System economics
Transmission distance (mileage), cost of, 124, 125, 127, 131
Transmission efficiency, 20, 46, 47, 55, 58, 61, 127, 138–152, 208, 217–218, 259, 264, 268, *269*, 270, 276, 277–278
Transmission facilities; services, 123–131, 207. *See also* Dial network; Transmission, voice-grade
Transmission impairments, 48–49, 104, 106, 107, 115–123, 126, 132, 138, 152, 172, 179–180
Transmission quality, 109, 110, 120, 202, 208
  improvement in, 109, 120, 208
Transmission sequence, 152–154, 223, 227, 234, 240
Transmission speed, 46, 47, 50, 52, 54, 55, 62, 121, 123, 124, 127, 164, 172, 212–216, 228, 276–277, 278. *See also* Facsimile, high-speed; System economics, trade-offs
Trends. *See* Facsimile, future of; Facsimile trends; Future
TRI-TAC program (U.S. military), 228
Trouble, isolation of, 62–63, 201
Troxel, Donald E., 224(fn)
TRT Telecommunications, 130, 218
Trunks, telephone, 109, 119, 120
Trucking. *See* Freight expediting
Tube, vacuum, 53, 111–112. *See also* Cathode ray tubes; Photomultiplier
TV. *See* Television; Video systems
TV-facsimile
  high-speed, 14, 277
  weather satellite, 54, 130–131
Two-dimensional (vertical) compression, 146–148
Two-level (black-white) signaling, 58, 75–76, 130, 136–139, 144, 149, *156*, 164, 179–183
TWX, 39, 261–262
Type sizes, 186–191
Typewriter, electric, 261–263, 271, 273

Ultrafax (RCA), 277
Unattended operation. *See* Facsimile automation
Unifax (UPI), 18
United Press Associations, 15
United Press International (UPI), 15, 18, 23, *24*, 130, 157, 195
U.S. Army Electronics Command, 226. *See also* Facsimile, military; Dept. of Defense
U.S. Government. *See* Dept. of Defense; Federal Communications Commission; Government regulations; Regulatory bodies
U.S. Navy. *See* Naval Electronic Systems
U.S. Postal Service. *See* Postal Service
University of California, 38–39, 152
University of Illinois, 253

University of Lund, 77
University of Nevada, 189-190, 198(fn)

Vacuum tubes, 53, 111-112. *See also* Cathode ray tubes; Photomultiplier
Value-added carriers, 127
Variable-length coding, 159(fn)
Variable velocity scanning, 141-143, 144, 150
Vector encoding; vectorization, 151-152
Venus Scientific, Inc., 258
Versatec (Inc.), 258
Vertical compression. *See* Two-dimensional compression
Victor Comptometer, 19, 247
Victor Graphic Systems, 19, 149
Video disc, 260, 279-280
Videofile (Ampex), 260
Videograph, 18, *84*
Video Hard Copy (Tektronix), 258
Video recording, 245, 258, 260, 279-280
Video systems, 171, *196*, 253-260, 278, 279-280. *See also* Television
Video tape, 260, 279
Videovoice (RCA), 258
Vidicon, 130
Viewdata, 274
Visual Sciences, Inc., 19
Visual Transmission (FCC definition), 225-226
Vocoder, 140-141, *142*
Voiceband channels. *See* Transmission, voice-grade
Voice communication, 46, 47, 49, 50, 103, 119, 121, 129, 134, 140-141, 151, 154, 207, 247, 266(fn), 267
Voice coordination, 152
Volume. *See* Transaction volume
Von Ardenne, Manfried, 14
Vydec (Corp.), 263

Wall Street Journal, the, 31, 32-33, 34, 131
Wang (Labs, Inc.), 263
Washfax III System, 228
Watson, Dr. H. A., 78, 279
Waveforms, 103, 104, 106, *149*, 150, 151, 162, 175
Weather charts. *See* Weather maps
Weather forecasting, 7, 24-27, 221
Weather maps, 7, 24, 26, 27, 41, 55, 58, 81, 130, 149, 157, 179, 193-194, 227-228, 231, 233-234, 236, 237-240, 243, 268
Weather networks, 24-27, 50, 229-230
Weather satellites, 24, 26, 27, 50, 54, 130-131, 201, 229
Wente, E. C., 3
Westar I satellite (W.U.), 32-33, 131
Western Electric Co., 6, *8*, 9, 15, 266(fn)
Western Michigan University, 39
Western Union (Telegraph Co.), 5, 6, 9, 14, 16, 17, 32-33, 35, 37, 60, 61, 80, 102(fn), 123, 124, 125, 131, 202, 216, 219
Western Union International, 130, 218
Westrex, 15
White-space skipping, 15, 54, 55, 97, 141-143. *See also* Redundancy reduction
Wide Area Telephone Service (WATS), 207, 263
Wideband channels. *See* Transmission, broadband
Wiltek, Inc., 261
Wirephoto (AP), 7, *8*, 19
Wire services. *See* Associated Press; United Press International
Word processing; word processors (WP), 261, 271
communicating, 245, 261-263, 270-271
Words, digital. *See* Code words
World Meteorological Organization (WMO), 27, 227, 230, 237-240, 243
WTMJ (Milwaukee), 10
W2XBF (N.Y.), 10
Wyle, H., 159(fn)

Xerography, 15, 82-83. *See also* Recording, electrostatic
direct, 82-83
transfer, 82-83, 202
Xerox (Corp.), 15, 17, *18*, 19, *28*, 54, 69, 81, 82, *83*, 125, 150, 154, 157-158, 198(fn), 216, 242, 263, 277
X-rays, transmission of, 41

Yellow Pages (advertisements), 147
Young, Chas. J., 10

Zero crossings, 149-150
"Zero frequencies," 103
Zoom capability, TV, 255, 256
Zworykin, V. K., 14